Animal Husbandry - Beliefs, Facts and Reality

Edited by Sándor Kukovics

Published in London, United Kingdom

Animal Husbandry - Beliefs, Facts and Reality
http://dx.doi.org/10.5772/intechopen.1003349
Edited by Sándor Kukovics

Contributors
Alfredo Pereira, Ana Geraldo, Arieh Brosh, Arnolfo M. Monleon, Asrat Guja Amejo, Aviv Asher, Basant Kumar Bhinchhar, Benjamin Adjei-Mensah, Cocou Claude Kpomasse, Cristina Conceição, Dănuţ Nicolae Enea, Eliane Gonçalves Gomes, Emanuel Carreira, Essodina Talaki, Flávio Silva, Francesca Camilli, Gary William Onan, Hitesh Muwal, Ioan Hutu, Jacopo Goracci, Jan Johannes Hlongwane, Jenny Potsiso Mokhaukhau, João Serrano, Kokou Tona, Kwena Mokoena, Lebo Trudy Rashijane, Livia Vidu, Madikadike Kagisho Molabe, Madumetja Cyril Mathapo, Mahlatsana Ramaesela Ledwaba, Maleke Dimpho Sebopela, Mamonene Angelinah Thema, Masindi Lottus Mphaphathi, Mmapatla Precious Senyolo, Mükremin Ölmez, Nkhumeleni Cathbert Negota, Paramveer Palriya, Paulo Infante, Rankotsane Victoria Hlokoe, Rolly Maapola, Roshan Riaz, Rui Charneca, Silvio Henrique Ribeiro Balduino, Tarkan Şahin, Thabang Luther Mashilo, Thabang Sako, Thivhilaheli Richard Netshirovha, Thobela Louis Tyasi, Urbano Gomes Pinto de Abreu, Vinod Bhateshwar, Özlem Karadağoğlu

First published in London, United Kingdom, 2025 by IntechOpen
IntechOpen is the global imprint of INTECHOPEN LIMITED, registered in England and Wales, registration number: 11086078, 167-169 Great Portland Street, London, W1W 5PF, United Kingdom

For EU product safety concerns: IN TECH d.o.o., Prolaz Marije Krucifikse Kozulić 3, 51000 Rijeka, Croatia, info@intechopen.com or visit our website at intechopen.com.

British Library Cataloguing-in-Publication Data
A catalogue record for this book is available from the British Library

Animal Husbandry - Beliefs, Facts and Reality
Edited by Sándor Kukovics
p. cm.

This title is part of the Agricultural Sciences Book Series, Volume 24
Topic: Animal Farming
Series Editor: W. James Grichar
Topic Editor: Vasileios Papatsiros

Print ISBN 978-1-83769-925-4
Online ISBN 978-1-83769-924-7
eBook (PDF) ISBN 978-1-83769-926-1
ISSN 3029-052X

If disposing of this product, please recycle the paper responsibly.

IntechOpen Book Series

Agricultural Sciences

Volume 24

Aims and Scope of the Series

The importance of agriculture cannot be overstated. It helps sustain life, as it gives us the food we need to survive and provides opportunities for economic well-being. Agriculture helps people prosper around the world and combines the creativity, imagination, and skill involved in planting crops and raising animals with modern production methods and new technologies. This series includes two main topics: Agronomy and Horticulture, and Animal Farming. This series will help readers better understand the intricacies of production agriculture and provide the new knowledge that is required to be successful. The success of a farmer in modern agriculture requires knowledge of events happening locally as well as globally that impact input decisions and ultimately determine net profit.

Meet the Series Editor

W. James Grichar has been employed with Texas A&M AgriLife Research for over 45 years with an emphasis on research in agronomy, plant pathology, and weed science. He obtained his BS from Texas A&M in 1972 and his Masters of Plant Protection in 1975. He has published 195 journal articles, over 330 research reports and briefs, 11 book chapters, and over 300 abstracts of profession meetings. He also directs research in many crops including corn, grain sorghum, peanuts, and sesame. He has held various positions in different professional societies including the American Peanut Research and Education Society, Southern Weed Science Society, and Texas Plant Protection Conference in addition to being Associate Editor for Peanut Science and Weed Technology. Significant accomplishments have included spearheading efforts to determine the optimum planting time for soybean production along the upper Texas Gulf Coast. These efforts have shown growers that soybean yields can be improved by 10 to 20% by following a late March to early April plant date. He also has been instrumental in developing a herbicide program for peanut production in the south Texas growing region. Through the development and use of herbicides that are effective against major weed problems in the south Texas region, peanut yields have increased by 25 to 30%.

Meet the Volume Editor

Prof. Dr. Sándor Kukovics spent 40 years in the Research Institute for Animal Breeding and Nutrition (Herceghalom, Hungary), being responsible for the small ruminants sector, edited 40 books, published more than 1,550 articles, and had licences for 4 products. He participated in the undergraduate and graduate education programs of various universities in Hungary. Since 1996, he has served as president of the Hungarian Sheep and Goat Dairying Public Utility Association and as executive manager of the Sheep and Goat Products' Board since 2010. Between 2015 and 2019, he served as vice president of the EU COPA-COGECA Working Party on Sheep and Goats. Since 2016, he has been a member of the Board of Directors within the International Goat Association. He organised more than 26 national and 16 international conferences.

Contents

Preface

The analyses, findings, and summaries contained in this volume present the reality and facts of animal husbandry, attempting to dispel the myths circulating in this field.

Over thousands of years of animal husbandry history, laws on ownership, registration, breeding, and trade have gradually developed, later supplemented by requirements for animal health and welfare.

These have been continuously tightened in response to the rapid increase in the number of people and the demand for animal protein, particularly over the last half-century. These changes have served to provide the human population with the safest possible supply. Over the last one and a half to two decades, the assessment of sustainability has played an increasingly important role in the process of animal husbandry and product production, so the requirements related to this have also appeared in the regulation of animal husbandry and the production of animal products.

Details of this process that took place in the Philippines were summarized by the authors of the first chapter.

Eggs are perhaps the oldest animal protein used by humans. Over the past half-century, significant progress has been made in egg production, including advancements in genetics, feeding, and housing, as well as the expansion of its human use.

As the human population grows, the demand for animal protein increases, and eggs are an excellent food source to meet this demand. Population growth has also been quite intense in Africa over the past half-century, a process that can be observed in the amount of eggs consumed and their production.

The authors analyze the changes in egg production and consumption in South Africa over the past decade in the second chapter and propose that the experiences necessitate the development and implementation of an intensive development strategy to meet the increasing needs.

The success of animal breeding can be influenced by many factors. Therefore, in past centuries, animal breeders have sought to identify those characteristics in each animal species, the knowledge of which has allowed them to predict individual production capabilities.

Morphological factors have a direct and indirect effect on body weight development, with the latter being a key determinant of meat production. Knowledge of the characteristics of local indigenous breeds also lays the foundation for the development of breeding and production characteristics in the case of the goat species.

The authors of the third chapter reported on the correlations observed between individual body measurements and body weight in the case of indigenous goats.

The possibility of animal husbandry and product production is also influenced by several environmental factors. Developing animals that can be successfully kept and bred under given environmental conditions is a continuous challenge for livestock keepers. This system of relationships is further complicated by the applicable breeding procedures, the needs of local and exotic breeds, and the possibilities of keeping them in production, as well as the new challenges posed by global warming.

Small-scale broiler chicken farming in West African agricultural areas is also influenced by several factors, which encompass every detail, from breed and genetic issues through compliance with external environmental conditions to economic efficiency keeping and the evaluation of fiscal solution options.

The authors describe this system of relationships in the framework of the fourth chapter.

The use of sperm extenders, antibiotics, and antioxidants is essential for preserving the quality of male reproductive material during cryopreservation. In recent years, certain plant extracts have been playing an increasingly important role in this process, helping to preserve the quality and motility of cryopreserved sperm.

The authors summarized the possible raw materials (plant species) for the preparation of such plant extracts in the fifth chapter. As a result of their study, they found that the extracts made from these presented plants can not only be used with excellent results in semen cryopreservation but also are much cheaper than the previously used artificial antibiotics and antioxidants.

The intensive growth of the human population necessitates the increasing use of animals in production to meet the demand for animal protein. In most cases, this is accompanied by an increase in the size of livestock farms and the number of animals kept on them. However, this growth also brings with it a necessary change in environmental factors. This concept primarily involves the emission of greenhouse gases (GHG), the regulation and reduction of which is also expected for livestock farmers in the context of livestock farming, although livestock farming contributes a negligible share to the production of these gases. In contrast to the share of energy production and supply in greenhouse gas (GHG) emissions, livestock farming accounts for less than 6% of the total GHG balance.

Nevertheless, researchers are seeking methods and materials that can reduce these emissions.

In the sixth chapter, the authors analyzed and summarized the effects of certain natural feed additives used in the broiler chicken production process on GHG emissions.

Beef production has faced considerable criticism in the past decade due to its high greenhouse gas (GHG) emissions per unit of meat and its land requirements; however, the situation is far from simple. On the one hand, although the GHG emissions of beef cattle farming per unit of product are the highest among the animal production processes, they are a fraction of the emissions of the energy sector, and on the other hand, many variants of beef production differ in fundamental aspects. Extensive, semi-intensive and intensive beef cattle farming are suitable for the production of beef of different qualities, and their relationship with the environment is also different.

The authors summarized the characteristics of the different variants of beef production in the seventh chapter.

The sustainability of beef production has been supported in numerous studies by various authors, who have tried to provide a clearer picture of the relationship between meat yield and the environment for those who attack it, primarily green organizations and other members of the human population. However, organic beef production deserves a separate chapter in this process, as it must meet strict requirements to achieve and maintain this certification. For beef production to be both sustainable and organic, it must also meet economic requirements. In Brazil, organic beef production is also supported by tax reductions.

The authors analyzed this production process and its income-generating ability in the eighth chapter.

Since beef cattle are ruminants, their most natural environment is grazing. However, the quality of the pasture determines the number of animals that can be kept on it, influences the grazing period necessary for the animals to be satiated, and also the period necessary for the animals to develop properly.

However, achieving the expected production level requires individual registration of the animals, detailed knowledge and monitoring of the daily behavioural process, and up-to-date analysis of the results. An integral part of this is the knowledge of the breeding and selection characteristics of the animals, as well as their utilization.

The authors present their experiences in developing and utilizing a more efficient method of pasture utilization applicable to beef cattle farming in the ninth chapter.

Forested areas and the pastures connected to them create a special system of conditions for livestock farming and product production. Especially in Mediterranean regions, there is a significant interaction between livestock farming and the use of forest areas. On the one hand, areas used in agroforestry serve as a source of feed and absorb a large proportion of the greenhouse gases produced, on the other hand, animals, through their grazing and defecation, help maintain the plant species richness of forest areas and replenish the energy and nutrients of the soil used by plants.

The authors of the tenth chapter have systematized knowledge of this process.

In Portugal and the neighbouring areas of Spain, a unique agro-sylvo-pastoral ecosystem has developed, which also bears the characteristics of dry Mediterranean pastures. Extensive animal husbandry, characterized by low input and output, is the typical form of production in this area. These characteristics provide livelihood opportunities for the rural human population living in this region.

The development of this unique system and the centuries-old process of product utilization are summarized and evaluated by the authors within the framework of the eleventh chapter.

Extensive animal husbandry and product production have existed since the beginning of human animal husbandry, although some modifications have occurred since then. A special version of this is the migratory grazing of animals, which involves the systematic utilization of mountain summer pastures and other fodder sources left in the valley - which also extends to the grazing of forests and wooded groves to a significant extent. This process, called transhumance, is still present in the case of sheep in the utilization of pastures dotted with high mountains in various countries today.

The authors summarize the details of this special animal husbandry in the twelfth chapter.

India is the second-largest sheep-producing country in the world after China. Sheep, which are an integral part of the Indian pastoral economy, traverse vast pastures, exhibiting complex migration patterns and behavioural adaptations to cope with various environmental stressors.

India's diverse topography, from the foothills of the Himalayas to vast plains and coastal regions, offers a wide range of habitats for sheep populations.

As a result, Indian sheep farmers follow a variety of sheep migration patterns to exploit different forage resources. These migration patterns also vary from region to region.

The authors have examined and summarized the characteristics of this diverse and complex system in the thirteenth chapter.

The livestock farming in the Sub-Saharan African Region, and Ethiopia in particular, which is based on small-scale private farms, must meet a diverse set of conditions. This requires an ever-increasing resilience of both the farms and the animals they raise.

Ethiopia encompasses a wide range of production opportunities, from challenging desert and semi-desert regions to high mountain areas and tropical conditions, to which farmers and their animals must adapt, considering the expected sustainability conditions.

The authors of the fourteenth chapter, analyzing the resilience of livestock farming in this diverse region, conclude that to improve the sustainability and resilience of livestock farming, developments, education and achieving a more programmable version of production will be necessary in the future, which cannot do without state interventions to achieve the goal.

Dr. Sándor Kukovics
Professor,
Hungarian Sheep and Goat Dairying Public Utility Association and
Sheep and Goats Product's Board,
Hungary

Section 1

Regulation of Animal Husbandry

Chapter 1

Futures of Livestock Industry from the Viewpoints of Jurisprudence and Regulatory Measures in the Philippines

Arnolfo M. Monleon

Abstract

The livestock industry, comprising the sources of protein foods (eggs, meat, and milk), contributing enormous income, balancing trade, and attaining the food requirements in the Philippines. Filipinos have unique food preferences, and traditional delicacies are served usually in important events showcasing the rich food cultures and gastronomic traditions. Thereby, consumption and demands for animal products will continue to rise associated with the increasing trends in the human population and economic status. The sustainability and progress of the livestock industry, however, are challenged continuously by disruptions in production and input supply associated to disease outbreaks, natural calamities, and regulatory measures on biosecurity, importation, and urbanization. This review provides outlooks and viewpoints considering the issues and issuances affecting directly or indirectly the future of the livestock industry. The most prominent issue is the public clamor on the protection of general welfare sending a "chilling effect" or "doubt" to those engaged in the livestock industry. Also, jurisprudence provides different viewpoints on genetically modified organisms (GMOs) thereby products of biotechnology and gene engineering have limitations or restrictions on applications and uses. These challenges and issues, however, have optimistic viewpoints for achieving the standards of perfection toward the new beginning of the future of the livestock industry.

Keywords: biotechnology, food safety, importation, urbanization, welfare

1. Introduction

The livestock industry is a dynamic, progressive, and vibrant component of the Agriculture Sector of the Philippines that is directly under the control, regulation and supervision of the Executive Department (Section 17, Article VII of the 1987 Philippine Constitution). In the first quarter of 2024, the livestock industry posted an output amounting PhP 128.22 billion (2.56 billion USD) accounting for about 30% of the total agricultural outputs [1, 2]. The poultry sub-sector contributed the most (53.63%) valued at PhP 68.76 billion (1.38 billion USD) with an increment of 5.9% from 2023 data [2, 3]. The hog sub-sector, which is a major contributor had 4.3% decline in the value of production, due to the tremendous effect of African swine fever (ASF) since 2019 [2–4]. The key players in the hog sub-sector, however, are seeing positive development with expected 3% increases in the production this 2024 [5].

Increments in outputs from livestock production are continuously sought even at the global scale to meet the market demands. The driving forces in consumption have been associated to human population growth, rising income and urbanization, and dietary preferences with pork, poultry (chicken), and beef as the usual protein source in the Filipino diet. The demands, nevertheless, are continuously not meet locally due to several pressures in production like disease outbreaks, production fluctuations from natural causes (e.g. heat stress, typhoons, and torrential rains) and high cost of inputs. In the 1990s or 34 years ago, all these factors and scenarios in supply and demand were identified and found still applicable to the same challenges facing the livestock industry [6–8]. The most profound pressures nowadays are the socio-political issues on the environment, public welfare, and public health [9].

In the recent review (2024) as forecasted in a review in 2010; three major issues were raised, (1) the need to feed additional human population at scarce resources competing for land and water, and food and feed, (2) the need to operate in a carbon-constrained economy directly focused on efficiency, productivity, and reproductive outputs, and (3) the need to satisfy socio-politico-economic calls for "clean, green, and ethical (CGE)" products and industries [6, 7]. Here, this review provides outlooks and viewpoints from the Philippines considering recent issues and issuances on biosecurity, environment, food safety, importation, and public welfare affecting directly or indirectly the futures of the livestock industry.

2. Methodology

The data and information were taken from several government, and private and/or commercial online resources that include the website of the Supreme Court (https://sc.judiciary.gov.ph/), the Lawphil Project of the Arellano Law Foundation (https://lawphil.net/), Chan Robles Virtual Law Library (https://chanrobles.com/), the Official Gazette of the Republic of the Philippines (https://www.officialgazette.gov.ph/), Department of Agriculture (https://www.da.gov.ph/), Bureau of Animal Industry (https://www.bai.gov.ph/), Philippine Council for Agriculture and Fisheries (https://pcaf.da.gov.ph/), Philippine Council for Agriculture, Aquatic, and Natural Resources Research and Development (https://www.pcaarrd.dost.gov.ph/) Philippine Statistics Authority (https://psa.gov.ph/). Some specific terms or information (e.g. per capita consumption, human population) were searched from the Microsoft Bing (https://www.bing.com/) for cross-checking purposes.

The relevant data and information were organized in a table format to provide stand-alone information and present the key facts for the discussion of the topic thereby providing a consolidation of all available and current resources. The findings and discussion were presented carefully and with due diligence in consonance with all laws, jurisprudence, and issuances from the Philippine government. Nothing in this review chapter is claimed for personal gain or profit but merely to consolidate the recent information and trend that have a direct influence on the future of the Philippine livestock industry.

3. Results and discussion

3.1 Historical jurisprudence on livestock and relevant matters

The cattle (*Bos sp.*) has been mentioned in 75 of 121 resolved cases (or 61.98%) from 1904 to 2022 (**Table A1**). In a landmark case of Compañia Agricola de Ultramar, a corporation established in Madrid, Spain in 1893 provides in its Article 3 of incorporation that "*In order to exploit and develop the agricultural industry in the Philippine Islands and other Spanish colonies, the partnership may acquire any land, canals, and irrigating marshes or runways, overflows, waterfalls, quarries, and other real estate, and such cattle as may be useful for agricultural exploitation; to exploit or alienate said property, and to rent, by way of a change, or underlease, as may be convenient for the interests of the partnership, the reality; to establish agricultural colonies and to invest capital at interest with a mortgage upon rural or urban property, and to acquire credits with such guaranties; to grant loans upon crops, cultivated lands, cattle, agricultural machines, and in turn to borrow money on mortgage guaranty; to lease rural or urban property* [10]." The cattle is valued to have the most utility and economic use considering its various uses as draft, and as source of meat and milk. Sadly, however, cattle were mentioned in several resolved cases (34 of 71) of crimes against persons and property (e.g. theft, robbery, larceny, and even murder).

Most notably, cattle were mentioned in several resolved cases or disputes concerning national interests that include food safety, taxation, biosecurity, and importation (**Table 1**). Issues on food safety have been resolved as early as 1910 prohibiting the slaughter of cattle for human consumption without permission from the authorities [11]. Tax is the lifeblood of the state that usually construed against the people and imposed as early as 1918 [12]. The issues of biosecurity and importation are almost synonyms prohibiting entry of livestock from other place (or other country usually) to protect the current population from the possible entry of dreaded diseases of economic importance [13].

Subject matter	Commodity or property	Reference year	Number of cases
Food safety	Carabao, cattle, fresh meat	1910, 1913, 1929, 1933, 1955, 1956	6
Taxation	Cattle; frozen meat; meat, bones, hides of cow and carabao; frozen butter and veal; fresh meat, cattle import; slaughterhouse fee	1918, 1927, 1930, 1934, 1956, 1959, 1963	7
Biosecurity, disease prevention	Horse, cattle, carabao	1911, 1922, 1931, 1984, 2009	5
Environmental protection, public health, public welfare	Hog, fowl	2008, 2011, 2022	3
Agrarian reform	Livestock business	1990, 2005, 2007, 2010, 2011, 2013, 2018, 2022	10
Import	Cattle	1908, 1909, 1910, 1918, 1922, 1959	7
Intellectual property protection	Trademark	1983, 2003, 2020	3
Protection of property	Cattle, carabao	1904–2021	71

Table 1.
Subject matter of national interests involving livestock in resolved cases or disputes from 1904 to 2022.

3.2 Supply, demand, and utilization of livestock and its products

3.2.1 Supply

The chicken (*Gallus gallus domesticus*), categorized into native or improved chicken, broiler, and layer, with a population inventory of 202.82 million heads is the prominent source of animal-based protein in the Philippines (**Table 2**). The native or improved chickens comprise of 43% of the total chicken inventory are grown intensively in the Visayas region. The broilers and layers are commonly grown in large-scale commercial operations usually located near the Metropolis or the consuming public. The majority of broiler inventory was recorded in the Central Luzon while the layer chickens (35.18%) are found in the Calabarzon region [2]. The chicken inventory and category distribution were estimates based on quarterly surveys providing loopholes in its true population status considering that the broilers have a short production lifespan of 28–35 days as compared to layer chickens (100 weeks) and native chickens (150–180 days) [14]. Thereby, an improved survey and estimates, and intelligent tools and strategies are seriously needed. A revised classification of livestock and poultry farm based on animal type and population scale was introduced and implemented which will guide in the survey to collect reliable data for decision and policy making [15].

Duck, commonly the mallard duck (*Anas platyrynchos*), is distant next to the chicken inventory with 14.49 million heads grown for eggs, particularly in the Central Luzon contributing 51.2% in duck egg production. There is limited production for the broiler duck or meat-type duck. The spent duck layers or culled from the egg production are then slaughtered and cooked as "kinulob na itik," a traditional delicacy in Laguna [16].

The swine (*Sus sp.*) industry, even tremendously affected by the African swine fever since 2019, had contributed 302.65 billion pesos (6.05 million USD) to the agriculture sector (**Table 3**). Optimistically, the Philippine Swine Industry Roadmap will gear the industry toward achieving greater status better than the pre-ASF outbreak [17]. The government and private sector partnerships are guided with the Philippine Industry Roadmap for carabao [18] dairy [19], poultry (broiler) [14], poultry (layer) [20], and small ruminants [21].

Commodity	**Inventory (In million head)**	**Classification, %**		
		Smallhold	**Semi-commercial**	**Commercial**
Chicken	202.82	Not specified	Not specified	Not specified
Native/improved chicken (43.0%)	87.20			
Broiler chicken (34.7%)	70.42			
Layer chicken (22.3%)	45.20			
Duck	14.49	64.80	32.70	2.60
Swine	9.86	67.50	3.30	29.20
Goat	3.86	99.20	0.39	0.37
Carabao	2.74	99.30	0.50	0.20
Cattle	2.59	82.30	14.30	3.40

Table 2.
Inventory distribution based on farm-holding classification as of September 2023 (https://psa.gov.ph/).

Commodity	Volume of production (In thousand metric tons, liveweight except eggs and dairy)	Value of production (At Current Prices, in million PhP)	Forecasted volume in production in 2032 (In million tons)
Chicken	1866.65	234,888.61	2.41
Chicken eggs	708.50	87,091.68	
Duck	26.25	4647.63	
Duck eggs	55.64	7149.28	
Hog	1737.12	302,656.81	2.21
Cattle	236.65	37,176.81	
Carabao	130.27	19,570.44	
Goat	70.38	14,431.39	
Dairy	30.28	1686.65	

Table 3.
Volume and value of production in 2022, and forecasted production (https://psa.gov.ph/).

3.2.2 Demand

Food animals or animal-sourced foods still remain as an important resource to meet the food requirements of the increasing human population. The chicken and hog sub-sectors with forecasted growth in the volume of production will still be the food resources affordable to Filipinos. Chicken is widely consumed having few restrictions as to custom and tradition, religion, and health concerns. With better technology and good husbandry, pork will become a nutritious food resource even for those health-conscious individuals.

In 2023, the meat consumption per capita was 22.03 kg from the chicken sub-sector (12.57 kg), pork (7.88 kg), and beef (1.58 kg) [1–3]. Other livestock species, namely; carabao, goat, sheep, horse, and duck are supplementary sources of meat being cooked for special Filipino dishes (e.g. adobo or cooked with cocomilk). With projected increases to 132.77 million human population in the Philippines in 2032, the meat production target must increase to 11–12% to meet the current per capita consumption. Egg, milk, and fish are the other sources of proteins but consumption is low thereby undernutrition is still prevalent in the Philippines [22].

3.2.3 Utilization

Livestock products (eggs, meat, and milk) have several uses, utilities, and preparations considering the diverse culture, tradition, and beliefs in the Philippines. The "lechon" (roast whole pork) has gained more prominence being served in grandiose events and occasions as well as in festivities honoring a patron saint [23]. The "lechon" is prepared from particular breed, the native pig called markaduke, that is grown to perfection for the best desired crispiness, juiciness, and sensory attributes [Video 1, Available from: https://www.youtube.com/watch?v=_pfpgn1_kWM/IEC#16: The Finest Breed for Lechon and Processed Meat Products]. Other traditional food products from pork are the "itag," "tapa," and "bagnet" having geographical methods of preparations and uses.

The street-food markets and night markets are the best hang-outs to taste and savor unconventional foods like the "balut" (hard-boiled embryonated duck egg), "kwik-kwik" (fried flour-coated egg and/or day-old chick), "abnoy" (baked eggs rejected at first candling), and "betamax" (roast chicken's blood). Moreover, several restaurants and fast-food chains are producing variations in food products like the fried chicken, which differs considerably from one restaurant to another.

3.3 Challenges and issues confronting the livestock industry

3.3.1 Limits in business operation

Section 16 of Republic Act No. 7160 (RA 7160. Local Government Code of 1991) grants the Local Government Units (LGUs) the power to enact ordinances and regulations for the promotion and protection of the general welfare of its constituents. The Municipality of Biñan in the province of Laguna (in Calabarzon region) is a LGU that invoked this delegated police power in RA 7160, in approving and implementing a Municipal Ordinance regulating urban control zones for agricultural use and gradual phase-out of large piggery, fowl, and other livestock farms located within their Municipality [9]. The affected farms by the Municipal Ordinance were large-scale commercial farms, had filed a Petition before the court assailing the validity of the Municipal Ordinance. In 2022, the Supreme Court upheld the validity of the Municipal Ordinance in regulating the business operation of livestock farms at specified population size (i.e. not more than 10 swine and 500 chickens) to promote the general welfare of its residents by preserving their comfort and convenience (**Table 4**). However, such Municipal Ordinance is contrary and cannot help

Commodity	Regulatory measures	Source
All animals	To protect and promote the welfare of all animals in the Philippines by supervising and regulating the establishment and operations of all facilities utilized for breeding, maintaining, keeping, treating or training of all animals either as objects of trade or as household pets. For purposes of this Act, pet animal shall include birds.	Section 1 of Republic Act No. 8485. The Animal Welfare Act of 1998 as Amended in Republic Act No. 10631 (2013)
All organic food and input	All organic food and input establishments must register with the Director of the Bureau of Agriculture and Fisheries Product Standards (BAFPS), registration under this section shall begin within ninety (90) days of the enactment of this Act.	Section 16 of Republic Act No. 10068. Organic Agriculture Act of 2010
Organic agriculture farms, livestock and poultry and its feeds	...the Bureau of Animal Industry (BAI) of the Department of Agriculture shall be responsible for the registration of organic agriculture farms, livestock and poultry and its feeds.	Section 13 of Republic Act No. 11511 (2020). An act amending Republic Act No. 10068 or The Organic Agriculture Act of 2010
Swine poultry	Must not be more than: 10 heads swine 500 birds	Municipality of Biñan, Laguna et al. vs. Holiday Hills Stock and Breeding Farm Corporation and Domino Farms, Inc. G.R. No. 200403 (2022)

Table 4.
Regulation on business operation and/or volume of production.

to the current program entitled "INSPIRE (Integrated National Swine Production Interventions for Recovery and Expansion)" to rehabilitate the Philippine hog industry from the tremendous effect of the African swine fever (ASF) [17].

3.3.2 Financing scandal

The Quedan and Rural Credit Guarantee Corporation (QUEDANCOR) is a government-owned and controlled corporation created under Republic Act No. 7393 in 1992 but was abolished for various reasons by Memorandum Order No. 13 in 2017. The QUEDANCOR Swine Program (CG-QSP) was tasked to support the swine industry through credit but became controversial due to questionable purchasing agreements [24].

The Development Bank of the Philippines (DBP) financing scheme for a poultry broiler project that had caused suffering to the debtor was declared unscrupulous. In lieu thereof, the Court ruled against the DBP in order to indemnify the debtor including payment for the loss of income [25].

3.3.3 Transboundary diseases

Classical swine fever (CSF) and African swine fever (ASF) are two of the many diseases confronting the swine sector. The CSF is endemic while the ASF occurred in 2019 only but with tremendous impact [26]. The government is implementing strict regulations on international trade to protect the livestock industry (**Table A2**).

3.3.4 Threats to the general welfare

In two landmark cases in 2008 [27] and 2011 [28], the Court had ordered to several government agencies to provide comprehensive plan and action and submit reports on progressive activities in upholding their mandates to preserve and protect the environment, particularly the Manila Bay. Specifically, the Department of Agriculture is tasked to report the contributed pollution from agricultural and livestock sectors into the waterways going to the Manila Bay. The Court ruled that mishandling of pig manure is a liability; thereby farms owner indemnified the affected party [29].

3.3.5 Precautionary principle

The Court ruled in favor of the constitutional right to a balanced and healthful ecology (Section 16, Article II, 1987 Philippine Constitution) thereby permanently preventing the field testing for *Bt talong*, a bio-engineered eggplant (*Solanum melongena*). Relatively, the Department of Agriculture Administrative Order No. 08, series of 2002 that authorizes the importation and use of modern biotechnology was declared null and void. Furthermore, the Court ordered to stop any application for contained use, field testing, propagation and commercialization, and importation of genetically modified organisms (GMOs) [30]. Hence, any legislation toward utilization of GMOs must be guided with careful and decisive collaboration and undertaking for meeting the societal and judicial scrutiny.

3.4 Opportunity and possibility at the crossroads

3.4.1 Standard of welfare

The Animal Welfare Act of 1998 (Republic Act No. 8485 as amended) imposes the standards for housing facilities for all animals either as trade or as household pets. Those engaged (or to engage) in the livestock rearing must secure a permit and registration from the government office having jurisdiction in such registration matters. Similarly, those engaged in organic production are required to register with designated government office as prescribed under the Organic Agriculture Act of 2010 (RA 10068) as amended by RA 11511 in 2020 (**Table 4**). The category or classification based on farm-holding size (smallholder, semi-commercial, and commercial) was refined for the monitoring of compliance with regulation and management of the livestock industry (**Table A3**).

3.4.2 Biotechnology and gene engineering

Recent review promotes the development and implementation of risk-proportionate regulatory approaches and policies on the potential of animal biotechnologies to improve the sustainability and security of global food systems. This initiative hinges on the application of genetic engineering and genome editing. Genetic engineering is modifying the genome to produce a modified phenotype, thus called genetically modified organisms (GMOs). Genome editing is a precise modification of the genetic constitution of an organism that leads to a modified phenotype. Both techniques are used to create cisgenic (within-species) or transgenic (between-species) modifications, with the minute distinction between "GMO" and "non-GMO" animals and their genotypes [31]. Gene engineering is foreseen to increase productivity, produce disease resistance, and intensify biomedicine [32]. Moreover, these techniques paved the way for understanding the genes that control susceptibility to stress, and other physiological processes involving fetal programming, nutrient metabolism, and better nutrition as well as minimizing methane emission [7].

3.4.3 Intellectual property protection

The Supreme Court upheld the decision granting protection to the rightful owners of an intellectual property who sold and distributed roasted pigs using the registered trademarks. The petitioner who was found in bad faith for copying the trademark was ordered to pay for exemplary damages and attorney's fee, and ordered to cease and desist [33]. In a similar case, the petitioner had claimed as the sole owner of the business mark and that the other marks as used by the respondent were confusing to the public considering that both marks refer to the same classification. Thus, the court ruled in favor of the petitioner [34].

4. Conclusion

The livestock industry is progressing through government and private sector initiatives and collaborations toward meeting the desired requirements of the increasing demands and satisfying the socio-politico-economic expectations. Cattle have a long history of economic utility but its production output was surpassed by the chicken and pork sub-sectors, which are the leading producers of meat for human consumption. The animal-protein sources (e.g. meat, milk, and eggs) had low consumption per capita thereby malnutrition and/or undernutrition are still widespread in the Philippines.

Several interventions on biosecurity, food safety, importation, and production processes were imposed to protect and secure the livestock industry as well as improve its production system to align with the international standards. However, public clamor on the protection of public health, public welfare, and comfort and convenience to the citizenry is sending a "chilling effect" or "doubt" to multi-national and large-scale business sectors to expand operation and/or investment toward improving their production system, which later can be stopped (or closed) through a *doctrine of res judicata* (claim preclusion) and *stare decisis* (precedent).

Nevertheless, the challenges and anticipations may have optimistic viewpoints for achieving the standards of perfection as harnessed with continuing quests for better procedures and strategies toward the new beginning of the future of the livestock industry.

Acknowledgements

The Marinduque State University, the author's employer, is hereby acknowledged.

Conflict of interest

The author declare no conflict of interest.

Appendix

Year	Commodity/property	Resolved case/dispute	Source
1904	Cattle	Civil partnership	G.R. No. 1184. April 22, 1904
1905	104 cow, 31 carabao, 32 sheep, 17 goat, 1 horse	Recovery of personal property	G.R. No. 1341. November 08, 1905
1905	20 cattle	Robbery	G.R. No. 2208. May 05, 1905
1906	1000 cattle	Breach of contract	G.R. No. 2391. April 28, 1906
1907	81 cattle	Recovery of personal property	G.R. No. 2980. January 02, 1907
1907	883 cattle, 200 calves, 83 carabao, 74 goat, 23 sheep, 21 horse	Recovery of personal property	G.R. No. 3787. November 14, 1907
1908	1 caraballa and her calf	Recovery of personal property	G.R. No. 3837. August 01, 1908
1908	200 cattle	Will and testament	G.R. No. 4029. January 25, 1908

Year	Commodity/property	Resolved case/dispute	Source
1908	25 carabao	Recovery of personal property	G.R. No. 4292. December 04, 1908
1908	150 cow, 52 carabao	Import: Breach of contract	G. R. No. 4395. September 09, 1908
1909	Cattle (import)	Partnership sharing	G.R. No. 4861. March 20, 1909
1909	Cattle valued at P400	Theft	G.R. No. 5004. November 19, 1909
1910	Carabao	Violation of Act No. 1147, An Act regulating the registration, branding, and slaughter of large cattle.	G.R. No. 5060. January 26, 1910
1910	677 cattle and carabao (import, welfare)	Violation of Act No. 55 as amended by Act No. 275 (animal welfare)	G.R. No. 5270. January 15, 1910
1910	20 carabao	Recovery of personal property	G.R. No. 5524. March 21, 1910
1911	Horse	Provincial resolution pursuant to Act No. 83, as amended by Act No. 138 (disease prevention)	G.R. No. 6016. March 26, 1911
1912	Cattle	Will and testament	G.R. No. 7075. March 25, 1912
1913	Bovine	Violation of Ordinance No. 12 pursuant to Act No. 1147 (food safety)	G.R. No. 7852. August 18, 1913
1913	33 cattle, 28 fictitious cattle	Misappropriation of public funds	G.R. No. 8029. December 29, 1913
1913	150 cattle (branded)	Violation of Act No. 1740	G.R. No. 8191. March 04, 1913
1913	Cattle	Violation of Act No. 2030 (Theft)	G.R. No. 8687. October 16, 1913
1913	2 cattle	Theft	G.R. No. 8856. November 21, 1913
1914	1 caraballa and her calf	Theft	G.R. No. 8931. March 14, 1914
1914	Cattle	Recovery of personal property	G.R. No. 8976. December 02, 1914
1914	Carabao	Theft	G.R. No. 9267. March 02, 1914
1915	101 cattle	Recovery of personal property	G.R. No. 9749. November 13, 1915
1916	1 caraballa and her calf	Theft	G.R. No. 10078. February 05, 1916
1916	Cattle	Property sharing	G.R. No. 10722. February 18, 1916
1916	2 cow with calves	Theft	G.R. No. 11750. November 24, 1916
1917	3 cattle	Recovery of personal property	G.R. No. 11476. March 15, 1917

Year	Commodity/property	Resolved case/dispute	Source
1917	Carabao	Theft	G.R. No. 11636. February 28, 1917
1917	Caraballa	Recovery for damages (killed by a train)	G.R. No. 11730. March 24, 1917
1917	Horse	Violation of Ordinance No. 13, series of 1914 (loose of horse in a plaza) pursuant to Act No. 1147	G.R. No. 11779. February 23, 1917
1918	Cattle and carabao (import)	Breach of contract	G.R. No. 10029. July 15, 1918
1918	1000 cattle	Act No. 2339 (taxation)	G.R. No. 11962. August 27, 1918
1918	Carabao	Larceny	G.R. No. 13081. March 20, 1918
1920	2000 cattle	Property sharing	G.R. No. 14807. March 17, 1920
1922	Draft cattle and bovine cattle (import, serum)	Implementation of Administrative Order No. 21 of the Bureau of Agriculture (immunized from rinderpest)	G. R. No. 19628. December 04, 1922
1926	Cattle (Permit fee)	Misappropriation of public funds	G. R. No. 24622. January 28, 1926
1926	Cattle	Contract of employment (labor)	G. R. No. 25151. October 04, 1926
1927	Frozen meat as beef, pork, lamb, mutton, deer	Ordinance No. 1264, the so-called Meat Ordinance (taxation)	G.R. No. 26124. February 14, 1927
1928	Carabao, cattle	Recovery of personal property	G.R. No. 29044. October 02, 1928
1929	435 cattle	Property sharing	G.R. No. 30427. August 12, 1929
1929	Pork	"Ordinance No. 12 was adopted for the purpose of securing pure food for the inhabitants of the municipality, thereby protecting their health and comfort."	G.R. No. 30903. September 24, 1929
1930	Meat, bones, and hides of cow and carabao	Section 1459 of the Administrative Code (taxation)	G.R. No. 33138. November 20, 1930
1931	Cattle	Theft	G.R. No. 34612. October 01, 1931
1931	Cattle	Theft	G.R. No. 34634. October 01, 1931
1931	10 cattle	Act No. 3155 (prohibits importation to prevent disease)	G. R. No. 34674. October 26, 1931
1932	Offspring of cattle	Partition of conjugal property	G.R. No. 34989. April 05, 1932
1932	Cattle	Larceny	G.R. No. 35522. April 05, 1932
1932	Cattle	Larceny	G.R. No. L-35578. April 05, 1932

Year	Commodity/property	Resolved case/dispute	Source
1933	Slaughterhouse, sale of meat	Act No. 2758, as amended by Act No. 3632	G.R. No. 38256. December 16, 1933
1934	Frozen butter and veal	Philippine Tariff Act of 1909 (customs duty)	G. R. No. 39679. March 13, 1934
1936	Cattle	Theft	G. R. No. 44355. January 03, 1936
1946	Cattle	Theft	G.R. No. L-121. March 14, 1946
1948	Cattle	Theft	G.R. No. L-1728. November 26, 1948
1951	Cattle	Partition of property	G. R. No. L-1866. May 30, 1951
1955	Meat	Ordinance No. 3563 in Manila to ban the sale of fresh meat outside city markets (business regulation)	G.R. No. L-6762. February 28, 1955
1955	200 cattle	Recovery of personal property	G.R. No. L-7046. August 31, 1955
1956	Fresh meat	Ordinance No. 3563 which was declared by this Court constitutional in G. R. No. L-6762	G.R. No. L-7453. May 11, 1956
1956	Hog, cattle	Ordinance No. 3 (taxation)	G.R. No. L-9171. August 27, 1956
1957	321 cattle	Recovery of personal property	G. R. No. L-11005. October 31, 1957
1959	Cattle (import)	Violations of Central Bank Circular No. 31 in connection with Section 34 of Republic Act No. 265 (Central Bank Act)	G.R. No. L-10829-30. May 29, 1959
1962	3 bulls	Agreement or contract	G. R. No. L-17474. October 25, 1962
1963	Slaughterhouse fees	Ordinance No. 24 (fees)	G.R. No. L-15807. April 22, 1963
1966	Cattle	Theft	G.R. No. L-18507. March 31, 1966
1968	Carabao	Recovery of personal property	G.R. No. L-24490. May 29, 1968
1970	Carabao	Recovery of personal property	G.R. No. L-24814. January 30, 1970
1974	Cattle	Partition of property	G.R. No. L-19872. December 03, 1974
1976	2 goat	Theft	G.R. No. L-40037. April 30, 1976
1977	Cattle	Theft	G.R. No. L-39229. December 29, 1977
1981	Cattle	Cattle rustling	G.R. No. 51368. November 06, 1981

Year	Commodity/property	Resolved case/dispute	Source
1983	Native lechon	Business name	G.R. No. 50437. February 28, 1983
1984	26 carabao and a calf	Presidential Executive Order No. 626-A (1980). confiscation and forfeiture by the government of carabaos transported from one province to another	G.R. No. L-64279. April 30, 1984
1986	Hog, cattle, carabao	Claim for unpaid passage and freight.	G.R. No. L-27134. February 28, 1986
1987	Carabao	Executive Order No. 626-A	G.R. No. 74457. March 20, 1987
1990	Hog feeds	Claim for payment	G.R. Nos. 84,572-73. November 27, 1990
1990	Swine	Agrarian (swine raising is industrial activity)	G.R. No. 86889. December 4, 1990
1991	Cattle	Cattle rustling	G.R. Nos. 66,401-03. February 13, 1991
1991	Cattle	Cattle rustling	G.R. No. 91721. July 31, 1991
1998	Cattle	Cattle rustling	G.R. No. 128517. September 10, 1998
1999	Veterinary quarantine clearance to import	Falsification of public document	G.R. No. 124140. November 25, 1999
2000	Hog manure	Claim for damages and lost of profit	G.R. No. 125018. April 6, 2000
2001	Cattle	P.D. No. 533, otherwise known as the Anti-Cattle Rustling Law	G.R. No. 120548. October 26, 2001
2001	Cattle	P.D. No. 533, otherwise known as the Anti-Cattle Rustling Law	G.R. No. 140937. February 28, 2001
2003	Parasiticide	Patent	G.R. No. 126627. August 14, 2003
2003	Cattle, hog	Contract growing agreement	G.R. No. 153126. September 11, 2003
2004	Veterinary biologicals	Contract of carriage	G.R. No. 150094. August 18, 2004
2004	Animal feeds	Sum of money and damages	G.R. No. 152219. October 25, 2004
2005	Cattle	Agrarian (Cattle breeding)	G.R. NO. 162070. October 19, 2005
2006	Cattle, hog	Termination of employment	G.R. NO. 154284. October 27, 2006
2007	1 cattle (male)	P.D. No. 533, otherwise known as the Anti-Cattle Rustling Law	G.R. NO. 154941. July 09, 2007
2007	Alabang Stock Farm	Recovery of public property	G.R. No. 161639, July 30, 2007

Year	Commodity/property	Resolved case/dispute	Source
2007	401 cattle, 20 horse, 8 carabao	Agrarian exemption	G.R. NO. 169277. February 09, 2007
2007	Cattle (trading)	Contract breach	G.R. NO. 171250. July 04, 2007
2008	Veterinary medicine (manufacturing)	Termination of employment	G.R. No. 151309. October 15, 2008
2008	Livestock pollution	Manila Bay clean up	G.R. Nos. 171,947-48 December 18, 2008
2009	Cattle (importation, production, fattening and distribution)	Employer-employee relationship	G.R. No. 153192. January 30, 2009
2009	Horse	Administrative Order No. 55 dated 28 March 1994 by the Department of Agriculture (biosecurity, Equine Infectious Anemia (EIA))	G.R. No. 175220. February 12, 2009
2010	Cattle	Agrarian (Cattle raising as defense)	G.R. No. 154094. March 09, 2010
2010	5477 cattle, 115 horse	Recovery of personal property	G.R. No. 162079. March 18, 2010
2010	Cattle	Agrarian (Cattle raising as defense)	G.R. No. 170623. July 07, 2010
2011	Cattle, pigs, and other livestock	Agrarian (Livestock raising as defense)	G.R. No. 182332. February 23, 2011
2011	Livestock pollution	Manila Bay clean up	G.R. Nos. 171,947-48 February 15, 2011
2011	Cattle-financing program	Violation of Republic Act (R.A.) No. 3019, otherwise known as the Anti-Graft and Corrupt Practices Act	G.R. No. 167219. February 08, 2011
2011	105 cattle, 5 carabao, 11 horse, 9 goat, 18 swine as defense	Agrarian (Livestock raising as defense)	G.R. No. 178895. January 10, 2011
2013	Cattle	Agrarian (Cattle raising as defense)	G.R. No. 170018. September 23, 2013
2014	Cattle	Recovery of personal property (cattle breeding and dispersal)	G.R. No. 191838. October 20, 2014
2015	Agoo Breeding Station	Land donation	G.R. No. 197115. March 23, 2015
2016	Carabao	P.D. No. 533, otherwise known as the Anti-Cattle Rustling Law	G.R. No. 212186. June 29, 2016
2016	Lechon sauce	Intellectual property (trademark)	G.R. No. 198889. January 20, 2016
2017	Swine	Farm rehabilitation (loan)	G.R. NO. 197313, September 06, 2017
2018	Cattle, carabao, goat	Agrarian (Livestock business as defense)	G.R. No. 228503. July 25, 2018

Year	Commodity/property	Resolved case/dispute	Source
2019	Baguio stock farm	Recovery of public property (land)	G.R. No. 201273. August 14, 2019
2019	Swine farm	Labor (illegal dismissal, other benefits)	G.R. No. 232989. March 18, 2019
2020	Hog parent stock	Claim of unfair competition	G.R. No. 205835. June 23, 2020
2020	Veterinary drug	Intellectual property	G.R. No. 211850. September 8, 2020
2020	QUEDANCOR swine program	Administrative	G.R. No. 232844. July 7, 2020
2021	4765 hog	Replevin with damages (chattel mortgage)	G.R. Nos. 201,044 & 222,691. May 05, 2021
2021	Broiler	DBP loan dispute	G.R. No. 224138. October 06, 2021
2022	Hog, fowl, and other livestock	Municipal ordinance 04 (2004). Gradual phase-out of large farm	G.R. No. 200403. October 10, 2022
2022	Cattle, goat, poultry	Agrarian (livestock business as defense)	G.R. No. 243310. August 15, 2022

Table A1.
Chronology of resolved cases or disputes involving livestock or relevant matters of interests.

Commodity	Country	Sector	Period	Source
Pork	Taiwan	Six (6) FMEs	Three (3) years (Jan 3, 2023–Jan 3, 2026)	D.O. No. 01 series 2023
Poultry meat (Turkey, Goose and Chicken)	Poland	Twelve (12) FMEs	Three (3) years (Jan 3, 2023–Jan 3, 2026)	D.O. No. 02 series 2023
Poultry meat (Duck, Goose and Chicken) and Pork	Hungary	Five (5) FMEs	Three (3) years (Jan 3, 2023–Jan 3, 2026)	D.O. No. 03 series 2023
Poultry meat (Turkey and Chicken)	Chile	Six (6) FMEs	Three (3) years (Apr 2, 2020–Apr 2, 2023)	D.O. No. 05 series 2020
Poultry meat (Turkey and Chicken), Pork and Beef	Brazil	Twenty four (24) FMEs	Three (3) years (Apr 2, 2020–Apr 2, 2023)	D.O. No. 06 series 2020
Pork and Beef	Spain	Not specified	Three (3) years (Apr 2, 2020–Apr 2, 2023)	D.O. No. 07 series 2020
Buffalo meat	India	Six (6) FMEs	Three (3) years (Oct 5, 2020–Oct 5, 2023)	D.O. No. 12 series 2020
Pork	Sweden	Not specified	Three (3) years (Until Oct 7, 2022)	D.O. No. 14 series 2019
Buffalo meat	India	Sixteen (16) FMEs	Three (3) years (Oct 1, 2018–Sep 30, 2021)	D.O. No. 09 series 2018

Table A2.
Regulation on international trade (export to the Philippines).

Commodity	Classification			Source
	Smallholder	**Semi-commercial**	**Commercial**	
Swine	Tending 1–10 sow level or 1–100 heads	Tending 11–50 sow level or 101–500 heads	Tending 51 sow level and above or 501 heads and above	PSA Board Resolution No. 11, series of 2023
Carabao	Tending not more than five (5) caracows	Tending 6–50 heads of caracows	Tending 51 caracows and above	PSA Board Resolution No. 04, series of 2022
Cattle (Dairy and Beef)	Tending not more than five (5) cows and five (5) yearlings or calves	Tending 6–50 heads of cows	Tending 51 cows and above	PSA Board Resolution No. 04, series of 2022
Goat	Tending not more than thirty five (35) does	Tending 36–100 does	Tending more than 100 heads of does	PSA Board Resolution No. 04, series of 2022
Broiler chicken	Raising 500 birds and below	Raising 501–10,000 birds	Raising 10,001 birds and above	PSA Board Resolution No. 04, series of 2022
Layer chicken and Native/improved chicken	Raising 250 birds and below	Raising 251–5000 birds	Raising 5001 birds and above	PSA Board Resolution No. 04, series of 2022
Duck	Raising 250 birds and below	Raising 251–5000 birds	Raising 5001 birds and above	PSA Board Resolution No. 04, series of 2022

Table A3.
Industry structure based on farm-holding classification as of September 2023 (https://psa.gov.ph/).

Author details

Arnolfo M. Monleon
College of Agriculture, Marinduque State University, Philippines

*Address all correspondence to: arnolfommonleon@gmail.com

References

[1] Unlocking Growth – Challenges and Opportunities in the Philippine Livestock and Poultry Sector [Internet]. Available from: https://blog.fast.com.ph/unlocking-growth-challenges-and-opportunities-in-the-philippine-livestock-and-poultry-sector/ [Accessed: June 16, 2024]

[2] The Value of Production in Agriculture and Fisheries Inches up by 0.05 Percent in the First Quarter of 2024 (at Constant 2018 Prices) [Internet]. Available from: https://psa.gov.ph/content/value-production-agriculture-and-fisheries-inches-005-percent-first-quarter-2024-constant [Accessed: June 16, 2024]

[3] Q4 Agricultural Value Production up by 0.7% — PSA [Internet]. Available from: https://www.philstar.com/business/agriculture/2024/01/30/2329661/q4-agricultural-value-production-07-psa [Accessed: June 16, 2024]

[4] From INQUIRER: African Swine Fever Infection Downtrend Reported [Internet]. Available from: https://www.da.gov.ph/from-inquirer-african-swine-fever-infection-downtrend-reported/ [Accessed: June 16, 2024]

[5] Despite ASF Threat, Philippine Hog Industry Sees Better 2024. Available from: https://business.inquirer.net/448349/despite-asf-threat-philippine-hog-industry-sees-better-2024#ixzz8dHcax2vi [Accessed: June 16, 2024]

[6] Thornton PK. Livestock production: Recent trends, future prospects. Philosophical Transactions of the Royal Society B. 2010;**365**:2853-2867. DOI: 10.1098/rstb.2010.0134

[7] Martin GB. Perspective: Science and the future of livestock industries. Frontiers in Veterinary Science. 2024;**11**:1359247. DOI: 10.3389/fvets.2024.1359247

[8] Philippines Livestock And Poultry Market (2024-2030) Outlook. Available from: https://www.6wresearch.com/industry-report/philippines-livestock-and-poultry-market-outlook [Accessed: June 16, 2024]

[9] Supreme Court. Municipality of Biñan, Laguna vs. Holiday Hills Stock and Breeding Farm Corporation. G.R. No. 200403, October 10, 2022. Available from: https://sc.judiciary.gov.ph/200403-municipality-of-binan-laguna-rogelio-v-lee-antonio-p-aguilar-and-roberto-hernandez-vs-holiday-hills-stock-breeding-farm-corporation-and-domino-farms-inc/ [Accessed: June 16, 2024]

[10] Supreme Court. The Compañia Agricola de Ultramar vs. Anacleto Reyes et al., G.R.No. 1184. April 22, 1904. Available from: https://chanrobles.com/cralaw/1904aprildecisions.php?id=131 [Accessed: June 16, 2024]

[11] Supreme Court. The United States vs. Luis Toribio. G.R. No. 5060. January 26, 1910. Available from: https://lawphil.net/judjuris/juri1910/jan1910/gr_l-5060_1910.html [Accessed: June 16, 2024]

[12] Supreme Court. Philip C. Whitaker vs. James J. Rafferty as Collector of Internal Revenue of the Philippine Islands. G.R. No. 11962. August 27, 1918. Available from: https://lawphil.net/judjuris/juri1918/aug1918/gr_l-11962_1918.html [Accessed: June 16, 2024]

[13] Supreme Court. Mauricio Cruz vs. Stanton Youngberg, Director of the Bureau of Animal Industry. G. R. No. 34674. October 26, 1931. Available from: https://lawphil.net/judjuris/juri1931/oct1931/gr_l-34674_1931.html [Accessed: June 16, 2024]

[14] DA-BAR (Department of Agriculture – Bureau of Agricultural Research through the UPLB Foundation, Inc. in collaboration with Philippine Council for Agriculture and Fisheries. The Philippine Poultry Broiler Industry Roadmap (2022-2040). Diliman, Quezon City, Philippines; 2022. 75 p. Available from: https://pcaf.da.gov.ph/index.php/cir-broiler/ [Accessed: June 16, 2024]

[15] PSA Board Resolution No. 04 Series 2022. Approving and Adopting the Revision in the Classification of Livestock and Poultry Farm from Backyard and Commercial to Smallhold, Semi-Commercial, and Commercial Farms, and the Definitions by Animal Type. Available from: https://www.psa.gov.ph/system/files/psa-board/PSA%2520Board%2520Reso%2520No.%252004%2520series%2520of%25202022_0.pdf [Accessed: June 16, 2024]

[16] Kinulod na itik. The Philippines Today. Available from: https://thephilippinestoday.com/kinulob-na-itik/ [Accessed: June 16, 2024]

[17] DA-BAR (Department of Agriculture – Bureau of Agricultural Research through the UPLB Foundation, Inc. in collaboration with Philippine Council for Agriculture and Fisheries. The Philippine Hog Industry Roadmap (2022-2026). Diliman, Quezon City, Philippines; 2022. 132 p. Available from: https://pcaf.da.gov.ph/index.php/cir-hog/ [Accessed: June 16, 2024]

[18] DA-BAR (Department of Agriculture – Bureau of Agricultural Research through the UPLB Foundation, Inc. in collaboration with Philippine Council for Agriculture and Fisheries. The Philippine Carabao Industry Roadmap (2022-2026). Diliman, Quezon City, Philippines. 108 p. Available from: https://pcaf.da.gov.ph/index.php/cir-carabao/ [Accessed: June 16, 2024]

[19] DA-BAR (Department of Agriculture – Bureau of Agricultural Research through the UPLB Foundation, Inc. in collaboration with Philippine Council for Agriculture and Fisheries. The Philippine Dairy Industry Roadmap (2020-2025). Diliman, Quezon City, Philippines. 107 p. Available from: https://www.da.gov.ph/wp-content/uploads/2023/05/Philippine-Dairy-Industry-Roadmap.pdf [Accessed: June 16, 2024]

[20] DA-BAR (Department of Agriculture – Bureau of Agricultural Research through the UPLB Foundation, Inc. in collaboration with Philippine Council for Agriculture and Fisheries. The Philippine Poultry Layer Industry Roadmap (2022-2040). Diliman, Quezon City, Philippines. 107 p. Available from: https://pcaf.da.gov.ph/index.php/cir-layer/ [Accessed: June 16, 2024]

[21] DA-BAR (Department of Agriculture – Bureau of Agricultural Research through the UPLB Foundation, Inc. in collaboration with Philippine Council for Agriculture and Fisheries. The Philippine Small Ruminants Industry Roadmap (2022-2040). Diliman, Quezon City, Philippines. 91 p. Available from: https://pcaf.da.gov.ph/index.php/cir-sr/ [Accessed: June 16, 2024]

[22] NVN M, Demombynes G, SFA P, AJV A. Undernutrition in the Philippines: Scale, scope, and opportunities for nutrition policy and programming. In: International Development in Focus. Washington, DC: World Bank; 2021. DOI: 10.1596/978-1-4648-1701-4.

License: Creative Commons Attribution CC BY 3.0 IGO

[23] Valdez MLA. Parada ng Lechon in Balayan, Batangas Philippines to honor St. John the Baptist: An ethnographic study. Asia Pacific Journal of Multidisciplinary Research. 2014;**2**(3). Available from: https://oaji.net/articles/2014/1543-1418699647.pdf [Accessed: June 16, 2024]

[24] Supreme Court. Heirs of Nelson Cabrera Buenaflor, Namely, Pura R. Buenaflor, Karina R. Buenaflor, Kenneth R. Buenaflor, Paul R. Buenaflor and Mark R. Buenaflor vs. Field Investigation Office, Office of the Ombudsman. G.R. No. 232844, July 7, 2020. Available from: https://chanrobles.com/cralaw/2020julydecisions.php?id=453 [Accessed: June 16, 2024]

[25] Supreme Court. Development Bank of the Philippines vs. Evelina Togle and Catherine Geraldine TogleG.R. No. 224138. October 06, 2021. Available from: https://lawphil.net/judjuris/juri2021/oct2021/gr_224138_2021.html [Accessed: June 16, 2024]

[26] Hsu C-H, Chang C-Y, Otake S, Molitor TW, Perez A. Strategies for transboundary swine disease management in Asian Islands: Foot and mouth disease, classical swine fever, and African swine fever in Taiwan, Japan, and the Philippines. Veterinary Sciences. 2024;**11**:130. DOI: 10.3390/vetsci11030130

[27] Supreme Court. Metropolitan Manila Development Authority, Department of Environment and Natural Resources, Department of Education, Culture and Sports, Department of Health, Department of Agriculture, Department of Public Works and Highways, Department of Budget and Management, Philippine Coast Guard, Philippine National Police Maritime Group, and Department of the Interior and Local Government vs. Concerned Residents of Manila Bay, Represented and Joined by Divina V. Ilas, Sabiniano Albarracin, Manuel Santos, Jr., Dinah Dela Peña, Paul Dennis Quintero, Ma. Victoria Llenos, Donna Caloza, Fatima Quitain, Venice Segarra, Fritzie Tangkia, Sarah Joelle Lintag, Hannibal Augustus Bobis, Felimon Santiaguel, and Jaime Agustin R. Oposa, G.R. Nos. 171947-48, December 18, 2008. Available from: https://lawphil.net/judjuris/juri2008/dec2008/gr_171947_2008.html [Accessed: June 16, 2024]

[28] Supreme Court. Metropolitan Manila Development Authority, Department of Environment and Natural Resources, Department of Education, Culture and Sports, Department of Health, Department of Agriculture, Department of Public Works and Highways, Department of Budget and Management, Philippine Coast Guard, Philippine National Police Maritime Group, and Department of the Interior and Local Government vs. Concerned Residents of Manila Bay, Represented and Joined by Divina V. Ilas, Sabiniano Albarracin, Manuel Santos, Jr., Dinah Dela Peña, Paul Dennis Quintero, Ma. Victoria Llenos, Donna Caloza, Fatima Quitain, Venice Segarra, Fritzie Tangkia, Sarah Joelle Lintag, Hannibal Augustus Bobis, Felimon Santiaguel, and Jaime Agustin R. Oposa, G.R. Nos. 171947-48, February 15, 2011. Available from: https://lawphil.net/judjuris/juri2011/feb2011/gr_171947_2011.html [Accessed: June 16, 2024]

[29] Supreme Court. Remman Enterprises v. Court of Appeals and Crispin E. Lat, G.R. No. 125018, April 6, 2000. Available from: https://chanrobles.com/cralaw/2000aprildecisions.php?id=481 [Accessed: June 16, 2024]

[30] Supreme Court. International Service for the Acquisition of Agri-Biotech Applications, Inc. vs. Greenpeace Southeast Asia (Philippines), Magsasaka at Siyentipiko sa Pagpapaunlad ng Agrikultura (MASIPAG), Rep. Teodoro Casiño, Dr. Ben Malayang III, Dr. Angelina Galang, Leonardo Avila III, Catherine Untalan, Atty. Maria Paz Luna, Juanito Modina, Dagohoy Magaway, Dr. Romeo Quijano, Dr. Wenceslao Kiat, Jr, Atty. Harry Roque, Jr., Former Sen. Orlando Mercado, Noel Cabangon, Mayor Edward S. Hagedorn and Edwin Marthine Lopez. G.R. No. 209271. December 8, 2015. Available from: https://lawphil.net/judjuris/juri2016/jul2016/gr_209271_2016.html [Accessed: June 16, 2024]

[31] Hallerman E, Bredlau J, Camargo LSA, Dagli MLZ, Karembu M, Kovich D, et al. Enabling regulatory policy globally will promote realization of the potential of animal biotechnology. CABI Agriculture and Bioscience. 2024;**5**:25. DOI: 10.1186/s43170-024-00221-6

[32] Kwon D-H, Gim G-M, Yum S-Y, Jang G. Current status and future of gene engineering in livestock. BMB Reports. 2024;**57**(1):50-59. Available from: www.bmbreports.org [Accessed: June 16, 2024]

[33] Supreme Court. Emzee Foods, Inc. vs. Elarfoods, Inc. G.R. No. 220558, February 17, 2021. Available from: https://elibrary.judiciary.gov.ph/thebookshelf/showdocs/1/67321 [Accessed: June 16, 2024]

[34] Supreme Court. UFC Philippines, Inc. (Now Merged with Nutri-Asia, Inc., as the Surviving Entity) vs. Fiesta Barrio Manufacturing Corporation. G.R. No. 198889, January 20, 2016. Available from: https://chanrobles.com/cralaw/2016januarydecisions.php?id=90 [Accessed: June 16, 2024]

Section 2

Relationships and Interaction among Animal Farming and Human Being

Chapter 2

Analyzing the Relationship between Egg Production and Population Growth in South Africa from 2013 to 2022

Jenny Potsiso Mokhaukhau, Mmapatla Precious Senyolo and Jan Johannes Hlongwane

Abstract

Eggs offer nutritional benefits and are a vital source of proteins for humans globally. Therefore, the study examined the relationship between egg production and population growth in South Africa using secondary time series data from the year 2013 to 2022. The number of observations amounted to 10 and the data was obtained from Statistics South Africa and the Department of Agriculture, Land Reform, and Rural Development in South Africa. The data was analyzed using various analytical models, such as the descriptive statistics and Augmented Dickey-Fuller (ADF) test to assess the unit root. In addition, the Johansen cointegration test, Vector Autoregressive (VAR) model to test for the relationship was employed. The results show that egg production displayed a significant variation with population growth. The average tons is between 445 and 594 per annum while the average population growth is 1.195 annually. The cointegration results confirmed a long-term relationship between the variables. The VAR model demonstrates that past population growth predicts egg production which suggests that population trends forecast the output of agricultural production in South Africa. Therefore, this emphasizes the need for strategic planning to meet the yearly demand for products such as eggs in South Africa as the population grows.

Keywords: egg production, population growth, Johansen cointegration test, vector autoregressive model, South Africa

1. Introduction

Eggs are an essential part of the human diet, providing valuable nutrients and serving as a versatile ingredient in various cuisines [1]. Eggs have long been recognized for their nutritious components and contributions to human dietary requirements across the globe. In South Africa, as in many other countries, eggs are a staple food, consumed by people across all socio-economic strata. Egg protein has demonstrated efficacy in reducing malnutrition, enhancing muscle health, and promoting feelings of fullness, thus aiding weight loss [2]. Furthermore, it offers added advantages, such as bolstering immunity, lowering blood pressure, and exhibiting potential anticancer properties [3]. Rich in essential amino acids and vitamins, eggs are utilized in various productions such as confectionery and household food preparations. Given the population growth, there is dire need to meet the dietary needs of humans and eggs offer such benefits. Therefore, understanding the relationship between egg production and population growth is crucial for ensuring food security, economic stability, and sustainable agricultural practices.

According to Tukur [4], egg production in Africa has increased since 1994, however, the per capita consumption remains low due to rapid population growth. The inefficiency of the poultry industry in Africa primarily stems from the prevalent extensive management system among farmers. Transitioning from traditional methods of production to a modern approach could improving the poultry output. Given this, the South African Poultry Association [5] reported the consumption of eggs in 2016 was 7.65 kg in 2017, a decrease from 8.39 kg in 2016. Therefore, there has been a decrease in the consumption of eggs by 9.4% in 2017 despite the population increase in 2017. Despite a 1.6% population growth, egg consumption per person dropped by 9.4%. This decline was attributed to the reduced supply and higher prices resulting from outbreaks of Avian Influenza. The highest egg consumption per person was recorded in 2012 at 152.5 eggs.

Food production has been affected by various challenges which include climate change and the growing population. Climate change poses significant challenges to food production because of extreme weather events such as drought and floods which can devastate food production. For instance, numerous studies focused on the relationship between food production and climate change [6, 7]. Noteworthy, other studies examined the role of population growth on food production [8, 9]. Furthermore, research on the context of egg consumption among households revealed that 74% of households in Nigeria consumed eggs once a week [10]. This indicates the importance of eggs and their nutritional benefit in meeting the daily nutritional requirements.

In South Africa, eggs are consumed almost every other day. For instance, during the first year of the Covid-19 pandemic, the consumption of eggs per individual in South Africa was 159 [11]. Similarly, the total egg production in 2020 amounted to over 50 billion ZAR whereas broiler meat accounted for over 1.9 million tons. In terms of rand value, broiler and egg production represent the largest segment of South African agriculture, accounting for 16.6% of all agricultural production (a decrease from 18.4% in 2020) and 39.9% of all animal products (down from 41.9%) [11]. From the animal products, about 16.6% comes from poultry (i.e., 13.6% poultry meat and 3.0% from eggs). Similarly, global egg production has significantly expanded over the decade with over 93 million tons produced in 2022 which is a 69% increase compared to the 2000 production [12, 13]. In addition, global data indicates a continuous rise in egg production worldwide, averaging a 3% annual increase over the past decade [12].

Broiler feed prices have remained unstable in South Africa. For instance, in 2017, 87% of broiler producers purchased their feed in bags, while the remaining 13% bought in bulk [5]. The average price for a 50 kg bag of laying mash, excluding VAT and transport, was R254.84 compared to 2016 when the price decreased by only 4.4%. A total of 1802.1 tons of bagged layer feed were purchased in 2017, averaging about 620 bags per farmer [5]. The average feed intake was 110.6 g per head. The total expenditure on bagged feed amounted to R8.649 million for the year. Farmers who bought in bulk purchased 3900 tons at a weighted average price of R4,263.20 per ton for layer mash. The average feed intake for the year was estimated at 131.1 g per head. The total expenditure on bulk feed was R15.869 million. In 2017, the average price for layer feed was R3,457 per ton, marking an annual decrease of 15.0%. This drop came after yearly increases of 23.7, 12.0, 4.8, 0.5, and 18.9% over the previous five years. The Animal Feed Manufacturers Association (AFMA) reported that national feed sales for the layer industry totaled 885,676 tons from April 1, 2016, to March 31, 2017, representing a 6.9% decrease compared to the previous year [5]. These change in prices affect the production and consumption of eggs, therefore, the objectives of the study are to:

i. Describe the egg industry in South Africa.

ii. To analyze the causal relationship between egg production and population growth.

iii. Analyze the cointegration between egg production and population growth.

iv. Determine the relationship between egg production and population growth.

2. Literature review

Selmi et al. [14] noted that chicken eggs are a staple food, rich in animal protein, and reasonably priced, making them accessible to all people. This leads to a high consumption of chicken eggs compared to other animal protein sources. However, current production levels do not meet public consumption demands, and prices are continually fluctuating. According to Botha [15], the nationwide lockdown implemented to curb the spread of the coronavirus disease (Covid-19) pandemic provided a temporary relief for egg producers, as there has been a surge in home-cooked meals and baking activities. In the past years, the egg industry faced persistently low prices due to increased egg production, while demand remained constant. The Covid-19 pandemic has impacted poultry producers, but the effects differ significantly across various regions. Factors such as local lockdown measures, supply chain disruptions, varying consumer demand, differences in government support, and regional economic conditions all contribute to the diverse impacts experienced by poultry producers worldwide. Egg and broiler producers in 2021, began facing sustained challenges due to increasing feed costs, limited disposable income, high unemployment rates, and decreased demand from traditional markets as a result of ongoing Covid-19 lockdowns.

During the Covid-19 lockdown in South Africa, the cost of eggs increased as consumers altered their consumption and buying behaviors as they anticipated stockpiling food, such as milk, eggs, and bread in preparation for the quarantines [16]. With dining out no longer an option, the demand for eggs shifted from the food service sector to retail, as more meals were prepared at home [17]. The average price of eggs per dozen in 2020 was R14,35. In July 2020, the graded and ungraded eggs were R15.25 and R13.44 per dozen, respectively. This was a decrease of 2.5% and a yearly increase of 2.4% for graded eggs and, a 5.3% decrease on month-to-month with an 8.9% year-to-year increase [18]. A decline in both production and consumption of eggs was in 2017. This could be attributed to the price of cull hen which was R29.08 in

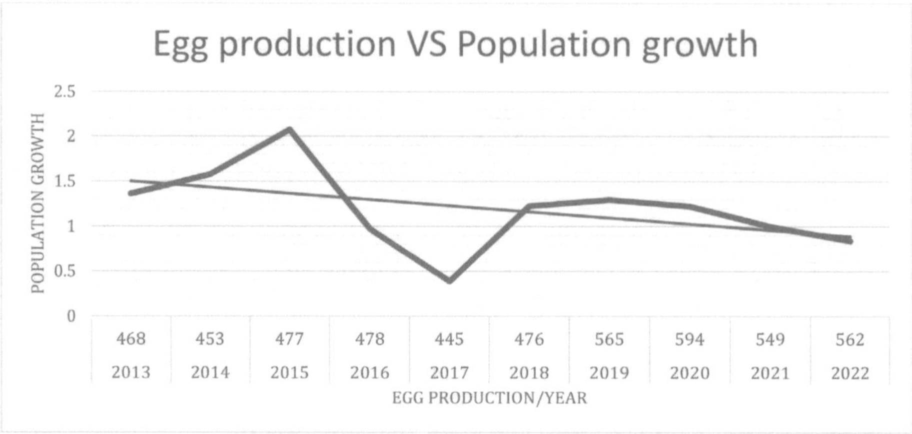

Figure 1.
Production of eggs vs. population growth. Source: Own compilation [2024].

2017. Although feed prices decreased in 2017, the price of eggs increased due to the scarcity of supply caused by the Avian Influenza outbreak [18].

Figure 1 presents the behavior of egg production relative to the population growth in South Africa from 2013 to 2022. The results showed that in 2015, the population growth was over 2% whereas the egg production was at 477 tons. According to the results presented in **Figure 1**, the population growth was low in 2016, and the production of eggs dropped to 445 tons in 2017. Furthermore, both egg production and population growth grew in 2018. This could be attributed to favorable economic conditions and increased demand for eggs. Therefore, higher population growth drives the increase in the demand for eggs. This ultimately boosts the economic performance of the egg industry while improving the livelihoods of farmers.

3. Methodology

The analysis of the relationship between egg production and population growth was based on the secondary time series data from South Africa spanning from 2013 to 2022 which accounts for 10 observations. The egg production data in tons was obtained from the Department of Agriculture, Land Reform, and Rural Development whereas the population growth data was obtained from Statistics South Africa. Descriptive statistics were employed to describe the egg industry in South Africa. Diagnostic tests such as the Augmented Dickey-Fuller (ADF) test were used to test for the unit root after which the Granger Causality test was used to analyze the causal relationship between egg production and population growth. The Johansen cointegration test was employed to analyze the cointegration between egg production and population growth. In order to determine the relationship between egg production and population growth in South Africa, the study employed the Vector Autoregressive (VAR) model.

3.1 Model description

The following models were employed to address the objectives of the study.

3.1.1 Descriptive statistics

The central measures of tendency from the descriptive statistics were used to describe the South African egg industry. These measures included the mean, standard deviation, minimum, and maximum.

3.1.2 Unit root: Augmented Dickey-Fuller (ADF) test

The unit root test is important in testing for stationarity among the dependent and independent variables. According to Guo [19], a unit root characterized by a nonstationary time series implies that the overall mean and the variance are not constant over time. Therefore, the time series data that is stationary suggests that the mean and the variance will not change over time.

The ADF is a common unit root test that is used to check whether a given time series is stationary or nonstationary. This test was first established by Dickey and Fuller in 1979 to demonstrate the existence of unit root with the primary emphasis of testing the null hypothesis that φ = 1 in the equation

$$Y_i = \varphi Y_i + U_i \tag{1}$$

Therefore, the hypotheses are given as:

H0: Time series data encompasses a unit root (φ = 1).

H1: Time series data is stationary (φ < 1).

The conditions are that if the data does not have a unit root problem, the null hypothesis will be rejected.

3.1.3 Granger causality test

The Granger Causality test was employed to analyze the causal relationship between egg production and population growth. The causality test is used to assess the relationship between a dependent variable and an independent variable in time series data. The test is mostly used to ascertain that one time series can be used to determine the other. Therefore, if X is important in predicting the values of the variable Y, then it is said to Granger-cause the variable Y.

The relationship between X and Y variables can be established by employing the following model:

$$Y_t = b_0 + a_0 X_t \sum m_j = 1\, a_j X_{t-j} + \sum n_i = 1 b_i Y_{t-i} + u_t \tag{2}$$

$$X_t = c_0 + d_0 Y_t \sum n_j = 1\, c_j Y_{t-j} + \sum m_i = 1 d_i Y_{t-i} + v_t \tag{3}$$

where the disturbance terms of the model is u_t and v_t. The correlation between X and Y can be analyzed by testing the null hypothesis that $a_j = d_j = 0$ for all $j\ (j = 0, 1 \ldots m)$ opposite to the alternative hypothesis that $a_j \neq 0$ and $dj \neq 0$ for some j.

3.1.4 Johansen cointegration test

The Johansen cointegration test was employed to analyze the cointegration between egg production and population growth. The Johansen cointegration test plays a pivotal role in understanding the relationship between variables in time series data.

Using the residuals which is the error term of the regression equation, the least-squares regression equation and the eigenvalues of data transformation, the Johansen cointegration test yields an estimate for all cointegrating vectors within the context of unit root analysis [20].

3.1.5 Vector Autoregressive (VAR) model

According to Loiboo et al. [21], the Vector Autoregressive model (VAR) is an econometric analysis model which involves estimating the regression equation in which the current value of each variable is expressed as a function of the lagged values of itself and each of the selected variables. Consequently, the dependent and independent variables are included in the autoregressive equation of the model and no variable is assumed to be exogenous. In VAR models, the dependent variable is a linear function of both lagged values of itself, therefore, the Ordinary Least Squares (OLS) can be used to calculate the Vector Autoregressive (VAR) model [22]. As a result, the VAR model entails calculating a regression equation where both current and historical values of each variable are evaluated. This approach assumes that no external factors and includes all variables in each autoregressive equation within the whole model. In this approach, no variable is pre-assumed to be exogenous, and all variables are included in each autoregressive equation within the system.

Following the general VAR model can be given by:

$$Y_t = \alpha + \beta_1 X_{t-1} + \beta_2 X_{t-2} + \beta_3 X_{t-3} + \ \dots\ + \beta_p X_{t-p} + u_i \tag{4}$$

where, Y_{t-1} represents the variable's value *ith* time period which is called the *ith lag* of the Y_t. X is the dependent variable. α represents the constant and u_i is the disturbance term.

4. Results and discussion

The results presented in **Table 1** show the outcome of the measures of central tendencies. These include the mean, minimum, maximum, and standard deviation for egg production and population growth. For egg production, the minimum value is 445 per 1000 tons and the maximum is 594 per 1000 tons. The standard deviation of the egg production and population growth amounted to 54.471 and 0.450 indicating significant variability in egg production and population growth, respectively. The average population growth for the period 2013 to 2022 in South Africa was 1.195%.

Properties	Egg production	Population growth (%)
Mean	506.700	1.195
Minimum	445.000	0.387
Maximum	594.000	2.074
Standard deviation	54.471	0.450
Source: Study results [2024].		

Table 1.
Descriptive statistics of egg production and population growth (N = 10).

4.1 Unit root test

To determine the stationarity of the data set, the study employed the Augmented Dickey-Fuller (ADF) unit root test. The conditions of rejecting the null hypothesis at unit root of 5% significant level is that the absolute value of the ADF statistics is higher than the critical value. As evidenced by the results in **Table 2**, both the population growth (GROWTH) and the egg production (TONS) are stationary at the first difference. The ADF for GROWTH and TONS is 4.08 and 3.7, respectively. At the 5% level of significance, the null hypothesis regarding the stationarity of the unit root test is rejected since the results proved that the ADF for D(TONS) is higher than the critical value of 4.08. Based on this outcome, the Johansen cointegration test can be used.

4.2 Johansen cointegration test results

To determine the existence of the relationship between egg production and population growth, the Johansen cointegration test was employed. The results of this cointegration are presented in **Table 3** for both Trace and Max-Eigen statistics.

	Population growth D(GROWTH) X		
Null hypothesis D(GROWTH) has a unit root			
Test critical values	**% level**	**t-statistics**	**Prob.**
	5% level	−4.450425	
ADF test statistics		−4.084211	0.0709
	Egg production D(TONS) Y		
Null hypothesis D(TONS) has a unit root			
Test critical values	**% level**	**t-statistics**	**Prob.**
	5% level	−3.403313	
ADF test statistics		−3.700926	0.0351

Source: Study results [2024].

Table 2.
Augmented Dickey-Fuller results.

Trace statistics results				
Hypothesized number of CE(s)	**Eigenvalue**	**Trace statistics**	**0.05 critical value**	**Prob.****
None*	0.991800	51.40786	25.87211	0.0000
At most 1	0.802570	12.97897	12.51798	0.0419
Max-Eigen statistics results				
Hypothesized number of CE(s)	**Eigenvalue**	**Max-Eigen statistics**	**0.05 critical value**	**Prob.****
None*	0.991800	38.42889	19.38704	0.0000
At most 1	0.802570	12.97897	12.51798	0.0419

**Denotes rejection of the hypothesis at the 0.005 significance level.*
***Mackinnon-Haug-Michelis p-values.*
Source: Study results [2024].

Table 3.
The cointegration test results using trace and Max-Eigen statistics.

The results of the Trace statistics showed that there is cointegration at a 5% level of significance, therefore, the null hypothesis is accepted because the egg production and the population growth are cointegrated. The Trace results further showed that there is one cointegration that is associated with egg production and population growth at a 5% level of significance. Given that the p-value of the Trace statistics is less than 5% significance, the null hypothesis is rejected, and it is assumed that at least one cointegration relationship exists between the variables.

The Max-Eigen statistics test in **Table 3** revealed that at least one cointegration exists between egg production and population growth at a 5% level of significance. The null hypothesis was tested, and it was revealed that as the rank was 0, the p-value for the Max-Eigen statistics was less than 5%. Therefore, the null hypothesis was rejected. The Trace and Max-Eigen statistics results suggest that there exists a long-run relationship between egg production and population growth.

4.3 Vector autoregressive correction results

The VAR equations are given by:

$$\begin{aligned} GROWTH = {} & C(1) * GROWTH(-1) + C(2) * GROWTH(-2) \\ & + C(3) * TONS(-1) + C(4) * TONS(-2) + C(5) \end{aligned} \tag{5}$$

$$\begin{aligned} TONS = {} & C(6) * GROWTH(-1) + C(7) * GROWTH(-2) \\ & + C(8) * TONS(-1) + C(9) * TONS(-2) + C(10) \end{aligned} \tag{6}$$

where C(1) to C(10) represent the coefficient of the lags. The estimates for the VAR are 10 estimates and each of the variable population growth and egg production (GROWTH and TONS respectively) has 5 estimates.

The VAR results were estimated, and the results are presented in **Table 4**. The adjusted R-square for the GROWTH model is given by 0.4009 which implies that 40.09% of the variation in the independent variable is explained by the variation in the dependent variable. The same applies to the adjusted R-square of the TONS model which implies that 91.14% of the variation in the independent variable is explained by the variation in the dependent variable.

Given that C(7) is the coefficient of the seventh lag of production of eggs (TONS), the results indicate that the production of eggs might not increase population growth ceteris paribus. This is true because the demand for eggs depends on the population growth and consumption. The negative and significant coefficient of −69.73771 suggests that past population growth is likely to predict egg production. This implies that demographic trends in South Africa might have a strong link with the agricultural output.

In addition, C(8) and C(10) represent the eight and the tenth lag of the coefficient of production of eggs with coefficients of 0.835724 and 266.6787, respectively. Both results are significant at a 5% level of significance. In addition, these results imply that holding other variables constant, as population growth increases, egg production is likely to increase in South Africa. This positive relationship suggests that an expanding population drives higher egg production to meet the growing demand ensuring food security and economic stability in the country.

Variable	Coefficient	Std error	t-statistics	Probability
C(1)	0.386244	0.509453	0.758155	0.4771
C(2)	−0.528573	0.469901	−1.124861	0.3036
C(3)	−0.003160	0.005736	−0.550910	0.6016
C(4)	0.000959	0.006241	0.153614	0.8830
C(5)	2.446736	2.292466	1.067294	0.3269
C(6)	3.892069	22.62330	0.172038	0.8691
C(7)	−69.73771	20.86691	−3.342024	0.0156**
C(8)	0.835724	0.254717	3.280994	0.0168**
C(9)	−0.175300	0.277128	−0.632561	0.5504
C(10)	266.6787	101.8016	2.619593	0.0396**
Model equation: GROWTH				
R-square: 0.3996			Adjusted R-square: 0.4009	
Model equation: TONS				
R-square: 0.9114			Adjusted R-square: 0.7932	

***Denotes significant levels at 5% level.*
Source: Study results [2024].

Table 4.
Vector autoregressive model results.

Null hypothesis	Observations	Lags	F-statistics	Prob.
TONS does not Granger cause GROWTH	8	2	0.21580	0.8174
GROWTH does not Granger cause TONS			5.76065	0.0939

Source: Study results [2024].

Table 5.
The Granger causality test results.

4.4 The granger causality test

The results presented in **Table 5** illustrate that the null hypothesis which stated that egg production (TONS) does not cause population growth (GROWTH), is accepted at the significant level of 10%. However, the null hypothesis implying that population growth does not cause egg production is rejected. This suggests that population growth (GROWTH) can forecast the production of eggs in South Africa. In addition, the results imply that there is a one-way causality between egg production and population growth. Furthermore, the results imply that in the long run, changes in egg production align with changes in population growth.

5. Conclusion and recommendation

The study focused on the dynamic relationship between egg production and population growth in South Africa over the period of 10 years that is, from 2013 to 2022.

While both egg production and population growth showed considerable fluctuation over the 10 years, there exists a significant long-term relationship among these variables. Therefore, population trends are likely to influence the production of eggs and predict the future demand. Thus, this drives the output in egg production within the country. Given these outcomes, opportunities for small-scale broiler producers exist to also venture into egg production to feed the growing nation. Not only will this meet the food security status and nutritional needs among households but also poses financial benefits for those in the egg industry.

Conflict of interest

The authors declare no conflict of interest.

Author details

Jenny Potsiso Mokhaukhau*, Mmapatla Precious Senyolo and Jan Johannes Hlongwane
Department of Agricultural Economics and Animal Production, University of Limpopo, South Africa

*Address all correspondence to: jenmkha@gmail.com

References

[1] Gunjan K, Pandey RP, Himanshu K, Mukherjee R, Chang CM. Navigating the global egg shortage: A comprehensive study of interconnected challenges. MedRxiv. 2024:24304250

[2] Puglisi MJ, Fernandez ML. The health benefits of egg protein. Nutrients. 2022; **14**(14):2904

[3] Barona J, Fernandez ML. Dietary cholesterol affects plasma lipid levels, the intravascular processing of lipoproteins and reverse cholesterol transport without increasing the risk for heart disease. Nutrients. 2012;**4**(8): 1015-1025

[4] Tukur HM. Egg Production in Africa. Improving the Safety and Quality of Eggs and Egg Products. Sawston, Cambridge: Woodhead Publishing; 2011. pp. 27-38

[5] The South African Poultry Association. Egg Industry Stats Summary for 2017 [Internet]. The South African Poultry Association; 2017. Available from: https://www.sapoultry.co.za/pdf-statistics/Egg-industry-stats-summary.pdf [Accessed: June 05, 2024]

[6] Ray DK, West PC, Clark M, Gerber JS, Prishchepov AV, Chatterjee S. Climate change has likely already affected global food production. PLoS One. 2019;**14**(5):e0217148

[7] Leal Filho W, Setti AF, Azeiteiro UM, Lokupitiya E, Donkor FK, Etim NN, et al. An overview of the interactions between food production and climate change. Science of the Total Environment. 2022;**838**:156438

[8] Pereira HC. Food production and population growth. Land Use Policy. 1993;**10**(3):187-190

[9] Ammani AA, Hassan AA. Population growth and food production: An investigation into the tenacity of the malthusian doctrine in Nigeria. Journal of Agripreneurship and Sustainable Development. 2022; **5**(3):33-41

[10] Olaniyi OZ, Adelayo A, Mary OO, Rasak A. Determinants of the household consumption of eggs in Oyo state a case study of Ibarapa central local government. Journal of Marketing and Consumer Research. 2015;**18**:53-59

[11] South African Poultry Association. 2021 Industry Profile [Internet]. South African Poultry Association; 2021. Available from: https://www.sapoultry.co.za/wp-content/uploads/2023/01/2021-Industry-Profile.pdf [Accessed: June 20, 2024]

[12] International egg commission. Global Egg Production Continuos to Grow [Internet]. International Egg Commission; 2024. Available from: https://www.internationalegg.com/resource/global-egg-production-continues-to-grow/#:~:text=According%20to%20data%20from%20the,growth%20of%20global%20egg%20production [Accessed: June 20, 2024]

[13] Food Agriculture and Organization. Agricultural Production Statistics 2000–2022 [Internet]. FAO; 2022. Available from: https://openknowledge.fao.org/server/api/core/bitstreams/dc1b16d6-44af-4d50-9544-10fcb8a3779f/content [Accessed: June 20, 2024]

[14] Selmi S, Putra A, Khairunnisa NP, Asriani PS. Fluctuations and movement patterns of race chicken egg prices in Bengkulu province. Jurnal Sain Peternakan Indonesia. 2023;**18**(1):57-63

[15] Botha L. Egg Demand Shoots through the Roof during Lockdown [Internet]. Farmers Weekly; 2020. Available from: https://www.farmersweekly.co.za/agri-news/south-africa/egg-demand-shoots-through-the-roof-during-lockdown/ [Accessed: June 05, 2024]

[16] Hafez HM, Attia YA, Bovera F, Abd El-Hack ME, Khafaga AF, de Oliveira MC. Influence of COVID-19 on the poultry production and environment. Environmental Science and Pollution Research. 2021;**28**: 44833-44844

[17] Animal Nutrition Competence (ANCO). How Are Egg Prices and Egg Producers Responding to Covid 19 [Internet]. ANCO; 2020. Available from: https://www.anco.net/how-are-egg-prices-and-egg-producers-responding-to-COVID-19/ [Accessed: June 05, 2024]

[18] South African Poultry Association. Monthly Egg Price Report: July 2020 [Internet]. SAPoultry; 2020. Available from: https://www.sapoultry.co.za/pdf-statistics/egg-price-report.pdf [Accessed: June 05, 2024]

[19] Guo Z. Research on the Augmented Dickey-Fuller Test for Predicting Stock Prices and Returns. Advances in Economics, Management and Political Sciences. 2023;**44**(1):101-106

[20] Dolado JJ, Jenkinson T, Sosvilla-Rivero S. Cointegration and unit roots. Journal of Economic Surveys. 1990;**4**: 249-273

[21] Loiboo D, Luvanda E, Osoro N. Population and economic growth in Tanzania. Tanzania Journal for Population Studies and Development. 2022;**28**(2):20-42

[22] Keating J. Structural approaches to vector autoregressions. Federal Reserve Bank of St. Louis Review. 1992;**1992**(74): 37-57

Section 3

Animal Breeding

Chapter 3

Application of Path Analysis for Direct and Indirect Effects of Morphometric Traits in Body Weight of Non-Descript Indigenous Goats of Polokwane Local Municipality

Madumetja Cyril Mathapo, Thobela Louis Tyasi, Kwena Mokoena, Rankotsane Victoria Hlokoe, Madikadike Kagisho Molabe, Lebo Trudy Rashijane and Thabang Sako

Abstract

The study was conducted to investigate the direct and indirect effect of morphometric traits on body weight of non-descript indigenous goats. A total of 139 goats aged between 1 and 5 years old were used in the study. Heart girth (HG), body length (BL), withers height (WH), rump height (RH) and body weight (BW) were measured in the study. Pearson correlation results indicated a positive ($P < 0.01$) correlation between BW and RH, WH, sternum height (SH), HG and BL with correlation coefficients of (r = 0,63**), (r = 0,50**), (r = 0,50**), (r = 0,33**) and (r = 0,31**), respectively. Path Analysis results indicated that RH (0,58) had the highest direct effect on the BW, while HG (−0,03) had a negative lowest effect on the BW. HG had the highest indirect effect on BW via RH. Simple linear regression model indicated that RH had the high coefficient of determination ($R^2 = 0.41$), adjusted coefficient of determination ($AdjstR^2 = 0.40$), and lowest mean square error (MSE = 107,38). This study suggests that RH can be used as a selection criteria to improve the BW of South African non-descript indigenous goats. The study further suggests that RH can be used to estimate BW of non-descript indigenous goats in the absence of weighing scale.

Keywords: path analysis, regression model, morphometric traits, non-descript indigenous goats, body weight

1. Introduction

Body weight is one of the most crucial economic traits as it plays a huge role in animal selection [1, 2]. Morphometric traits have been used as a predictor of body weight of the non-descript indigenous goats using different models. However, it has been highlighted that most the methods that have been used such as the Pearson correlation may be inadequate and not cover all aspects of the relationships involved between body weight and the morphometric traits [3]. Path analysis is a form of statistical multiple regression analysis that looks at the links between a dependent variable and two or more independent variables with the aim of evaluating the causal models [4]. It breaks down the correlation coefficients into components of direct and indirect compound path and effects, while measuring the direct and indirect effects of one variable on another [5]. Non-descript indigenous goats are crossbred populations that commonly raised under traditional systems [6–8]. They are one of the most prominent goat breeds in South Africa, characterized by a slow growth rate and small mature body weight. However, the relationship between body weight and morphometric traits of non-descript indigenous goats have not been studied using any multivariate analysis technique that involves the use of path analysis. Therefore, this study is aiming at investigating and establishing the direct and indirect effects between body weight and some of the morphometric traits through the application of path analysis.

2. Methods and materials

2.1 Study site

The current study was carried out in Madiga villages of Polokwane Local Municipality in Capricorn District of Limpopo Province, South Africa. Madiga is situated 12 km west of the University of Limpopo. The mean annual rainfall is less than 400 mm, with temperatures ranging between 5–28°C in winter and 10–36°C in summer [9, 10].

2.2 Experimental animals, management, and design

Non-descriptive indigenous goats of different sexes from Madiga villages within Polokwane Local Municipality aged between 1 and 5 years were used in the study. A traditional management grazing system was used, where goats were allowed to graze on the communal land during the day and, returned to the kraal in the afternoon and provided with water. The non-pregnant and healthy goats were randomly selected to form part of the study [11].

2.3 Sampling technique and sample size

A cross-sectional study design was used in the current study, where Madiga village was visited to measure all the biometric traits of interest on each indigenous goat once without administering any treatment. A purposive sampling procedure was employed; Polokwane Local Municipality was purposively selected as the first stage since it had the highest population of indigenous goats, as indicated by the

Department of Agriculture, Land Reform, and Rural Development in Limpopo. Madiga villages were randomly selected. A total of 139 goats were randomly selected, with 30 bucks and 109 does.

2.4 Data collection

The BW of goat was measured using a weighing scale calibrated in kilograms (kg), whereas biometric traits were measured using a wooden ruler and a tailor measuring tape (cm) [12]. The following biometric traits were taken as described by Yakubu [13], briefly, withers height (WH), sternum height (SH), body length (BL), heart girth (HG) and rump height (RH). To avoid false results when measuring, it was made sure that the animal stands stationary on the ground to evade anatomical distortions [14]. Furthermore, the goats were measured in the morning before grazing to avoid inaccurate measures such as the effect of water and feeds on structure and size of the goats. **Figure 1** below displays the biometric traits measured during data collection.

2.5 Statistical analysis

Statistical package for social sciences [12] v.29.0 was used to analyze the collected data (IBM Corp., NY, USA) [15]. Pearson's correlation was employed to determine the association between body weight and biometric traits, whereas path coefficient analysis determined the direct and indirect effects of collected traits on body weight. A simple linear regression model was used to establish a model to predict body weight from biometric traits.

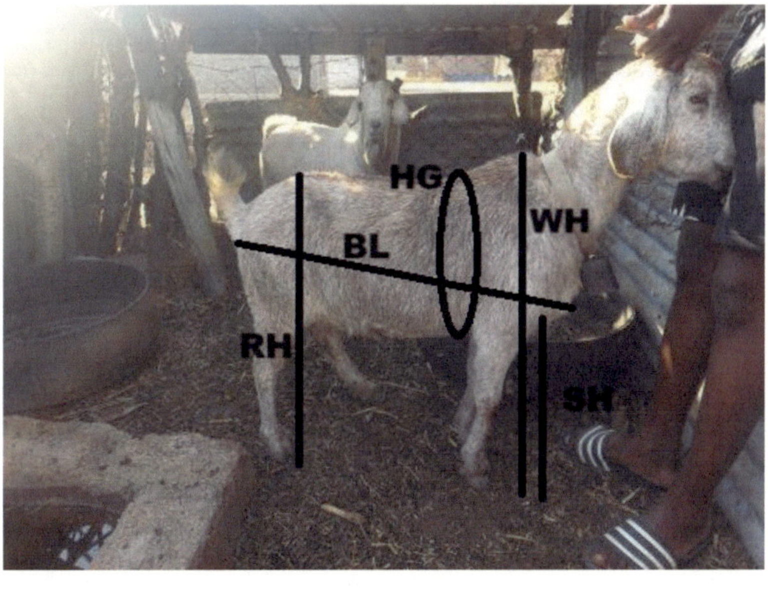

Figure 1.
Showing the measured biometric traits of non-descript indigenous goat WH: Withers height; HG: Heart girth; BL: Body length; RH: Rump height; SH: Sternum height. The picture was taken by Mr MC Mathapo (Source: own picture).

3. Results

Person correlation displayed in **Table 1** shows that the relationship between body weight and biometric traits ranged from 0,31 to 0,63. All biometric traits studied in the present study showed a positively high statistically significant correlation ($P < 0.01$) with BW. RH showed the highest positive significance correlation with BW, whereas BL showed the lowest positive significance correlation with BW.

The path analysis coefficient showing the direct and indirect effect of biometric traits on body weight is shown in **Table 2**. The RH had the highest direct effect on BW while HG had the lowest direct contribution on BW. Indirect path analysis results displayed that the HG showed highest indirect contribution on BW through RH.

Table 3 indicates the simple linear regression model established from positive direct biometric traits. RH model indicated R^2 = 0.41, AdjstR^2 = 0.40, MSE = 107,38 and a highly positive correlation of r = 0.63. The model further indicated that improving RH by 1 cm will increase BW by 0.99 kg. WH model indicated R^2 = 0.25, AdjstR^2 = 0.25, MSE = 135.38 and a highly significant positive correlation of r = 0.50. The model indicated that improving WH with 1 cm will increase BW with 0.71 kg.

Traits	BW	HG	RH	BL	WH	SH
BW		0,33**	0,63**	0,31**	0,50**	0,500**
HG	0,33**		0,56**	0,85**	0,66**	0,473**
RH	0,63**	0,56**		0,52**	0,75**	0,759**
BL	0,31**	0,85**	0,52**		0,71**	0,516**
WH	0,50**	0,66**	0,75**	0,71**		671**
SH	0,50**	0,47**	0,75**	0,51**	0,67**	1

*BW: Body Weight, HG: Heart Girth, RH: Rump Height, BL: Body Length, WH: Withers Height, SH: Sternum Height, *: Correlation is significant at the 0.05 level (two-tailed), **: Correlation is significant at the 0.01 level (two-tailed) (Source: Own tests).*

Table 1.
Correlation coefficient between biometric traits and body weight of goats.

Traits	Correlation coefficient with BW		Indirect effect				
		Direct effect	HG	RH	BL	WH	SH
HG	0,33**	−0,03		0,33	−0,05	0,07	0,02
RH	0,63**	0,58	−0,02		−0,03	0,08	0,02
BL	0,31**	−0,06	−0,03	0,30		0,07	0,02
WH	0,50**	0,10	−0,02	0,44	−0,04		0,03
SH	0,50**	0,04	−0,01	0,44	−0,03	0,07	

BW: Body Weight. HG: Heart Girth. RH: Rump Height. BL: Body Length. WH: Withers Height. SH: Sternum Height (Source: Own tests).

Table 2.
Path coefficient analysis of biometric traits and body weight.

Traits	Model	r	R^2	AdjstR^2	MSE
RH	−17,03 + 0.99RH	0.63	0.41	0.40	107.38
WH	−2.33 + 0.71WH	0.50	0.25	0.25	135.38
SH	−9.99 + 1.29SH	0.50	0.25	0.24	135.66

RH: Rump Height. WH: Withers Height. SH: Sternum Height (Source: Own tests).

Table 3.
Simple linear regression model.

SH model indicated R^2 = 0.25, AdjstR^2 = 0.24, MSE = 135.66 and a highly significant positive correlation of r = 0.50. The model further indicated that by improving SH with 1 cm will increase BW with 1.29 kg.

4. Discussion

The availability of computer packages in animal breeding leads to the greatest increase in the use of path analysis techniques [16]. Pearson's correlation was used to examine the association between the biometric traits and body weight. The outcomes revealed that all the biometric traits studied in the current study displayed a positively high statistically significant correlation with body weight. The rump height and withers height showed the highest positive significant correlation with the body weight. The findings of the study are consistent with those of Berhe [17], who found that body weight had a positively high significant correlation with rump height and body length in Maefur Goat Population in Tigray, Northern Ethiopia. Tyasi et al. [16] reported a positive significant correlation between body weight and sternum height, rump height and body length in Dorper sheep. Studies reported a positively high statistically significant correlation between body weight and body length in Assam Hill goats Khargharia et al. [18], Ganjam goats of Odisha Karna et al. [19], and yearling female Boer goats Mathapo and Tyasi [20], which is in line with the present study. The outcomes of the study imply that improving the rump height, body length, heart girth, withers height and sternum height might enhance the body weight in goats. Associated traits are assumed to be regulated by similar genes [21].

Tyasi et al. [22] detailed that the correlation coefficients assist in identifying the magnitude of relationships without assessing the cause-effect association between the traits. Thus, the use of path coefficient analysis helps to determine the direct and indirect influence of the biometric traits on goat's body weight. The path analysis was secondly used to determine the direct and indirect effects of biometric traits on body weight in goats. The findings showed that rump height had the highest direct effect on body weight, while heart girth had the lowest direct contribution to body weight. The results also displayed that heart girth had the highest indirect contribution to body weight through rump height. Nonetheless, more studies were done in chickens using path coefficient analysis for improvement of body weight in indigenous Nigeria chickens Egena et al. [23], Taishan chickens Qiao et al. [24], Chinese indigenous Dagu male chickens Tyasi et al. [25] and Zambian indigenous free-range chickens Liswaniso et al. [26].

Other studies of path analysis in cattle focused on milk yield improvement in Triveni crossbred cattle Gosavi et al. [27], Brown Swiss cattle Tahtali et al. [28], and Jersey dairy cattle Gorgulu [29]. The study further used simple linear regression to establish a model for predicting body weight from direct biometric traits in goats. The outcomes revealed that the model with rump height had a higher coefficient of determination and correlation coefficient and a lower mean square error. The findings of the present study differed with other studies in Etawah Peranakan goats Dahklan et al. [30], Kacang goats Depison et al. [31] and Saburai goats Dakhlan et al. [32], who found that heart girth was the best predictor of body weight. Rather et al. [33] reported that a model involving the withers height is the best to use when predicting body weight in Kashmir merino sheep. Kumar et al. [34] found a model consisting of heart girth gave a higher coefficient of determination and was recorded as a good model for prediction of body weight in sheep. The findings of the study suggest that rump height can be used as a predictor of body weight in goats. According to Ashwini et al. [35], increasing the heart girth by 1 cm resulted in an increase of 2.048 kg of body weight in crossbred cattle.

5. Conclusion

Pearson's correlation was used to determine the relationship between body weight and biometric traits in goats. It is concluded that rump height, body length, heart girth, sternum height, and withers height had a positively high significant correlation with the body weight in goats. Furthermore, Path analysis revealed that rump height had the highest direct effect on body weight while heart girth had the lowest direct contribution to body weight. The results also displayed that heart girth had the highest indirect contribution to body weight through rump height. More studies need to be conducted in path analysis with the major aim of improving body weight in other goat breeds. The regression findings displayed that the rump height made the highest contribution to the variation of body weight, therefore, it can be used as the predictor of body weight in goats. More studies need to be conducted in path analysis with the major aim of improving body weight in other goat breeds.

Acknowledgements

We would like to acknowledge the non-descript indigenous goat farmers from Madiga village for allowing us to work with their animals.

Conflict of interest

The authors declare no competing interests.

Declarations

We declare that this is our work.

Author details

Madumetja Cyril Mathapo*, Thobela Louis Tyasi, Kwena Mokoena,
Rankotsane Victoria Hlokoe, Madikadike Kagisho Molabe,
Lebo Trudy Rashijane and Thabang Sako
Department of Agricultural Economics and Animal Production, University of Limpopo, South Africa

*Address all correspondence to: madumetja.mathapo@ul.ac.za

References

[1] Zhang L, Wang F, Gao G, Yan X, Liu H, Liu Z, et al. Genome-wide association study of body weight traits in Inner Mongolia cashmere goats. Frontiers in Veterinary Science. 2021;**8**:1-9

[2] Ouchene-Khelifi NA, Ouchene N. Statistical models based on morphometric traits for live body weight estimation in goats. Agricultural Science and Technology. 2021;**13**(2):134-140

[3] Yakubu A. Path coefficient and path analysis of body weight and biometric traits in Yankasa lambs. Slovak Journal of Animal Science. 2010;**43**(1):11-25

[4] Chaitanya G, Prabhuling T, Hanumanthappa D. Path analysis: An overview and its application in social sciences. International Journal of Agriculture Extension and Social Developments. 2024;**7**(4):299-303. DOI: 10.33545/26180723.2024.v7.i4d.556

[5] Keskin A, Kor A, Karaca S, Mirtagioglu H. A study of relationships between milk yield and some udder traits by using of path analysis in Akkeci goats. Journal of Animal and Veterinary Advances. 2005;**4**:547-550

[6] Mdladla K, Dzomba EF, Muchadeyi FC. The potential of landscape genomics approach in the characterization of adaptive genetic diversity in indigenous goat genetic resources: A south African perspective. Small Ruminant Research. 2017;**150**:87-92

[7] Ramukhithi FV, Lehloenya KC, Kotze A, Nephawe KA, Chokoe TC, Seshoka MM, et al. Phenotypic characterisation of south African unimproved indigenous and Tankwa goats. American Journal of Animal and Veterinary Sciences. 2020;**14**(4):207-220

[8] Tyasi TL, Mathapo MC, Mokoena K, Maluleke D, Rashijane LT, Makgowo KM, et al. Assessment of relationship between body weight and morphological traits of south African non- descript indigenous goats. Journal of Animal Health and Production. 2020;**8**(1):32-39

[9] Kutu FR, Asiwe JAN. Assessment of maize and dry bean productivity under different intercrop systems and fertilization regimes. African Journal of Agricultural Research. 2010;**5**:1627-1631

[10] Maswana M, Mphahlele R, Mathapo MC, Hlokoe VR, Tyasi TL. Evaluation of socio-economic status and breeding practices for goat farmers in Madiga Village of Capricorn District, Limpopo Province, South Africa. Genetics and Biodiversity Journal. 2022;**6**(3):72-79. DOI: 10.46325/gabj.v6i3.212

[11] Yakubu A, Ladokun AO, Adua MM. Bio-prediction of body weight from zoometrical traits of non-descript goats using linear and non-linear models in north Central Nigeria. Journal of Livestock Research for Rural Development. 2011;**23**(6):1

[12] Rashijane LT, Mokoena K, Tyasi TL. Using multivariate adaptive regression splines to estimate the body weight of Savanna goats. Animals. 2023;**13**:1146. DOI: 10.3390/ani13071146

[13] Yakubu A. Fixing collinearity instability in the estimation of body weight from morpho-biometrical traits of west African dwarf goats. Trakia Journal Science. 2009;**7**:67-73

[14] Okpeku M, Yakubu A, Peters SO, Oozoje M, Ikebi OA. Application of multivariate principal component analysis to morphological characterization of indigenous goats in southern Nigeria. Acta Agriculture Slovenica. 2011, 2011;**98**(2):101-109

[15] Frey F. IBM SPSS: Statistical Packages for Social Sciences for Windows: Base System user's Guide, IBM Statistics. Chicago: SPSS Inc; 2023. p. 29. DOI: 10.2527/jas.2013-6967

[16] Tyasi TL, Mathye ND, Danguru LW, Rashijane LT, Mokoena K, Makgowo KM, et al. Correlation and path analysis of body weight and biometric traits of Nguni cattle breed. Journal of Advanced Veterinary and Animal Research. 2020;**7**(1):148-155

[17] Berhe WG. Relationship and prediction of body weight from morphometric traits in Maefur goat population in Tigray, northern Ethiopia. Journal of Biometrics and Biostatics. 2017;**8**:370

[18] Khargharia G, Kadirvel G, Kumar S, Doley S, Bharti PK, Das M. Principal component analysis of morphological traits of Assam Hill goat in eastern Himalayan India. Journal of Animal and Plant Science. 2015;**25**(5):1251-1258

[19] Karna DK, Acharya AP, Das BC, Nayak GD, Dibyadarshini MR. Morphometry of Ganjam goats of Odisha and age specific body weight prediction from linear body measurements. The Pharma Innovation Journal. 2020;**9**(4):171-175. Available from: https://www.thepharmajournal.com/archives/2020/vol9issue4S/PartD/S-9-4-48-605.pdf

[20] Mathapo MC, Tyasi TL. Prediction of body weight of yearling Boer goats from morphological traits using classification and regression tree. American Journal of Animal and Veterinary Sciences. 2021;**16**(2):130-135

[21] Hlokoe VR, Mokoena K, Tyasi TL. Using multivariate adaptive regression splines and classification and regression tree data mining algorithms to predict body weight of Nguni cows. Journal of Applied Animal Research. 2022;**50**(1):534-539. DOI: 10.1080/09712119.2022.2110498

[22] Tyasi TL, Qin N, Jing Y, Mu F, Zhu H, Liu D, et al. Assessment of relationship between body weight and body measurement traits of indigenous Chinese Dagu chickens using path analysis. Indian Journal for Animal Research. 2017;**51**:588-593. DOI: 10.18805/ijar.v0iOF.6990

[23] Egena SSA, Ijaiya AT, Kolawole R. An assessment of the relationship between body weight and body measurements of indigenous Nigeria chickens (*Gallus gallus domesticus*) using path coefficient analysis. Livestock Research for Rural Development. 2014;**26**:51

[24] Qiao X, Wang N, Yang Y, Yang A, Zhang Q, Fan X. Establishment of optimum regression model and path analysis between liveweight and body size of Taishan chickens. China Poultry. 2017;**39**:54-56

[25] Tyasi TL, Xu R, Chen X, Niu X, Qin N, Sun X, et al. Prediction of carcass weight from body measurement traits of Chinese indigenous Dagu male chickens using path coefficient analysis. Indian Journal Animal Science. 2018;**88**:744-748

[26] Liswaniso S, Tyasi TL, Quin N, Sun X, Xu R. Assessment of the relationship between body weight and linear body measurement traits of Zambian indigenous free-range

chickens using path analysis. Sylwan. 2020;**164**:465-485

[27] Gosavi SL, Deokar DK, Hadawale NP, Bhagat AA. Correlation and path analysis of 300 days milk yield in Phule Triveni crossbred cattle. Research in Veterinary Science. 2010;**1**(2):78-82

[28] Tahtali Y, Şahin A, Ulutaş Z, Şirin E, Abacı SH. Determination of effective factors for milk yield of Brown Swiss cattle using by path analysis. Kafkas Universitesi Veteriner Fakultesi Dergisi. 2011;**17**(5):859-864

[29] Gorgulu O. Path analysis for milk yield characteristics in Jersey dairy cows. Asian Journal of Animal and Veterinary. 2011;**6**(2):182-188. DOI: 10.3923/ajava.2011.182.188

[30] Dakhlan A, Saputra A, Hamdani MDI, Sulastri S. Regression models and correlation analysis for predicting body weight of female Ettawa grade goat using its body measurements. Advances in Animal and Veterinary Sciences. 2020;**8**:1142-1146. DOI: 10.17582/journal.aavs/2020/8.11.1142.1146

[31] Depison D, Putra WPB, Gushairiyanto G, Alwi Y, Suryani H. Morphometric characterization of Kacang goats raised in lowland and highland areas of Jambi Province, Indonesia. Journal of Advanced Veterinary and Animal Research. 2020;**7**:734-743. DOI: 10.5455/javar.2020.g475

[32] Dakhlan A, Hamdani MDI, Putri DR, Sulastri QA. Short communication: Prediction of body weight based on body measurements in female Saburai goat. Biodiversitas. 2021;**22**:1391-1396. DOI: 10.13057/biodiv/d220341

[33] Rather MA, Bashir I, Hamdani A, Khan NN, Ahangar SA, Nazki M. Prediction of body weight from linear body measurements in Kashmir merino sheep. Advances in Animal and Veterinary Sciences. 2021;**9**(2):189-193. DOI: 10.17582/journal.aavs/2021/9.2.189.193

[34] Kumar S, Dahiya SP, Malik ZS, Patil CS. Prediction of body weight from linear body measurements in sheep. Indian Journal of Animal Research. 2018;**52**(9):1-5. DOI: 10.18805/ijar.B-3360

[35] Ashwini A, Ramya HN, Ramkumar C, Reddy KR, Kulkarni RV, Abinaya V, et al. Reactive mechanism and the applications of bioactive prebiotics for human health. Journal of Microbiological Methods. 2019;**159**:128-137. DOI: 10.1016/j.mimet.2019.02.019

Chapter 4

Challenges of Small-Scale Broiler Production in Rural West Africa

Benjamin Adjei-Mensah, Cocou Claude Kpomasse, Essodina Talaki and Kokou Tona

Abstract

Small-scale broiler production in rural West Africa faces significant challenges that impact its sustainability and economic viability. The primary obstacles include limited market access, insufficient financial support, inadequate technical expertise and equipment, and high climatic variability. These issues are exacerbated by climate change, which threatens to increase average temperatures by 2 to 6°C, further affecting broiler production. The economic and environmental challenges are compounded by high feed costs, diseases, and expensive day-old chicks, all of which reduce productivity and profitability. To address these challenges, it is crucial to improve productivity and resilience through sustainable practices, better market access, and financial support. Breeding programs should focus on improving the genetic potential of local breeds while maintaining their adaptation to tropical climates. The introduction of high-performance hybrid broilers, though beneficial, requires careful management to avoid issues like thermal stress. Financial constraints also hinder the growth of small-scale broiler farms, limiting their ability to invest in necessary inputs and technologies. Innovative solutions like the use of renewable energy sources and improved ICTs can help overcome these limitations. By addressing these challenges holistically, small-scale broiler production in West Africa can become more robust, contributing to food security and economic development in the region.

Keywords: broiler production, sustainability, climate change, heat tolerance, small scale

1. Introduction

Small-scale broiler production in West Africa plays a significant role in the region's agriculture. Across the region, these operations typically involve a few hundred to a few thousand birds per cycle, contributing significantly to local food security and economies [1, 2].

Numerous obstacles prevent small-scale broiler production in rural West Africa from being sustainable and from making a positive economic impact on the community. The broiler meat-producing industry in the region faces several challenges, including limited market accessibility, insufficient financial support services, insufficient capital, insufficient technical expertise and equipment, and high climatic variability [3–6]. These difficulties have an impact on the general economic development of West African rural regions in addition to limiting the expansion potential of small-scale broiler farms. It is critical to adequately address the constraints encountered by small-scale broiler producers, as they are made more difficult by the anticipated rise in average temperatures in the region as a result of climate change [5].

Furthermore, issues with the economy and environment are not the only constraints affecting small-scale broiler production in rural West Africa. Insufficient funding, pest and disease outbreaks, costly feed, vaccine costs, and day-old chick expenses are among the challenges faced by small-scale broiler producers in the area [7–9]. These problems affect the industry's overall productivity and sustainability in addition to the income of the farmer. In addition, a vicious cycle of low productivity and vulnerability is created when small farmers in the Global South are neglected amid climate change challenges, especially for those who depend on rainfed agriculture.

To overcome these issues, efforts must be made to increase the productivity and resilience of small-scale broiler production in rural West Africa. To overcome the obstacles small-scale broiler producers face, it is imperative to create favorable conditions for farmers to increase their output using current technologies, enhance market accessibility, offer financial support services, and provide technical expertise and equipment [4]. Additionally, encouraging sustainable methods such as raising organic chickens and looking at alternate feed sources might make small-scale broiler farms more resilient to changes in the weather and the economy.

Regarding the performance of broiler chickens, local chickens express low feed efficiency and growth rates. Their production performance is influenced by genetics, nutrition, and health. Improving productivity without changing their inherent morphological and environmental traits is suggested, emphasizing the preservation of their genetic pool while identifying key productivity factors. Most indigenous chicken breeds constitute about 80% of the global poultry population [10].

This chapter explores the various challenges faced by small-scale broiler producers in rural West Africa. By examining economic limitations, infrastructural deficiencies, technical knowledge gaps, environmental impacts, social factors, and policy-related issues, this analysis aims to provide a comprehensive understanding of the obstacles these farmers encounter and offer recommendations to help them overcome these challenges, with the ultimate goal of fostering a more robust and resilient broiler production sector in rural West Africa.

2. Overview of broiler production in West Africa

West African broiler production faces many difficulties, especially in light of climate change. There is a serious risk to sustained broiler production because of the high heat and unpredictable local environment, which is expected to increase by 2 to 6°C over the next century [5]. The effects of climate change are very great in West Africa due to over high reliance on rainfed agriculture [11]. Because the economy and institutions are unable to adapt to these changes, this vulnerability is made worse. West African agriculture is very susceptible to the impacts of climate change while being a significant economic sector. This is mostly due to its dependence on rainfall [12].

Moreover, the effects of climate change on agriculture in West Africa go beyond broiler production. The region's mostly rainfed agriculture is in danger because of the shifting patterns of the climate (**Figure 1**) [12]. The considerable climatic uncertainty

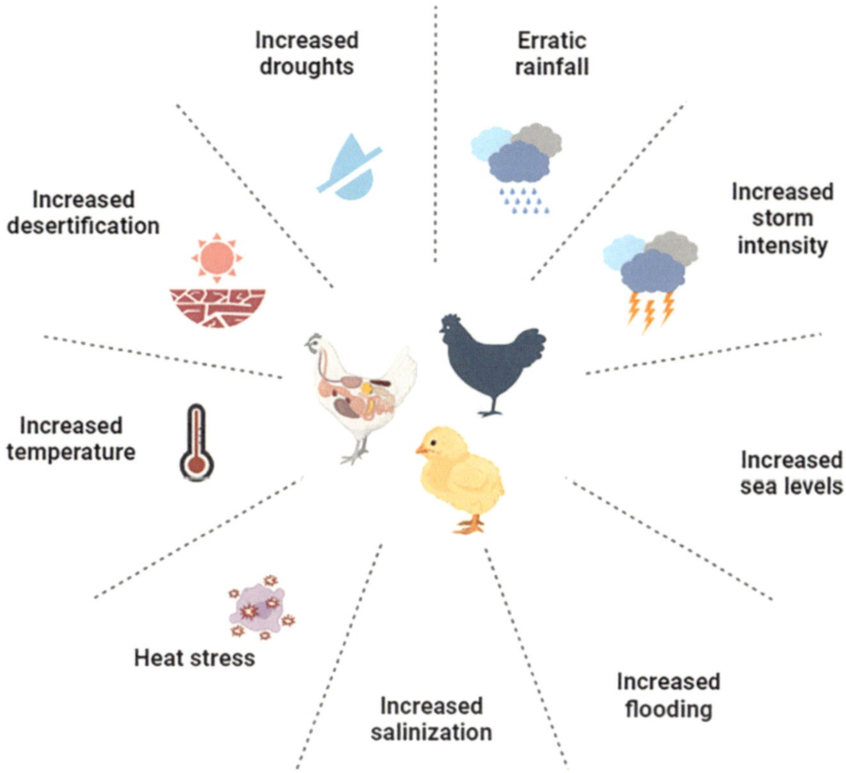

Figure 1.
Environmental impacts of climate change. Source: Developed by the authors.

in West Africa increases this vulnerability [11]. Because the area depends heavily on rainfed agriculture, adapting to climate change is difficult and threatens livelihoods and food security. Additionally, the probable increase in extreme weather events brought on by climate change may exacerbate the disruption of broiler production systems in West Africa, which would affect productivity and profitability.

Examining viable adaptation techniques for West African broiler production is essential given these difficulties. Adopting sustainable methods could be one strategy to lessen the effects of climate change on broiler farms. For example, the demand from consumers for sustainable products has led to a notable growth in the production of organic poultry [13]. The adoption of organic farming methods in broiler production has the potential to improve climate change resistance and satisfy consumer demand for eco-friendly products. Furthermore, looking for alternatives from non-traditional feed sources such as seed oils and fats from West African legumes may offer broilers more sustainable and alternate dietary options [14].

It is also crucial to make improvements to the region's broiler farms' productivity and profitability. Numerous factors have been found through studies to influence the profitability of broiler farms, such as the size of the flock, the expertise of the farmers, and membership in poultry associations [15]. Comprehending and refining these variables may enhance the financial viability of broiler farming in West Africa. Furthermore, investigating options like smallholder commercial broiler production could provide insights into improving revenue generation and the general viability of broiler farms in the area [16].

2.1 Objectives of the chapter

This chapter's objectives on the problems of small-scale broiler production in rural West Africa include providing a detailed understanding of the constraints faced by the small-scale broiler farmers in the area. The first goal is to pinpoint and investigate the major obstacles to small-scale broiler production in rural West Africa's long-term sustainability and growth. This entails looking at factors that impact small-scale broiler producers, such as restricted market access; a lack of funding, equipment, and technical know-how; as well as considerable climatic unpredictability. Second, we examine how climate change affects the region's small-scale broiler production and how it aggravates existing problems for farmers, such as the anticipated rise in average temperature and its consequences for sustainable broiler production.

Besides, other goals include assessing the financial barriers to small-scale broiler production in rural West Africa, including the availability of loans, the high cost of feed, vaccinations, and day-old chicks, all of which affect the productivity and profitability of broiler farms. The research also aims to explore the dominance of large businesses in the management of the commercial broiler value chain and the broiler meat sector, which poses challenges for small-scale producers in terms of competitiveness and market access. Additionally, the study intends to look at how local differences in maize prices affect the performance of small-scale broiler farming and how these economic dynamics impact the sector.

This chapter also captures the possible remedies and strategies to increase the productivity and resilience of small-scale broiler farming in rural West Africa. In order to address the challenges that small-scale broiler farmers have identified, it is necessary to explore avenues for expanding market access, providing financial support, providing technical skills and equipment, and promoting sustainable practices. In addition, the chapter reviews the wider consequences of overcoming these obstacles, including potential effects on livelihoods, food security, and regional economic development. In addition to guiding sector policy and practice to help develop a more resilient and sustainable broiler business in the region, the research aims to address these objectives and contribute to the body of information currently available on small-scale broiler production in rural West Africa.

3. Methodology

The literature review on the challenges of small-scale broiler production in rural West Africa has utilized a thorough technique to unearth and address the multifaceted issues facing this industry. To begin, an exhaustive search across renowned databases such as Scopus, Web of Science, PubMed, and Google Scholar has been done. This search has been supplemented by accessing regional databases like those supplied by the FAO and World Bank, as well as country-specific internet repositories. Keywords containing essential features of small-scale broiler production issues in the region drove the search, including terms such as small-scale broiler production, obstacles, rural West Africa, feed cost, disease outbreaks, market access, and infrastructure restrictions [17, 18].

Selected research, publications, and reports have been evaluated for relevancy, focusing primarily on issues important to small-scale broiler production within the rural West African context. Data taken from these sources have subsequently been combined to highlight common issues, including barriers relating to feed accessibility, disease control, market entry, and socioeconomic considerations. Building upon the aggregated findings, practical approaches are provided to solve the highlighted issues. These treatments are designed to be practicable and adaptable within the setting of rural West Africa, given the resource limits and particular circumstances of small-scale broiler farmers in the region [19].

4. Breeding and selection of broiler breeds suitable for rural West Africa

In rural West Africa, breeding and selecting broiler breeds suitable for small-scale production brings with it a host of challenges and difficulties that impact farmers and communities at large. The need to keep local chicken breeds acclimated to tropical climates, which can be endangered by shifting climatic conditions and disease pressures, is highlighted by the reliance on indigenous genetic resources for breeding initiatives in West Africa [20]. In order to fully realize the potential of the poultry industry in West Africa, appropriate breeding programs and strategies are required. This underscores the need to create sustainable breeding techniques that cater to the particular difficulties faced by regional poultry farmers [21, 22].

Native chickens are popular in tropical climates because they require less care, are resistant to common poultry diseases and variations in feed amount and quality, and show how native breeds can withstand environmental stressors (**Table 1**) [23]. For example, in Nigeria, the prevalence of coccidiosis in exotic chickens is higher than in local breeds, indicating the impact of farming practices on disease susceptibility and the requirement for biosecurity measures in poultry production [24]. This situation is a true representation of sub-Saharan. It is important to use breed selection strategies that consider local farming practices and environmental conditions. The introduction of improved breeds in Tanzania is in line with efforts to identify dual-purpose breeds suitable for local free-range scavenging conditions [25, 26].

The prevalence of avian tuberculosis in exotic breed chickens in Ethiopia in contrast to native breeds highlights the significance of disease surveillance and preventive methods in the poultry industry [27]. The distinct qualities and benefits of regional chicken breeds in Asian nations draw attention to the possibility of providing

Country	Types of breed	Purpose
Ghana	Naked neck, Frizzle, Peacomb	Dual purpose
Nigeria	Fulani chicken, Yoruba chicken, Noiler	Dual purpose
Togo	Naked neck, Frizzle	Dual purpose
Benin	Goliath chicken	Dual purpose
Senegal	Naked neck, Frizzle	Dual purpose
Burkina Faso	Konde	Dual purpose
Mali	Wassa-Che, Semba-Che	Dual purpose
Niger	Kolonto	Dual purpose
Cameroon	Dzaye, Dongwe, Zarwa, Poule Brahman	Dual purpose

Source: Developed by the authors.

Table 1.
Some common indigenous chickens in West Africa.

consumers with high-quality, healthful chicken meat, underscoring the importance of native breeds in environmentally friendly poultry production systems. The significance of investigating local genetic resources to improve breed performance and sustainability in tropical regions is highlighted by the genetic fingerprints and mechanisms of tolerance to climatic extremes found in indigenous livestock and poultry.

For high-performance broiler breeds, hybrid lines are suitable for rural West Africa, although their husbandry may be more demanding than for indigenous chickens [28]. Feathers on the head and neck of most broilers are removed in West Africa during the growing period due to hot temperatures, a practice that should be discouraged. This treatment is believed to be marasmus as there will be entanglement, which can lead to unattractive patches on the skin, friction injuries, and infections. On the other hand, helping the birds to cool down is part of the physiological regulation of the metabolism to maintain the normal body temperature of 41°C [29]. This removal of feathers can also cause thermal stress, after which the birds huddle close together to keep warm, disheveled feathers [30], which can result in cooling in the event of rainfall, wicking inflammation in the skin and remigration, which can be recognized by unattractive patches on the skin.

To sum up, a thorough approach that considers genetic variety, disease resistance, and local adaptation is required to address the challenges and concerns that farmers and communities face when creating and choosing broiler breeds appropriate for rural West Africa. Stakeholders may aid small-scale farmers in the area while also enhancing food security and economic growth by utilizing sustainable breeding practices and local genetic resources to increase the resilience, productivity, and sustainability of chicken farming.

4.1 Sources of breeding stock

Indigenous birds were reared mainly for family consumption of meat and eggs, while exotic chickens and hybrids were established for meat, eggs, and income generation. In rural areas, despite the significant increase of these types in the last five to ten years, livestock farming remains mainly for home consumption of eggs and the occasional sale of stocks, before females reach an approximate mass of 1 to 2 kg to the age of 2 to 5 months [31]. Besides, there was a slightly lower proportion of chicken mixes of exotic genotype for the male categories, compared to that of local types, although they were almost equally significant at the regional level. Overall, the results of this study suggest that the breeding and selection of local chickens and the source of breeding material used in West Africa are influenced by both socioeconomic and technical considerations.

Managing and breeding appropriate local chickens is perceived to contribute to promoting the welfare of rural people's livelihoods in developing countries [28]. Informally, in West Africa, farmers usually depend only on the local chicken ecotype for their production but use conventional medication and vaccination [32]. Breeding stock of exotic types was more mixed compared to local ecotypes, which are sourced from poultry product marketers. The exotic breeds mainly imported to the sub-Saharan region include Cobb 500, Sasso, Arbor Acre, and Hubbard (**Figure 2**) [33].

4.2 Genetic improvement and local adaptation

Achieving the best possible results in poultry farming in rural West Africa requires carefully selecting and breeding broiler breeds that are suitable for small-scale

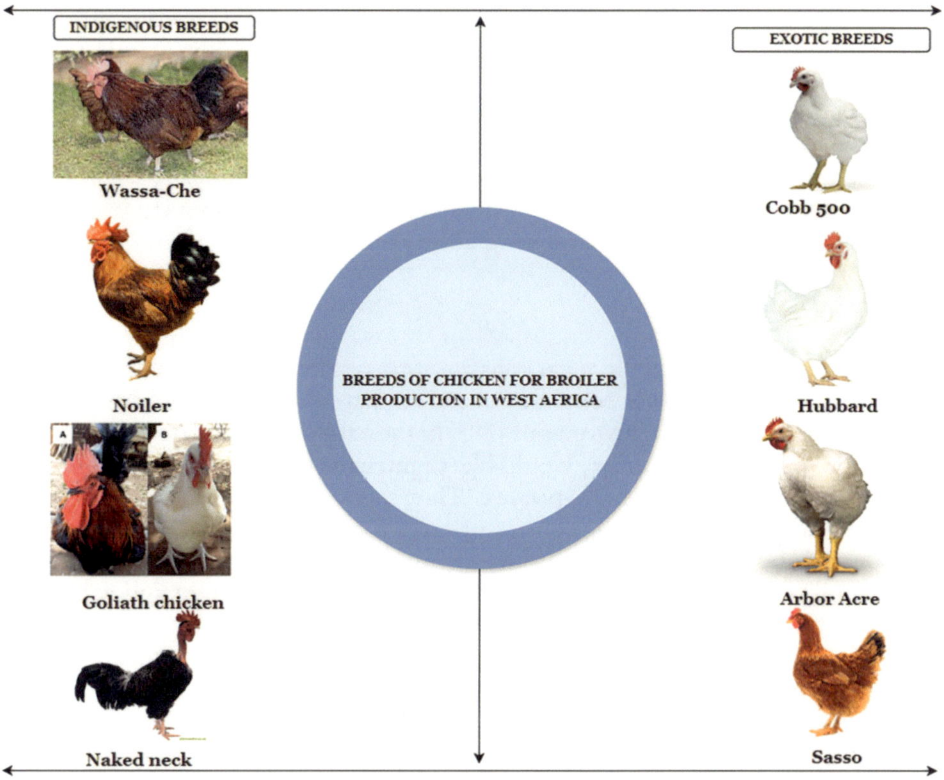

Figure 2.
Some common breeds of broiler chickens reared in West Africa. Source: Developed by the authors.

production. Even if there are successful examples of native chicken breeds that have adapted to tropical environments, it is important to assess their genetic diversity and adaptation. For instance, studies have shown that Naked Neck and frizzle-feathered chickens have adapted well to tropical temperatures exceeding 25°C, highlighting the need to preserve native breeds resilient to environmental conditions [34]. Additionally, studies conducted in Ethiopia have highlighted the importance of a chicken's capacity to adapt to its production environment and the significance of using the region's genetic resources to support sustainable breeding programs [28].

Additionally, the potential for incorporating locally adapted breeds into broiler production systems is shown by the popularity of frizzled feather chickens in Western Africa because of their capacity to adapt to tropical environments [35]. Additionally, studies have demonstrated the adaptability of Vanaraja chickens to local tropical circumstances, indicating their efficacy as home poultry keepers in tropical climates [36]. Stakeholders may identify genes and traits that support native chickens' adaptability to a range of climatic conditions by utilizing the genetic fingerprints of adaptations to tropical stress and domestication [37].

Moreover, the significant genetic variation and population division observed in native chickens in tropical regions offer opportunities for effective breed advancement and preservation strategies [20]. By comprehending the specific signals and genes linked to heat adaptation in native chickens from different tropical climate zones, researchers can enhance the adaptability and yield of native breeds when faced with environmental challenges [37]. The conservation and utilization of these genetic resources are crucial for the sustainable production of poultry, as demonstrated by the raising of native chickens, which are fundamental to smallholder farming systems in tropical regions [37].

Lastly, the resilience, productivity, and sustainability of poultry farming in rural West Africa may be increased by incorporating locally adapted chicken breeds into broiler production systems. Stakeholders may facilitate genetic conservation, improve breed performance, and provide food security for small-scale farmers in tropical areas by utilizing the genetic diversity and adaptability of native breeds.

4.3 Financial constraints

Limited resources due to financial constraints result in less creation of employment opportunities and fewer contributions to impacts on income inequality, poverty reduction, and livelihood improvement [38]. In the case of broiler farming, people belonging to the local community would like to intensively live with each other and always share their problems and happiness. They spend heaps of their time together. Sometimes, they share jointly the consumption of food and beverage, work together, and enjoy their festival parties. In all these cultural practices, they make use of poultry products, and thus, they contribute to nonmonetary employment and economic development.

Small farmers in West Africa face severe financial constraints to diversify their income through agricultural activities. The reasons are primarily limited access to capital, lack of information about modern farming methods, high prices for agricultural inputs, and a lack of financial and non-financial support from governments and development organizations [39]. For poultry farming, the major problems lie in the lack of capital for procuring the inputs, inexperience with the broiler farming system and feed management, financial constraints for proper feeding, limiting target markets with high risk in productivity, and lack of consulting services.

4.4 Limited access to inputs and technology

The use of Information and Communication Technologies (ICTs) for agriculture has brought about so many new solutions for maintaining food security, ensuring sustainable agricultural development, and generating employment in rural areas [40]. However, the fact that the involvement of ICTs in the agriculture sector is fairly low compared to other sectors is an important issue to be addressed. Training activities to be realized to fill this gap must be planned according to a fast, effective, and permanent education model related to diversified criteria of broiler farm management for smallholder farmers [38, 41]. For instance, fliers, E-books, computer programs, and videos can be created and prepared with a fast, effective, and permanent education model and technological packages developed for rearing broilers using rural physiognomies. Additionally, the use of renewable energy (RE) and relevant training activities for diversified criteria of broiler farm management for smallholders in West Africa seems to be especially important to resolve the low level of efficiency problems of broiler production [38, 41]. Due to a lack of capital and limited access to agricultural inputs and technology, small-scale broiler production in rural West Africa is characterized by extensive use of locally available feedstuffs and protein sources [42].

Consequently, production performance, in terms of growth and feed conversion rate, is generally compromised, with production parameters lower than recommended levels. This contributes to inefficient production and increased production costs due to longer fattening periods and higher levels of feed consumed to reach marketable live weight. In particular, the lack of access to appropriate feeding technology for broiler production is a major challenge at the level of small-scale rural farmers in West Africa [8, 43]. On-farm and scientific research demonstrated the disadvantages of using traditional methods in feeding broilers, in terms of both economic (higher feed cost) and ecological (waste disposal, manure management) aspects [44]. Solar and wind energy can alleviate the lack of access to inputs and technology. Both solar and wind power can be used in broiler facilities. Ventilation can be enhanced by integrating solar systems and distributed well through evaporative cooling. Solar wind power is especially useful in addressing the limitations associated with the national grid and power distribution systems. Consequently, remote areas in sub-Saharan Africa can be enabled to integrate solar and wind power in their agricultural facilities and to adapt and implement photovoltaic and wind electrical hybrid power ventilator technology in broiler houses.

4.5 Market access and pricing

One of the major constraints faced by smallholder broiler farmers is dominance and unfair competition from the large commercial poultry producers. They think that it is not beneficial to start such an economic activity, and therefore, they engage in other economic activities that are profitable [45]. Additionally, beyond the unfair competition posed by large poultry producers, their dominance extends to being primary value chain participants and providers of chicks or eggs. They offer loans to these farmers, which often leads to losses as chicks or eggs are sold on their behalf.

The main source of farming products for small-scale broiler farmers in rural West Africa are feed shops, poultry-producing companies, and veterinary shops [46]. However, they face unequal competition from large poultry producers who sell chicks, fully-grown broilers, and eggs at lower prices. These companies dominate the range of production stages, including feed, chickens, vaccinations, and veterinary services, ultimately lowering small-scale poultry farmers' incomes. Their dominance makes it difficult for small-scale farmers to sustain their livelihoods. It was also established that the lack of pre-sold chicken collection centers, processing facilities, and properly functioning and competing export subsidiaries makes small-scale rural broiler farmers unable to link up with both national and international customers [47].

5. Nutrition

5.1 Addressing feed challenges in small-scale broiler production: opportunities and alternatives in rural West Africa

Rural small-scale farmers often lack the resources to buy commercial poultry feeds. They therefore seek alternative diets for raising broilers, such as using by-products from government-owned rice mills available in the villages, soybean-maize diets, feeding the chicks with termites, leaves and gizzards, trapped insects, and entrails and other waste products from meals taken within the households [48]. The availability of feed resources and the expertise in feed formulations rarely satisfy the nutritional recommendations for broilers, hence the need for research to determine alternatives. This will promote the development of appropriate evidence-based replacements and supplements for standard diets. Feeds for broilers in rural West

Africa can be formulated using appropriate proportions of local feed resources available within the regions to guarantee a balanced ratio of the essential amino acids and energy that meet the biological needs of the broiler.

The high cost and unavailability of quality feedstuffs constitute major challenges to small-scale broiler production [42] in rural areas, particularly in West Africa [49]. In the rural areas where broiler production is often constrained by poor feeding and rearing management, fewer finishing houses, inappropriate equipment, and inadequate veterinary services, there is little scientific information documented about the composition of local feedstuffs, their nutritional worth, and the animals' response to them [10]. There is little attention to data on how feedstuffs are commercially formulated by local feed millers.

Broiler chickens require 23% crude protein and an energy content equivalent to 3000 kcal ME/kg to optimize their performances (widely used protein and energy values for poultry feed) (**Table 2**) [50]. The high-performance genetic strains require

	Starter (1–21 days)	**Finisher (22–42 days)**
Ingredients	**Composition (%)**	
Corn	57	60
Wheat bran	7	9
Soybean cake	18	16
Soybean	12.6	10
Oyster shell	2	1.5
Lysine	0.2	0.2
Methionine	0.2	0.3
Dicalcium Phosphate	1.1	1.2
NaCl	0.4	0.1
Vitamin premix*	1.5	1.7
Total	100	100
Nutrient composition		
Metabolizable energy (Kcal/kg)	2903	3059
Crude protein (%)	23.3	20.6
Crude fiber (%)	4.2	3.9
Crude ash (%)	2.8	2.9
Crude fat (%)	5.6	5.0
Calcium (%)	1.1	1.1
Available phosphorus (%)	0.6	0.7
Lysine (%)	1.2	1.3
Methionine (%)	0.5	0.6
Methionine + cystine (%)	0.9	0.8

*Source: Developed by the authors.**Vit. A, 12000 IU, vit. E, dl-α-tocopheryl acetate) 20 mg, menadione 2.3 mg, Vit. D3, 2200 ICU, riboflavin 5.5 mg, calcium pantothenate 12 mg, nicotinic acid 50 mg, Choline 250 mg, vit. B12 10 μg, vit. B6 3 mg, thiamine 3 mg, folic acid 1 mg, d biotin 0.05 mg. Trace mineral, mg/kg of diet): Mn 80 Zn 60, Fe 35, Cu 8, Selenium 0.1 mg.*

Table 2.
A typical example of the composition of a standard diet adopted by small holder broiler farmers.

more energy for growth. Modern broilers are bred to have rapid growth rates, particularly in the first three weeks post-hatch. The modern high-yielding broiler can reach market weight in 42 days and is ready for slaughter [32]. In rural settings of many poor countries like Sierra Leone, the sector is distressingly underdeveloped. The birds are selected according to the resource situation, local conditions, disease problems, and bird adaptability to the local environment and mostly for survival.

5.2 Availability of feed ingredients in rural areas

The majority of poultry farmers in rural areas are more concerned about feed access compared to other factors, as high feed costs affect the profitability of small-scale broiler production [41]. The limited alternative feed resources in rural areas force farmers to continuously purchase feed at prices they sometimes cannot afford. Non-availability of bagged feed also discourages potential smallholder farmers from taking up livestock farming in rural areas. Insufficient availability of feed additives and feed supplements, including suitable annual vaccine programs, has some implications for small-scale broiler production.

It is well known that rural areas in Western Africa are facing food-feeding challenges [38]. For most small-scale rural poultry production systems, access to feeds is the most important bottleneck in the generation of income and is closely linked to the availability and ease of access to feed raw materials. In Ghana and Burkina Faso, for example, farmers face many difficulties with access to feeds, and alternative feed ingredients are rarely available [51]. Constraints include poor transportation infrastructure in some areas, monopolistic pricing policies on the part of the few large

Figure 3.
Some common alternative feed resources for broiler production in the rural settings of West Africa. Source: Developed by the authors.

commercial feed producers, high transportation costs in relation to buying prices, and low purchasing power [51]. Ordinarily, only a few feed ingredients are available to small domestic poultry farmers, and new feed formulations are not brought to the available ingredients. The most frequently available poultry feed ingredient is maize and some premixes of incomplete feed for layers and broilers. Sometimes, small quantities of commercially manufactured concentrate are available, but only in cities or near major roads.

Farmers prefer to use locally available and cheap organic feed sources, which may influence the feed conversion efficiency [5]. However, the nutrient requirements for chicken can essentially be supplemented with leftover food, kernel cakes, spent grains, cereal hulls, and forages generated within the rural communities. The use of such locally available feed resources in small-scale broiler production reduces production costs, promotes a sustainable environment, and supports the local economy (**Figure 3**) [52].

6. Optimizing housing conditions for enhanced small-scale broiler production in rural settings

Though the local chicks expressed a remarkable quality of adaptation to extremely harsh environmental conditions, heat-induced physiological heat stress reduced growth rates and live weight gain and decreased the feed conversion ratio [53, 54]. Housing situations for small-scale broiler production in rural farming are critical for ensuring the most useful broiler health, growth, and productivity. In conventional confined systems, broilers are normally housed in well-constructed systems that provide safety from environmental extremes and predators. These houses are designed to hold foremost temperature, humidity, and ventilation, which are important for preventing warmth pressure and ventilation problems. Proper insulation, adjustable airflow structures, and heating mechanisms are important components that help create strong internal surroundings conducive to boosting feed utilization and growth [55].

In these limited housing structures, area allocation is another critical consideration. Overcrowding can result in elevated competition for assets, higher stress stages, and a higher risk of ailment transmission. Adequate space in keeping with chicken (10 birds per meter square), based totally on recommended stocking densities, guarantees that broilers have sufficient floor space, get free access to feed and water, and exhibit normal behaviors. Litter management inside those houses is also critical because it affects both welfare and the environment. Regular removal of litter and litter management can save the buildup of unsafe pathogens and ammonia gases, thereby enhancing bird welfare and performance [56].

In comparison, free-range structures permit broilers to get free access to the environment wherein they can forage and engage in natural behaviors along with pecking, scratching, and dirt bathing. These systems regularly contain an aggregate of indoor housing for the middle of the night refuge and safety, and precise outside spaces in the day. Free-variety broilers normally show off lower strain levels and improved welfare indicators due to the enriched surroundings and improved opportunities for exercise. However, free-variety production also provides challenges, which include increased vulnerability to predators, exposure to damaging climate situations, and ability to biosecurity dangers. Effective management practices, including predator management measures and the provision of shaded areas, are vital to mitigate those dangers and ensure the well-being of the birds.

The implications of housing situations on small-scale broiler production are great. Confined housing systems, with controlled environmental conditions, regularly result in better feed efficiency and rapid growth rate due to the solid, optimized environment. However, those structures can be more resource-extensive in terms of construction and maintenance expenses. The free-range system, while probably enhancing chicken welfare and farmers' perceptions of product quality, may bring about a slow growth rate and lower feed efficiency due to exposure to an uncontrolled environment [57].

7. Design considerations for rural environments

Existing house structures are used as chicken houses, which protect birds from the movements of human beings and help to avoid theft [32]. The pen layout seems to be adapted to the climate. In tropical and subtropical areas and during drought seasons, a roof system above the poultry pen is important, but floor ventilation is inadequate. There is an associated risk of heat accumulation, especially if floor insulation is inadequate and thermal mass is outside of the poultry pen. Structural modifications are very limited due to the generally small-scale production.

Good wall and roof materials, well-laid floors, raised structures, other measures to lower the indoor temperature, and subterranean construction are not often affordable or feasible. This is also true with modifications to alter the height or slope of the roof or to provide perches [41]. Moreover, the materials applied are quite different and can also reflect an individual or cultural background. The tradition of avoiding painting wood should be respected because there are health, fire, and other risks. Some materials, like a painted wooden wall, should be placed outside the poultry pen and should not be rearranged within the pen. Although most materials do not have a major effect on production, they contribute to the durability of the poultry house.

Water is a crucial resource for poultry production and must be stored safely and kept clean, which is challenged by the low cleanliness of many water sources. The use of water tanks, rainwater harvesting, and filters cannot always be recommended in an animal welfare-friendly manner due to their seasonal distribution and seasonal dry-ups [42]. For this reason and depending on location, a local and independent water system employing deep hand pumps can be used to avoid the risk of dehydration of the chickens in periodical increasing water demand and at water absence due to failure of the municipal water system.

Ventilation, temperature control, and lighting are three key functions for successful chicken rearing in rural West Africa, particularly in countries such as Nigeria and Mali. Ventilation systems help to ensure stable air exchange rates in the broiler house in order to keep the microclimate at an appropriate level for the chickens and to avoid a dangerous buildup of harmful gases. An ideal ventilation system should create a safe climate for the chickens while also reducing the smallholder's expenses for fuel used in cooling the farm, helping to mitigate the effects of rising temperatures [38].

8. Biosecurity measures to prevent disease

Although the use of vaccines is considered the golden rule in disease management because of the high price of vaccines, engaging in biosecurity practices will result in

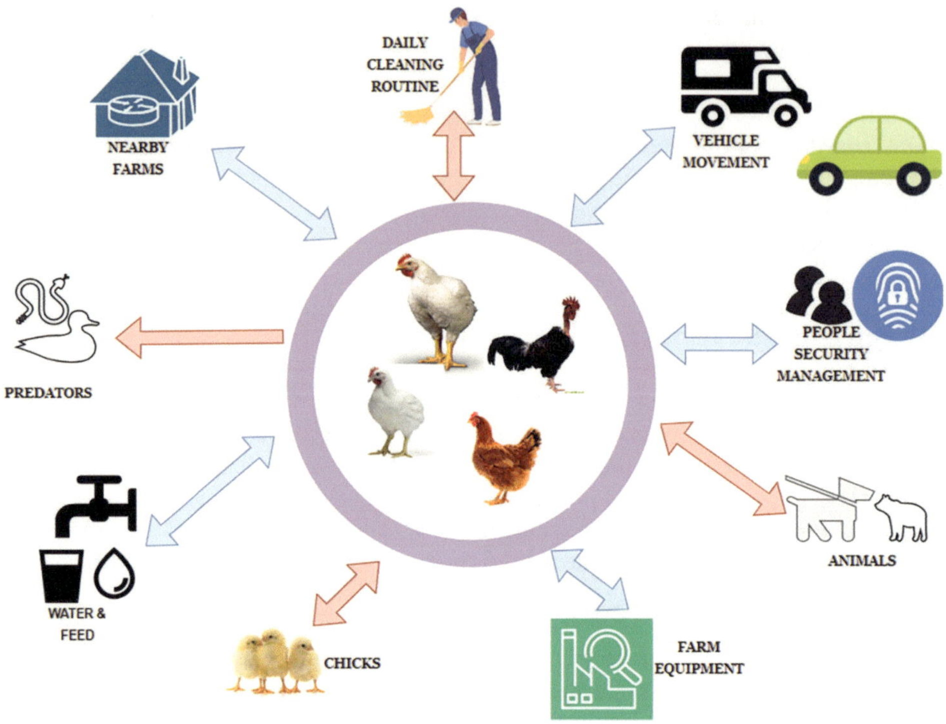

Figure 4.
Biosecurity measures. Source: Developed by the authors.

improved chicken health and production for the rural system. Farmers should ensure that no potential risk factors for disease transmission are present inside the brooder house and apply a single-direction flow of air into and out of the building (e.g., have inlet and outlet points of the brooder opposite one another). By restricting any source of infection, the resultant airflow will not prevent harmful pathogens from entering the building or from being transmitted within the unit [38].

Biosecurity or the implementation of a series of practices is used to prevent the introduction and spread of disease agents into a poultry flock (**Figure 4**) [58]. In Ghana, and generally in West Africa, farmers engage in low-input rural poultry production, based on local chickens and scavenging feeding systems. The role of these birds is very significant in food security and poverty alleviation, but the sector has been compromised by increases in the incidence and spread of diseases and egg and broiler losses [59].

9. Management practices

9.1 Effective management strategies for confined and free-range systems

Using the right management techniques is essential for optimizing broiler production in both limited and free-range systems in rural agricultural settings. Careful nutrition management is required to provide a balanced diet that meets the nutritional needs of broilers at different stages of growth [60].

Water quality and regular feeding schedules are essential for metabolic and hydration processes. Health management practices contribute to the flock's overall well-being by preventing disease outbreaks. Vaccination schedules, routine physical examinations, and biosecurity measures are necessary [61]. Maintaining the right temperature, humidity, and ventilation as well as cleaning up litter depend on environmental management. Crucial elements of free-range systems are pasture quality control, shaded areas, and shelters. Welfare management techniques provide broilers more freedom to move about and exhibit their natural behaviors, which enhances their overall quality of life. Using welfare management techniques improves the general well-being of broilers by giving them adequate space to roam about and exhibit their natural behaviors. Effective management of broilers in rural environments relies on striking a balance between the unique requirements and difficulties of restricted and free-range systems [62].

9.2 Daily care and routines for broilers

In large commercial farms, technical persons are appointed to look after such care and routine work, but it is still a critical issue on small broiler farms, where the responsibility lies with very few people. In these circumstances, it is difficult for the flock manager/caretaker to manage traditional and technical poultry activities simultaneously, which seriously impacts production performance. Here, lack of infrastructure, time, and knowledge on technology adoption hinder better care of a broiler flock. Aggregated guidance material or improved collaborative efforts can better acquaint caregivers with day-to-day practices and the needed technology [63]. Bulky and non-standardized literature about modern broiler management may also complicate better care. Care management with automation and smart technologies is prevalent in developed countries, while such facilities are quite less in poor, underdeveloped, and remote areas of the world. Open-litter housing in commercial chicks is of major importance for developing countries because it involves less capital investment and running costs [64]. High levels of care and smart technologies can further minimize the cost of production. Smart technologies to ascertain water levels, clean and dry excreta, monitoring of birds within the shed, and controlled environmental parameters will undoubtedly ensure better care for chicks.

Modern broiler chickens are highly productive commercial animals with the need for regular management and care. Broiler management includes feed, water and litter management, proper biosecurity and vaccination, and proper house cleaning [65]. Poultry production is much more dependent on proper housekeeping and management than any other animal production. Any loophole in care will directly affect the economics of production. For the successful rearing of broilers, a daily care routine is necessary to maintain the proper environment for the growth of birds. Farm staff are trained to follow proper daily management practices, such as timely removal of dead birds, moist litter, and drinking water availability, and environment control must be monitored regularly. In addition, other infrastructural and equipment-related care, machine and table hygiene, and proper record maintenance must be implemented daily [66].

9.3 Vaccination and veterinary services

In some villages, only some breeders and layers of female broilers were vaccinated against AI; all farm breeds reported being vaccinated at least once in the last 12 months; they are sexually transmitted from laying hens to broilers. Researchers and frontline disease control workers are cookers and can perform vaccine preparation and effective vaccination for their work. However, there is an identified need for better training of front-line workers and veterinarians in implementing effective vaccination campaigns. In summary, the different challenges in the villages reviewed are almost similar in this aspect: improving the human and economic capacity of the veterinary services.

On all sites and in all producer categories, vaccination with Newcastle Disease Virus (NDV) was common, but in addition to the presented statistics, for most of the various villages, there were also fowl pox, Gumboro, and other vaccinations done in single cases (**Table 3**) [51]. There is a special concern for vaccination against the avian influenza disease; since the October 2005 outbreak of avian influenza H5N1 in Nigeria, many small and large holders have been aware of AI and the importance of vaccinating against AI, but because of cost or other reasons, only 54% of broiler keepers have been working against avian influenza [42]. Despite the increased avian influenza vaccine coverage, coordination and oversight of the avian influenza vaccination camps and the quality and reliability of vaccines have been poorly managed [67].

9.4 Record-keeping and performance monitoring

High-feed prices are one of the constraints behind broiler production in the region, making broiler production unattractive to small-scale producers. In West Africa and sub-Saharan Africa in general, smallholder broiler production depends on conventional feed-milling operations, where feed prices are usually higher than the commensurate cost of the feed on the global market. However, expenses that are not recorded are hard to optimize [68].

More than two-thirds of the households do not keep records of daily feed consumption or weekly weight gain (**Tables 4** and **5**) [42]. Feeding generally starts early in the morning (7 am) and ends late in the evening (7 pm) [42]. However, this feeding regime resulted in high variation in feed intake, which can compromise growth performance. Farmers need to understand growth performance and the financial benefits of different feeding regimes so that they can make informed decisions on optimizing feeding to maximize profit. Robust systems therefore need to be developed for collating hybrid growth performance data on feed intake, weight gain, and carcass and meat quality attributes, which should both articulate and quantify the

Age of birds	Disease	Vaccine	Administration route
Day old	Infectious bronchitis, Newcastle Disease	Avinew, Vitapest	Eye drop/Course spray (administered at hatchery)
Day 14	Infectious bursal disease	Gumboro	Drinking water
Day 21	Newcastle Disease	Avinew or Lasota type (if birds will be kept for more than 42 days	Drinking water/fine spray
Day 28	Infectious bursal disease	Gumboro	Drinking water

Source: Developed by the authors.

Table 3.
Typical example of vaccination schedule for broiler rearing in the rural settings.

Date	Feed type	Feed Quantity (kg)	Number of birds	Mortality (number of dead birds)	Observation/ Comments
YY-MM-DD					
YY-MM-DD					
YY-MM-DD					
YY-MM-DD					
YY-MM-DD					
YY-MM-DD					
YY-MM-DD					

Source: Developed by the authors.

Table 4.
Typical example of record sheet for daily records of feeding and weighing of broilers.

Week	Total feed intake (kg)	Average daily feed intake	Start weight (kg)	End Weight (kg)	Weight gain (kg)	Mortality (number of dead birds)	Observation/ Comments
1							
2							
3							
4							
5							
6							

Source: Developed by the authors.

Table 5.
Weekly summary of records of broilers.

impact of these different feeding regimes on commercially relevant exotic strains and explore the wider implications of gene in queering consequences for commercial broiler welfare and production systems [31].

10. Economic and social impacts

10.1 The socioeconomic impact of small-scale broiler production in rural West Africa

Poultry contributes a significant percentage to the agricultural economy of developing countries [42]. In low-income countries, only small-scale rural poultry units are observed, and over large tons of chicken are produced annually. In Africa, there has been a general increase in poultry consumption over the years, as the demand for quality animal protein is on the rise.

A survey conducted among small-scale broiler production farmers in different small communities in the Ashanti region of Kumasi in Ghana revealed that the workers hired were mainly females and juveniles and that the production rate of the broilers was 92.2% [41]. This implies that poultry production is creating ample job opportunities in the rural populace, especially for females. Besides, it is also improving the standard of living in rural environments [31]. In the absence of formal employment, and with occupation options limited mostly to farming and small-scale trading, poultry farming becomes a critical aspect of rural livelihoods. The poultry farming sector is led by women and employs mainly women and young people, with the majority of households keeping broilers and layers.

Small-scale broiler production leads to efficient and stable income generation for rural people in West Africa [45]. High input prices, low access to inputs, unpredictable markets, and diseases are among the constraints faced by producers. Optimization of market access will lead to sustainable production and profitability [39]. For example, it was reported that the poultry farmers of Asante Akyem North District, Ghana, were challenged by a lack of access to capital [39]. Additionally, the net benefits of broiler production can substantially decrease even with just a small increase in feed prices. Broiler production produces net benefits, but these statistics need to improve significantly for it to be ranked as the biggest source of employment in the subregion.

10.2 Policy support and existing government initiatives

Broiler farmers in rural West Africa often lack access to quality feed, veterinary services, and extension services [42]. Initiatives for training and capacity building are two more significant areas of government support. To address the importance of technical knowledge and skills in broiler production, governments must establish extension services and training programs. These courses provide farmers with the best practices for managing feed, preventing illness, housing, and overall farm management. By assisting farmers in developing their technical skills, these programs boost production efficiency and improve the health of broiler chicks.

Additionally, improving access to financial resources is a significant issue for politicians. Governments must be moved to improve the financial and investment opportunities available to small-scale broiler farmers. For example, in Nigeria, the government has taken steps to encourage various stages of the production and marketing process [69]. The government has made significant investments in feed mills and hatcheries through contracts with major companies. Commercial hatcheries have collaborated with the government of Kano State to employ and train young people in the state and beyond. The government of Nigeria also guarantees for broiler farmers by providing support and credit finance through the Bank of Industry.

Governments must also recognize the need to promote gender parity and female involvement in broiler farming. In an effort to provide women with more access to resources, opportunities for education, and employment, policies and programs have been created. Women's empowerment programs may aim to create an environment where women may thrive in the broiler production industry by addressing difficulties that are exclusive to one gender. By encouraging gender-inclusive policies, governments may aid rural communities in achieving greater socioeconomic growth and reducing poverty [70].

Encouraging small-scale broiler farming in West Africa also requires regional cooperation and alliances with foreign agencies. To carry out development programs and exchange best practices, governments collaborate with groups such as the World Bank, the Food and Agriculture Organization (FAO), and other non-governmental organizations. These partnerships tackle comparable problems and encourage sustainable agricultural growth by pooling resources, funding, and technical expertise.

10.3 Training and capacity building

This context is because those services are often more concerned with cropping activities, which have specific seasonal schedules, and they are only involved in the duplication of poultry production technical sheets and advice on typical issues such as the use of local veterinary products against common diseases [71]. These farmers' pity also goes against the laws maintaining the chicken in broad daylight, which is part of the customs of some local communities, which somehow prevents measures to control the predators. As a priority, the project cycle should be properly designed and implemented to correlate with the sociology of the target population. More emphasis should be put on practical formation in addition to theoretical training. This project should involve sustainable funding for follow-up on populations and the results of the different batches produced. To this end, the continuation of such activity and the setting up of a demonstration site with the participation of the interns could be very relevant, if the alliances are properly built.

Extension services and training are crucial for fostering innovation adoption, ensuring success and profitability in broiler production, and ultimately sustainably adding value to chicken production systems in rural West Africa [42]. Training principally focuses on breed, feeding programs (ration calculation and feed formulation), housing construction, breed management, animal health practices, and biosecurity measures, among others. However, only 35% of farmers have acquired knowledge of various aspects of broiler management from formal education or previous training. This can be explained by the lack of support from the state, the limited availability of extension services, the proximity of demonstration sites but with very little relevance, and an elitist approach to the extension that favors large-scale producers. This can be explained by the fact that in agropastoral activities, breed management, feeding, and disease control are not sufficiently integrated into extension programs led by local agricultural services [72].

11. Conclusions

The sustainability of small-scale broiler farming in rural West Africa is in danger from a number of factors. A high degree of climate unpredictability, limited market access, inadequate financial support, a lack of technical competence, and capital shortages are some of these limitations. The output and profitability of small-scale broiler farms are also impacted by economic and environmental factors including disease outbreaks, high feed prices, and vaccine expenses. These problems are made worse by climate change, which highlights the necessity of improving resilience and adaptation. The region's small-scale broiler production must benefit from increased market accessibility, sustainable practices, and financial and technical support if it is to increase productivity and long-term sustainability.

Prioritizing several important areas is necessary to address these problems. Prioritizing the selection and breeding of suitable broiler breeds is the first step. Controlling financial constraints and limiting access to resources and technology are also essential. Although native breeds are more adapted to their surroundings, they frequently develop more slowly and with worse feed efficiency than exotic species. However, maintaining genetic diversity and adapting to regional conditions are essential for sustainable chicken farming. Farmers are unable to invest in necessary materials and technology due to financial constraints, which leads to the use of inefficient farming methods. By expanding access to renewable energy sources and bolstering market conditions, small-scale broiler farms may significantly increase their productivity and profitability.

The production of broilers on a small scale in Western Africa is significantly impacted by gender, especially the participation of women. Numerous aspects of chicken farming, including breeding, feeding, marketing, and financial management, are actively participated in by women. Enhancing women's involvement through targeted policies and initiatives can improve the sector's overall sustainability and productivity. Government programs already in place and policy support are essential for addressing the problems that small-scale broiler farmers face. The broiler industry in rural West Africa will expand and thrive if these policies are strengthened and effectively implemented.

Acknowledgements

The authors would like to express their gratitude to the Regional Excellence Centre on Avian Sciences (CERSA), University of Lome, Togo.

Conflict of interest

The authors declare no conflict of interest.

Author details

Benjamin Adjei-Mensah[1*], Cocou Claude Kpomasse[1], Essodina Talaki[1,2] and Kokou Tona[1]

1 University of Lome, Regional Centre of Excellence for Poultry Science, Lome, Togo

2 University of Lome, School of Agronomy, Lome, Togo

*Address all correspondence to: kamben86@gmail.com

References

[1] Wong JT, de Bruyn J, Bagnol B, Grieve H, Li M, Pym R, et al. Small-scale poultry and food security in resource-poor settings: A review. Global Food Security. 2017;**15**:43-52. DOI: 10.1016/j.gfs.2017.04.003

[2] Erdaw MM. Contribution, prospects and trends of livestock production in sub-Saharan Africa: A review. International Journal of Agricultural Sustainability. 2023;**21**(1):2247776. DOI: 10.1080/14735903.2023.2247776

[3] Carron M, Alarcon P, Karani M, Muinde P, Akoko J, Onono J, et al. The broiler meat system in Nairobi, Kenya: Using a value chain framework to understand animal and product flows, governance and sanitary risks. Preventive Veterinary Medicine. 2017;**147**:90-99. DOI: 10.1016/j.prevetmed.2017.08.013

[4] Okwara MO, Umebali EE, Agu-Aguiyi FN, Ukoha II, Anyanwu UG. Resource-use efficiency in small scale broiler production in Owerri municipal Council of Imo State, Nigeria. Asian Journal of Agricultural Extension Economics and Sociology. 2019;**36**(2):1-7. DOI: 10.9734/ajaees/2019/v36i230238

[5] Kpomasse CC, Oke OE, Houndonougbo FM, Tona K. Broiler production challenges in the tropics: A review. Veterinary Medicine and Science. 2021;7(3):831-842. DOI: 10.1002/vms3.435

[6] Dhillon R, Moncur Q. Small-scale farming: A review of challenges and potential opportunities offered by technological advancements. Sustainability. 2023;**15**(21):15478. DOI: 10.3390/su152115478

[7] Röling NG, Hounkonnou D, Offei SK, Tossou R, Van Huis A. Linking science and farmers' innovative capacity: Diagnostic studies from Ghana and Benin. NJAS-Wageningen Journal of Life Sciences. 2004;**52**(3-4):211-235. DOI: 10.1016/s1573-5214(04)80015-0

[8] Chiekezie NR, Nwankwo EC, Ozor MU. Analysis of small scale broiler poultry production in south East Nigeria, West Africa. International Journal of Animal and Livestock Production Research. 2022;**6**(1):1-6. DOI: 10.37745/ijahlpr.15/vol6n1116

[9] Mbuli CS, Fonjong LN, Fletcher AJ. Climate change and small farmers' vulnerability to food insecurity in Cameroon. Sustainability. 2021;**13**(3):1523. DOI: 10.3390/su13031523

[10] Kpomasse CC, Kouame YA, N'nanle O, Houndonougbo FM, Tona K, Oke OE. The productivity and resilience of the indigenous chickens in the tropical environments: Improvement and future perspectives. Journal of Applied Animal Research. 2023;**51**(1):456-469. DOI: 10.1080/09712119.2023.2228374

[11] Sultan B, Gaetani M. Agriculture in West Africa in the twenty-first century: Climate change and impacts scenarios, and potential for adaptation. Frontiers in Plant Science. 2016;**7**:1262. DOI: 10.3389/fpls.2016.01262

[12] Roudier P, Sultan B, Quirion P, Berg A. The impact of future climate change on west African crop yields: What does the recent literature say? Global Environmental Change. 2011;**21**(3):1073-1083. DOI: 10.1016/j.gloenvcha.2011.04.007

[13] Souillard R, Répérant JM, Experton C, Huneau-Salaun A, Coton J,

Balaine L, et al. Husbandry practices, health, and welfare status of organic broilers in France. Animals. 2019;**9**(3):97. DOI: 10.3390/ani9030097

[14] Lykke AM, Gregersen SB, Padonou EA, Bassolé IH, Dalsgaard TK. Potential of unconventional seed oils and fats from west African trees: A review of fatty acid composition and perspectives. Lipids. 2021;**56**(4):357-390. DOI: 10.1002/lipd.12305

[15] Adeyonu AG, Odozi JC. What are the drivers of profitability of broiler farms in the north-central and south-west geo-political zones of Nigeria? SAGE Open. 2022;**12**(1):21582440211071076. DOI: 10.1177/21582440211071076

[16] Gill T, Urban E, Ader D, Nisengwe R, Garner E. Piloting a commercial broiler production model for smallholders in Rwanda. African Journal of Food, Agriculture, Nutrition and Development. 2020;**20**(2):15660-15676. DOI: 10.18697/ajfand.90.18925

[17] Cronin P, Ryan F, Coughlan M. Undertaking a literature review: A step-by-step approach. British Journal of Nursing. 2008;**17**(1):38-43. DOI: 10.12968/bjon.2008.17.1.28059

[18] Ely C, Scott I. Essential Study Skills for Nursing. In: Maggie Nicol. Serie Editor. 1st ed. Elsevier Health Sciences, Edinburgh; 2007. p. 228

[19] Sukhera J. Narrative reviews: Flexible, rigorous, and practical. Journal of Graduate Medical Education. 2022;**14**(4):414-417. DOI: 10.4300/jgme-d-22-00480.1

[20] Otecko NO, Ogali I, Mauki DH, Ogada S, Moraa GK, Lichoti J, et al. Phenotypic and morphometric differentiation of indigenous chickens from Kenya and other tropical countries augments perspectives for genetic resource improvement and conservation. Poultry Science. 2019;**98**(7):2747-2755. DOI: 10.3382/ps/pez097

[21] Brahi OH, Xiang H, Chen X, Farougou S, Zhao X. Mitogenome revealed multiple post domestication genetic mixtures of west African sheep. Journal of Animal Breeding and Genetics. 2015;**132**(5):399-405. DOI: 10.1111/jbg.12144

[22] Gebremariam B, Girma F. Review work on body weight and egg production performance of chickens in Ethiopia. JBAH. 2019;**9**(8):16-18. DOI: 10.7176/jbah/9-8-03

[23] Lawal JR, Gulani IA, Ali AM, Bello AM, Abadam FA, Mustapha M, et al. Dry season prevalence of avian Coccidia infection in domesticated chickens (Gallus domesticus) in jere council, Borno state, Nigeria. Journal of Animal Science and Veterinary Medicine. 2016;**9**(6):653-659. DOI: 10.31248/jasvm2016.018

[24] Alemu SW, Hanotte O, Kebede FG, Esatu W, Abegaz S, Bruno JE, et al. Evaluation of live-body weight and the number of eggs produced for introduced and local chickens in Ethiopia. Acta Agriculturae Scandinavica, Section A—Animal Science. 2021;**70**(2):71-77. DOI: 10.1080/09064702.2021.1891278

[25] Alabi O, Shoyombo A, Jegede S, Oluba O, Akpor O. Rural production of tropically adapted breeds of chickens in rural areas of Kwara state, Nigeria. Potravinarstvo Slovak Journal of Food Sciences. 2019;**13**(1):1013-1017. DOI: 10.5219/1199

[26] Mpenda F, Lyantagaye S, Buza J. Immune response following Newcastle disease immunization and growth performance of kuroiler, broiler and

local Tanzanian chickens. International Journal of Livestock Production. 2020;**11**(1):1-7. DOI: 10.5897/ijlp2019.0628

[27] Debelu T, Abunna F, Mamo Kassa G, Ameni G. Epidemiology of avian tuberculosis in selected districts of Oromia region, Ethiopia. Veterinary Medicine International. 2022;**2022**(1):6933701. DOI: 10.1155/2022/6933701

[28] Dana N, Van der Waaij LH, Dessie T, van Arendonk JA. Production objectives and trait preferences of village poultry producers of Ethiopia: Implications for designing breeding schemes utilizing indigenous chicken genetic resources. Tropical Animal Health and Production. 2010;**42**:1519-1529. DOI: 10.1007/s11250-010-9602-6

[29] Fathi M, Martín Á, McClements DJ. Nanoencapsulation of food ingredients using carbohydrate based delivery systems. Trends in Food Science & Technology. 2014;**39**(1):18-39. DOI: 10.1016/j.tifs.2014.06.007

[30] Mota-Rojas D, Titto CG, Orihuela A, Martínez-Burnes J, Gómez-Prado J, Torres-Bernal F, et al. Physiological and behavioral mechanisms of thermoregulation in mammals. Animals. 2021;**11**(6):1733. DOI: 10.3390/ani11061733

[31] Ozenturk U, Chen Z, Jamone L, Versace E. Robotics for poultry farming: Challenges and opportunities. arXiv e-prints. 2023;**226**:109411. arXiv-2311. DOI: 10.48550/arXiv.2311.05069

[32] Alarcon P, Fèvre EM, Muinde P, Murungi MK, Kiambi S, Akoko J, et al. Urban livestock keeping in the city of Nairobi: Diversity of production systems, supply chains, and their disease management and risks. Frontiers in Veterinary Science. 2017;**4**:171. DOI: 10.3389/fvets.2017.00171

[33] Bolatito OA, Adebambo AO, Ilori BM, Folorunso O, Ajayi DA, Hasan T, et al. Utilisation and registration assessment status of chicken genetic resources in Nigeria. Asian Journal of Research in Animal and Veterinary Sciences. 2022;**9**(3):27-35

[34] Manyelo TG, Selaledi L, Hassan ZM, Mabelebele M. Local chicken breeds of Africa: Their description, uses and conservation methods. Animals. 2020;**10**(12):2257. DOI: 10.3390/ani10122257

[35] Ballantyne M, Woodcock M, Doddamani D, Hu T, Taylor L, Hawken RJ, et al. Direct allele introgression into pure chicken breeds using sire dam surrogate (SDS) mating. Nature Communications. 2021;**12**(1):659. DOI: 10.1038/s41467-020-20812-x

[36] Panigrahy KK, Behera K, Mohapatra LM, Acharya AP, Sethy K, Panda S, et al. Sex-related differences in hemato-biochemical indices of adult Vanaraja chickens during summer and winter seasons. Veterinary World. 2017;**10**(2):176. DOI: 10.14202/vetworld.2017.176-180

[37] Xu NY, Liu ZY, Yang QM, Bian PP, Li M, Zhao X. Genomic analyses for selective signatures and genes involved in hot adaptation among indigenous chickens from different tropical climate regions. Frontiers in Genetics. 2022;**13**:906447. DOI: 10.3389/fgene.2022.906447

[38] Tria D, Alghorbany A, Bin Muhamad AI, Alam MM. Government policies, financial scopes and technological usages for agricultural development and post-harvest loss

reduction in Algeria. International Journal of Postharvest Technology and Innovation. 2021. 2020;**7**(4):335-352

[39] Teye ES, Quarshie PT. Impact of Agriculture Finance in Modern Technologies Adoption for Enhanced Productivity and Rural Household Economic Wellbeing in Ghana: A Case Study of Rice Farmers in Shai-Osudoku District. Charlottesville, Virginia, USA: Center for Open Science (COS); 2021. pp. 1-15. DOI: 10.31235/osf.io/9ue2k

[40] El Bilali H, Allahyari MS. Transition towards sustainability in agriculture and food systems: Role of information and communication technologies. Information Processing in Agriculture. 2018;**5**(4):456-464. DOI: 10.1016/j.inpa.2018.06.006

[41] Mensah C, Enahoro D. Modeling poultry and maize sector interactions in southern Africa under a changing climate. SocArXiv ehd3j, Center for Open Science. 2022;**7**:10. DOI: 10.31219/osf.io/ehd3j

[42] Mbuza F, Manishimwe R, Mahoro J, Simbankabo T, Nishimwe K. Characterization of broiler poultry production system in Rwanda. Tropical Animal Health and Production. 2017;**49**(1):71-77. DOI: 10.1007/s11250-016-1160-0

[43] Etuah S, Ohene-Yankyera K, Liu Z, Mensah JO, Lan J. Determinants of cost inefficiency in poultry production: Evidence from small-scale broiler farms in the Ashanti region of Ghana. Tropical Animal Health and Production. 2020;**52**:1149-1159. DOI: 10.1007/s11250-019-02115-6

[44] Boumans IJ, Schop M, Bracke MB, de Boer IJ, Gerrits WJ, Bokkers EA. Feeding food losses and waste to pigs and poultry: Implications for feed quality and production. Journal of Cleaner Production. 2022;**378**:134623. DOI: 10.1016/j.jclepro.2022.134623

[45] Kruseman G, Dermawan A, Diagne M, Enahoro D, Frija A, Gatto M, et al. Foresight for income and employment: What can we learn for agricultural research for development. SocArXiv Papers. 2021:12-30. No. 783rw. DOI: 10.31219/osf.io/783rw

[46] Mankhomwa J, Tolhurst R, M'biya E, Chikowe I, Banda P, Mussa J, et al. A qualitative study of antibiotic use practices in intensive small-scale farming in urban and peri-urban Blantyre, Malawi: Implications for antimicrobial resistance. Frontiers in Veterinary Science. 2022;**9**:876513. DOI: 10.3389/fvets.2022.876513

[47] Wilson WC, Slingerland M, Oosting S, Baijukya FP, Smits AJ, Giller KE. The diversity of smallholder chicken farming in the southern highlands of Tanzania reveals a range of underlying production constraints. Poultry Science. 2022;**101**(10):102062. DOI: 10.1016/j.psj.2022.102062

[48] Karunatilake KE. Prospects of Feed Crops in Sri Lanka: The Role of CGPRT Crops. Bogor, Indonesia: CGPRT Centre; 2003

[49] Amole T, Augustine A, Balehegn M, Adesogoan AT. Livestock feed resources in the west African Sahel. Agronomy Journal. 2022;**114**(1):26-45. DOI: 10.1002/agj2.20955

[50] Arabi SA. The effects of different protein and energy levels on broilers performance under hot climatic conditions. Journal of Innovative Agriculture and Biological Research. 2015;**3**(1):19-28

[51] Enahoro D, Sircely J, Boone RB, Oloo S, Komarek AM, Bahta S, et al. Feed

biomass production may not be sufficient to support emerging livestock demand: Model projections to 2050 in southern Africa. SocArXiv. 2021. No. qmdrt. DOI: 10.31219/osf.io/qmdrt

[52] Chisoro P, Jaja IF, Assan N. Incorporation of local novel feed resources in livestock feed for sustainable food security and circular economy in Africa. Frontiers in Sustainability. 2023;**4**:1251179. DOI: 10.3389/frsus.2023.1251179

[53] Wasti S, Sah N, Mishra B. Impact of heat stress on poultry health and performances, and potential mitigation strategies. Animals. 2020;**10**(8):1266. DOI: 10.3390/ani10081266

[54] Biswal J, Vijayalakshmy K, Bhattacharya TK, Rahman H. Impact of heat stress on poultry production. World's Poultry Science Journal. 2022;**78**(1):179-196.DOI:10.1080/00439339.2022.2003168

[55] Jia LR, Han J, Chen X, Li QY, Lee CC, Fung YH. Interaction between thermal comfort, indoor air quality and ventilation energy consumption of educational buildings: A comprehensive review. Buildings. 2021;**11**(12):591. DOI: 10.3390/buildings11120591

[56] Sugiharto S. Dietary strategies to alleviate high-stocking-density-induced stress in broiler chickens–a comprehensive review. Archives Animal Breeding. 2022;**65**(1):21-36. DOI: 10.5194/aab-65-21-2022

[57] Zhao ZG, Li JH, Li X, Bao J. Effects of housing systems on behaviour, performance and welfare of fast-growing broilers. Asian-Australasian Journal of Animal Sciences. 2014;**27**(1):140. DOI: 10.5713/ajas.2013.13167

[58] Otte J, Rushton J, Rukambile E, Alders RG. Biosecurity in village and other free-range poultry—Trying to square the circle? Frontiers in Veterinary Science. 2021;**8**:678419. DOI: 10.3389/fvets.2021.678419

[59] Mujyambere V, Adomako K, Olympio SO, Ntawubizi M, Nyinawamwiza L, Mahoro J, et al. Local chickens in east African region: Their production and potential. Poultry Science. 2022;**101**(1):101547. DOI: 10.1016/j.psj.2021.101547

[60] Moss AF, Chrystal PV, Cadogan DJ, Wilkinson SJ, Crowley TM, Choct M. Precision feeding and precision nutrition: A paradigm shift in broiler feed formulation? Animal Bioscience. 2021;**34**(3):354. DOI: 10.5713/ab.21.0034

[61] Dhaka P, Chantziaras I, Vijay D, Bedi JS, Makovska I, Biebaut E, et al. Can improved farm biosecurity reduce the need for antimicrobials in food animals? A scoping review. Antibiotics. 2023;**12**(5):893. DOI: 10.3390/antibiotics12050893

[62] Dawkins MS, Cook PA, Whittingham MJ, Mansell KA, Harper AE. What makes free-range broiler chickens range? In situ measurement of habitat preference. Animal Behaviour. 2003;**66**(1):151-160. DOI: 10.1006/anbe.2003.2172

[63] Lu G, Li S, Mai G, Sun J, Zhu D, Chai L, et al. Agi for agriculture. arXiv preprint arXiv:2304.06136. 2023. DOI: 10.48550/arXiv.2304.06136

[64] Abukhattab S, Kull M, Abu-Rmeileh NM, Cissé G, Crump L, Hattendorf J, et al. Towards a one health food safety strategy for Palestine: A mixed-method study. Antibiotics. 2022;**11**(10):1359. DOI: 10.3390/antibiotics11101359

[65] Ohashi T, Saijo M, Suzuki K, Arafuka S. Deciphering the drivers

of smart livestock technology adoption in Japan: A scoping review, expert interviews, and grounded theory approach. arXiv preprint arXiv:2307.03338. 2023. DOI: 10.48550/arXiv.2307.03338

[66] Oloyo A, Ojerinde A. Poultry Housing and Management. London, UK, London, UK: IntechOpen; 2019. DOI: 10.5772/intechopen.8381

[67] Yang Y, Rajgopal J. Outreach strategies for vaccine distribution: A multi-period stochastic modeling approach. In: Operations Research Forum. Vol. 2, No. 2. Cham: Springer International Publishing; 2021. p. 24

[68] Laitinen A, Lindholm O, Hasan A, Reda F, Hedman Å. A techno-economic analysis of an optimal self-sufficient district. Energy Conversion and Management. 2021;**236**:114041. DOI: 10.1016/j.enconman.2021.114041

[69] Kassali R, Bakker-Edoh D, Kanake AM. Fundamentals in Effective Marketing. United States: Cari Journals USA LLC; 2022

[70] Heckert J, Martinez EM, Sanou A, Pedehombga A, Ganaba R, Gelli A. Can a gender-sensitive integrated poultry value chain and nutrition intervention increase women's empowerment among the rural poor in Burkina Faso? Journal of Rural Studies. 2023;**100**:103026. DOI: 10.1016/j.jrurstud.2023.103026

[71] Bettridge JM, Psifidi A, Terfa ZG, Desta TT, Lozano-Jaramillo M, Dessie T, et al. The role of local adaptation in sustainable production of village chickens. Nature Sustainability. 2018;**1**(10):574-582. DOI: 10.1038/s41893-018-0150-9

[72] Ali MS, Hossain MM. Factors influencing the performance of farmers in broiler production of Faridpur District in Bangladesh. World's Poultry Science Journal. 2010;**66**(1):123-131. DOI: 10.1017/s0043933910000127

Section 4

Reproduction

Chapter 5

The Use of Plants Extracts during Cryopreservation of Livestock Epididymal Semen

Masindi Lottus Mphaphathi, Maleke Dimpho Sebopela, Mahlatsana Ramaesela Ledwaba, Mamonene Angelinah Thema, Rolly Maapola, Nkhumeleni Cathbert Negota, Thabang Luther Mashilo and Thivhilaheli Richard Netshirovha

Abstract

The plants' extracts have recently emerged as a cheap and natural source of additives to preserve and enhance sperm functions during the cryopreservation process. Extracts from plants are reported to be the main sources of natural antioxidants with lower cytotoxicity as compared to therapeutic antioxidants. The numerous and valuable plant properties, such as antioxidants, antimicrobial, antiviral, anti-inflammatory, and antiparasitic, have been linked to plant extracts. Numerous plant extracts from different species have been used as preservatives for livestock semen cryopreservation. During these processes, the extracts are added to the semen extender in different doses to keep the sperm motile and viable throughout the cryopreservation, including during pre-freezing, equilibration, and post-thawing. The sperm quality parameters can be improved by adding these natural compounds that exert protection against bacteria and reactive oxygen species scavengers in semen extender. Although the application of plant extracts as preservatives for semen storage is still emerging, their potential is promising. They represent natural and economically viable alternatives to the conventional additives currently employed for semen preservation. This review focuses on the aspects related to plant extracts and their role during cryopreservation of semen from domesticated livestock species.

Keywords: plant extracts, cryoprotectants, testicles, cryopreservation, semen

1. Introduction

Normally, semen is collected from live male animals with the use of an artificial vagina or electro ejaculator method. Both methods have advantages and disadvantages during collection of semen from a live male animal. However, once the semen donor (male) died, both methods cannot be used to retrieve the semen. The only option left is the retrieval of epididymal semen and leads to the increase in the use of assisted reproductive technologies (ARTs), such as cryopreservation, *in vitro* fertilization, and artificial insemination [1]. However, there are difficulties in preserving the genetic material of domesticated animals through the use of ARTs. According to studies conducted previously, cryopreservation of semen causes sperm membrane oxidative, biochemical, and physical changes that reduce sperm viability and fertilizing ability [2]. Sperm motility, plasma membrane functioning, and acrosome integrity are all primarily affected by these cryoinjuries [3–5]. Therefore, in order to successfully preserve the genetic material of valuable animals, it is imperative that freezing procedures be improved [6].

Although egg yolk has been utilized for decades to preserve semen, there are biosecurity concerns when using it as an extender for cryopreservation [7, 8]. Plant extracts have emerged as a cheap and natural source of supplements used in cryopreservation procedures to protect and enhance sperm functionality. Most of the plant species have antioxidants, which potentially lessen the negative effects of oxidative stress on sperm function since they are reactive oxygen species (ROS) scavengers [9]. Furthermore, a number of antioxidant enzymes may become more active, and these organic substances may possess antibacterial qualities. Bacterial contamination of the semen before cryopreservation is another factor that lowers sperm motility during *in vitro* liquid semen storage [10, 11]. Consequently, during the cryopreservation of semen, the plant extract serves as a source of antioxidants and antimicrobials.

Plant extracts have been utilized by humans for dietary, cosmetic, medical, and pharmaceutical purposes throughout history. Alkaloids, phenolic chemicals, and compounds containing sulfur are examples of terpenoids, which are the primary source of the plant's pharmacological effect [12]. The protective mechanism of these chemical compounds is well known for shielding plants from possible infections. Phytochemicals also have many other qualities, including anti-inflammatory, analgesic, antiviral, antiparasitic, anticancer, and antimutagenic qualities [13]. Antioxidant-rich plant consumption or extracts from these plants have improved animal and human reproductive performance indices and shown some encouraging results for patients with infertility issues [14].

There are numerous elements that can be categorized as endogenous or environmental that affect the sperm quality. One of these is semen-borne bacterial contamination, which comes from both sources. According to Cojkic et al. [11], the existence of bacteria in semen samples has a detrimental effect on sperm quality in a number of ways, including reducing sperm motility and viability, triggering an early acrosome reaction, and causing sperm aggregation. However, these effects are linked to certain bacteria that are present in semen samples. Even though severe hygiene precautions are used during semen collection, these bacteria in semen samples are inevitable because the procedure is not sterile. Once more, the standard protocol for semen extender stipulates that commercial semen dosages for artificial insemination must contain antibiotics in order to suppress microbial growth. Antibiotics can not only reduce bacterial growth, but they also negatively impact sperm viability [15]. As a result, it is advised to use a variety of antibacterial agents in small amounts.

Several studies have assessed how natural antioxidants affect sperm during cryopreservation of semen. According to Santos-Sánchez et al. [16], antioxidants are thought to have the ability to neutralize the free radicals in lipid chains by providing a hydrogen atom, typically from a phenolic hydroxyl group. This transforms phenolic groups into stable free radicals that do not start or spread additional lipid oxidation. Malo et al. [17] demonstrated that adding rosemary aqueous extract to a semen extender improved post-thawed ram sperm characteristics without impairing the motility of boar and ram sperm. Due to their high polyphenol content, pomegranates have been utilized for their antibacterial and antioxidant qualities [18]. When added to semen extender, pomegranates have been shown to improve sperm quality [11]. It has been observed that adding ginger to semen extenders improved the frozen-thawed motility and fertility of ram epididymal semen [19].

Curcumin was previously tested as a supplement when added to semen prior to cryopreservation, based on its potent protective and antioxidative properties. It showed notable dose-dependent protective and antioxidative properties in those investigations using sperm samples from several animal species [20–22]. The studies on *Arctuim lappa* root [23], *Alnusincana* bark, and *Albiziaharveyi* leaf [24] in bull and ram, and Moringa oleifera leaf [25, 26] in buffalo and ram demonstrated that adding this extract to semen improved all post-thaw semen characteristics and biochemical levels. These results support the inclusion of these additives in a semen extender for cryopreservation process. Therefore, the primary focus of this review is on the characteristics of plant extracts and their role in the cryopreservation of epididymal semen from domesticated livestock species.

Higher production of ROS during livestock semen preservation led to oxidative stress, which affects sperm fertilizing ability [27]. One of the main contributory factors to low sperm quality is due to oxidative stress, which develops and causes oxidative damage to sperm glycoproteins and nucleic acids [28]. Livestock sperm is susceptible to this damage because of the high concentration of membrane [29]. All these factors have deleterious effects on sperm motility, viability, and morphology. Focusing on the challenges of disease transmission when using various animal protein sources in semen cryopreservation protocols [30], plant extracts have gained interest for use as cryopreservatives of livestock epididymal sperm [31].

The phenolic compounds from *Moringa oleifera* leaf extract (MOLE) are critical because of their potency to serve as antioxidants [10]. As such, the natural antioxidants of MOLE contribute to the requirements of sperm survival [32]. The MOLE extract had effective scavenging of free radicals when used in *in vitro* [33]. These flavonoids are mostly distributed in leaves [34] and are classified according to Flavonols and flavanols which has higher free radical-scavenging ability as compared to the synthetic antioxidant [32].

In ages, *Aloe vera* plant has been used for its medicinal purpose , and it has recently been employed as cryoprotectant alternative to egg yolk [10] during livestock sperm cryopreservation [35]. This is because egg yolk has xenobiotic contamination risks and is potent to introduction of exotic diseases through semen distribution around the globe. The endotoxins produced compromise the livestock sperm to fertilize the ovum during artificial insemination [36]. *Aloe vera* gel extract has pharmacological activities, such as anti-inflammatory, antifungal [27], and antimicrobial [37]. Moreover, it contains antioxidants, which are more crucial in semen cryopreservation, such as fat-soluble vitamins (A & C) and water-soluble vitamins (B2 & B12) [38], which play a role in inhibiting oxidative stress during preservation and freezing [39, 40].

2. Materials and methods review

A comprehensive electronic literature search was conducted across several databases, including Google Scholar, Web of Science, and PubMed. The search utilized a thorough combination of keywords and phrases, including "plant extracts", "cryoprotectants", "testicle", "semen", and "cryopreservation". This initial search resulted in a total of 1334 articles, as detailed in the accompanying flow diagram. Additionally, the reviewers manually examined the reference lists of relevant articles and reviews to identify other potentially eligible studies. Titles and abstracts of the articles were screened manually to exclude those not related to the topic, with a focus exclusively on studies involving plants. The remaining articles were further evaluated to determine their suitability based on the eligibility criteria for this review (**Figure 1**).

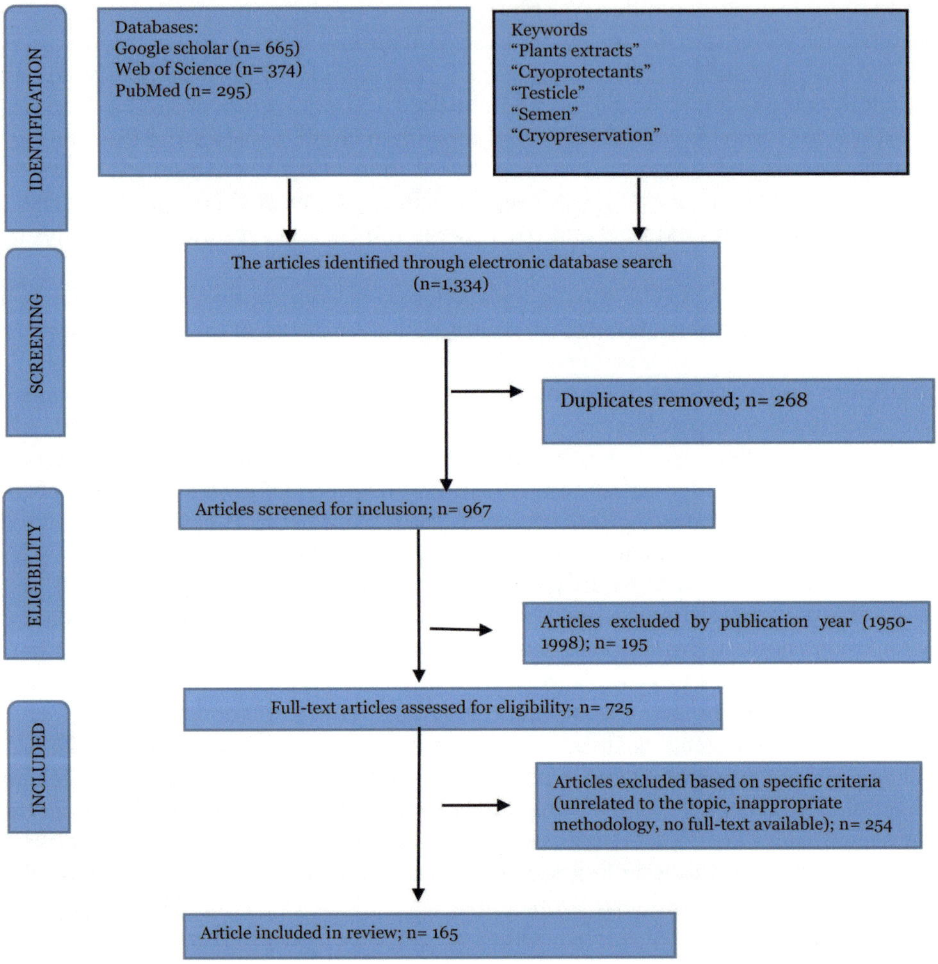

Figure 1.
Representation of inclusion and exclusion criteria, with corresponding results from the literature search.

3. Current challenges during semen cryopreservation from the livestock

3.1 Decreased sperm quality and functionality

Temperature fluctuations can affect sperm quality and functionality after thawing [2]. During the freezing and thawing processes, sperm is subjected to a variety of challenges, including ice formation, cold shock, chemical toxicity, osmotic stress, and oxidative damage. Cooling is a key stressor because membrane-bound phospholipids reorient themselves, disrupting membrane function and permeability [41]. Cryopreservation has been linked to decreased sperm motility, membrane integrity, and DNA integrity after thawing [42]. Cryopreservation stresses not only the physical properties of sperm, but also the chemical components required for tasks such as energy metabolism, which supports motility [43].

3.2 Cryoprotectant toxicity

To survive the cryopreservation procedure, sperm often requires the presence of a cryoprotectant, a type of substance that particularly acts to sustain the viability of cryopreserved cells [44]. Cryoprotectants are substances that prevent ice crystal formation while freezing the sperm [45]. Cryoprotectants are necessary for safeguarding sperm during freezing; however, their toxicity can have a negative impact on sperm quality [46]. Cryoprotectant may be considered harmful if cell membranes are breached or destroyed, enzyme performance is hindered, cell or embryo development is hampered, sperm motility is impeded, mitochondrial function is also reduced, or DNA, protein, or other macromolecules are disrupted [47]. Cryoprotectant toxicity is a significant barrier to realizing the full potential of artificial cryoprotection, but it is still a poorly understood phenomenon [48, 49]. Cryoprotectant toxicity has been identified as the single most limiting factor in the development of successful cryopreservation techniques [44]. Discovering non-toxic or less toxic cryoprotectant/s that preserve sperm viability is an ongoing issue [50]. Therefore, understanding the mechanisms of cryoprotectant toxicity and how to mitigate it could lead to successful semen cryopreservation from livestock [51].

3.3 Unimproved protocols for semen cryopreservation

Owing to cryobiology's significance for reproductive technologies, cryoprotectant drugs are being studied and new methods are being created to improve sperm cryo-survival [52]. Nevertheless, because a significant percentage of sperm die during the freezing-thawing process, new protocols developed have not yet reached the intended level [53]. As a result, creating the best freeze-thaw procedures for various livestock breeds and species continues to be difficult. Optimizing variables including freezing and thawing rates as well as the kind and quantity of cryoprotectants is crucial for achieving the best possible post-thaw sperm quality [45]. Universal cryopreservation techniques that are successful for all livestock species are difficult to create since different breeds and species have different reproductive physiologies and sperm properties. Specific characteristics of farm animal sperm, including size, shape, and lipid composition, influence their cryotolerance in different ways [54]. According to Layek et al. [36], various livestock species have distinct reproductive physiologies, and their sperms react differently to cryopreservation procedures. Therefore, creating cryopreservation procedures tailored to a species' needs is essential to preserving sperm quality in many species [55]. According to Lv et al. [54], the development of a successful technique for semen cryopreservation will enable the extension of artificial insemination in animal production and the long-term preservation of genetic resources.

3.4 Genetic damage and quality control

Reductions in fertility and offspring quality can result from genetic harm caused by cryopreservation [56]. Research into the mechanisms of cryo-damage and the creation of mitigation measures for genetic harm during cryopreservation are underway [57]. Successful artificial insemination programs depend on the quality and uniformity of frozen semen batches to use [58]. Low-quality batches must be identified and discarded using quality control techniques, such as sperm concentration, motility, and viability evaluations [59].

4. History of livestock semen cryopreservation using plant extract

Livestock semen cryopreservation dates for some centuries [60], with the current decades employing the supplementation of semen extenders with plant extracts to boost semen qualities [61]. The use of liquid nitrogen began when animal breeders started using modern cryobiological processes, owing to its advantages for long-distance transportation [62]. Researchers have noticed that there was an outline of research focus, addressing issues only relating to animal feeding and diseases, but with little focus being directed to animal reproduction and semen cryopreservation of livestock. It was reported that there is an effect of feeding and environment on the reproductive performance of livestock which subsequently has a deleterious effect on semen qualities [63]. Regarding this, there is a growing research focus on supplementing semen cryopreservation media with herbal plant extracts that are the source of non-enzymatic antioxidants [10].

Livestock farmers have used plant extracts to improve animal reproduction [64]. Plants amount to more than 90% of the total medicinal preparations in Africa and other countries around the world [65]. This is due to long trust as a source of health care that has no side effects and that is easily accessible to many farmers. Plant extracts contain a variety of nutrients, such as vitamins and phytochemicals that are important for semen cryopreservation [66]. Nutrients, such as zinc, vitamins, and antioxidants, are key players in the minimization of oxidative stress that subsequently influences fertility.

Due to the problems of the transmission of diseases through the use of various animal protein sources in semen extenders, most of plant extracts are being employed as semen preservatives for the preservation of semen for livestock [67]. Furthermore, plant extracts have recently emerged as a cheap and natural source of additives to preserve and improve sperm function during semen storage [68]. Most plant species are considered powerful sources of antioxidants, which can act as ROS scavengers for palliating the deleterious effects of oxidative stress on sperm function [69]. Moreover, these natural compounds can have antimicrobial properties and increase the activity of several antioxidant enzymes [10].

In numerous studies, the inclusion of plant extracts has been reported to have improved post-thaw sperm quality and fertilizing capacity [19] because of the significant reversal role of deleterious effects during cryopreservation and also plays a role in protection against ROS attack and growth of microbial substances [70]. For thousands of years in the history of medicine, plant extracts have been used to treat numerous diseases in animals, being the basis of modern technology in animal reproduction. Plant extracts contain a wide range of phytochemicals that safeguard against the potential growth of ROS in semen cryopreservation.

5. Plants

The effect of medicinal plant extracts on sperm motility and viability traits tested in different animal species, are shown in **Table 1**. Nevertheless, the sub-section below a provide an explanation of the important of these medicinal plants.

Plant species	Animal species	Extraction method	Plant part	Toxicity	Observed	References
Urtica dioica (Urticaceae)	Cattle	Water	Not specified	—	Improving sperm count, motility, morphology, and DNA integrity	[71]
Camellia sinensis (tea plant)	Pig	Water	Leaves, powder	DD	• Increased sperm motility and viability • Increased DNA integrity	[72, 73]
(Leucaenaleucocepala)	Sheep	Oil	Not specified	+	Improved sperm morphology	[74]
Syzygium cumini (Myrtaceae)	Cattle	Juice	Fruit	—	Improved percent sperm plasma membrane integrity	[75]
Schisandra chinensis (Schisandraceae)	Cattle	Ethanol	Fruit	—	Improved the sperm count, motility, and morphology	[76]
Sambucus nigra (Adoxaceae)	Cattle	Ethanol	Fruit	DD	Increased sperm motility, vitality, and oxidative profile	[77]
Salvia officinalis (Lamiaceae)	Pig	Water	Leaves	—	Protected boar epididymal sperm	[78]
Salvia miltiorrhiza (Lamiaceae)	Pig	Several steps	Not specified	—	Improved sperm quality	[79]
Punica granatum (Lythraceae)	Cattle and sheep	Juice	Fruit	DD	Protect ram sperm during cryopreservation without any adverse effects	[80]
Psidium guajava (Myrtaceae)	Goat	Several steps	Leaves	DD	Protective covering on the sperm	[81]
Portulaca oleracea (Portulacaceae)	Goat	Ethanol, methanol, water	Whole plant	DD		
Opuntia ficus-indica (Cactaceae)	Sheep	Acetone, oil	Cladodes, seeds	DD		
Echinacea purpurea (Asteraceae)	Sheep	Water	Leaves	—	Improvement in the number of viable sperm	[19]
Ilexparaguariensis (yerba mate)	Pig	Water	Leaves	—	Reduced sperm lipid membrane peroxidation, and protected DNA against oxidative damage during cryopreservation	[82]

Plant species	Animal species	Extraction method	Plant part	Toxicity	Observed	References
Zingiber officinale (Zingiberaceae)	Sheep	Water, ethanol	Rhizome	—	Improve the quality and fertility potential of frozen thawed	[19]
Cnidoscolus aconitifolius (Chaya)	Sheep	—	—	—	Improved sperm quality	[83]

DD = Dose-dependent and DNA = Deoxyribonucleic acid.

Table 1.
The effect of medicinal plant extracts on sperm motility and viability traits tested in different animal species.

5.1 *Aloe vera* (aloe)

Aloe vera (Aloe barbadensis miller) is an evergreen perennial plant that grows slowly to a height of 0.8 to 1 meter. It is a well-known herbal preparation with a long history of use [84]. It has been used as an alternative cryoprotectant to egg yolk for the cooling and cryopreservation of peccary semen [85]. The previous authors concluded that *A. vera* extract at a concentration of 20% can be used as an egg yolk substitute in the cooling and freezing of semen. High numbers of secondary abnormalities have been reported to impair sperm storage in the epididymis extended by *Aloe vera* extract [86]. Fakhrildin and Sodani [87] reported that the highest significant increase in sperm concentration, motility rate, progressive motility score, and normal sperm morphology as well as viability were recorded with a low dose gel extracts of fresh *Aloe vera* leaf of 5 μL/1 mL semen. In addition, little is known about how *Aloe vera* affects epididymal sperm morphology and viability in livestock animals.

5.2 *Aspalathus linearis* (rooibos)

Rooibos, scientifically known as *Aspalathus linearis*, is a woody legume plant that is native to the Cedarberg region in the Western Cape of South Africa. Rooibos has been traditionally drunk as a health beverage for over a century in South Africa and Europe [88–90]. The unfermented rooibos product has a green hue and is known as "green" rooibos. During fermentation, the color changes from green to red due to the oxidation of the polyphenols present, and it is then referred to as fermented or "red" rooibos [91]. Rooibos tea includes distinctive phenolic chemicals, with aspalathin and nothofagin being the most prevalent β-hydroxy-dihydrochalcone glucosides [92].

The plant was utilized in swine semen samples that were held for 96 hours at a temperature of 17°C, as documented by Ros-Santaella and Pintus [10] in 2017. Two variants of rooibos, one fermented and one unfermented, were utilized. The *Aspalathus linearis* concentration of the extract used ranged from 20 to 500 g/mL. Despite the fact that the unfermented version exhibited more antioxidant capacity and total phenolic content compared to the fermented form, the latter demonstrated superior benefits on sperm function. While storing semen, the use of this botanical extract maintained the speed, movement, and structural integrity of sperm, including the plasma membrane and acrosome. The greatest quantities employed resulted in a considerable reduction in overall sperm motility, while maintaining the integrity of the acrosome and membrane [10]. The optimal concentrations for boar semen were determined to be 20 and 60 g/mL.

5.3 *Camellia sinensis* (tea plant)

This plant has been widely used for semen preservation in a large number of animal species, such as boar, mouse, and ram [10]. Addition of seminal fluid with an extract concentration of 0.5–1 mg/ml increased sperm motility and viability during storage [10]. White and green tea have been tested on mouse epididymal sperm [72] stored at room temperature for 72 hours. However, Ahmed et al. [73] reported that the deceased occurrence of abnormalities with supplemented *Camellia sinensis* on semen extenders revealed better semen quality as compared to control in bulls. Khan et al. [93] reported that the addition of green tea extract in semen extender showed a significant response in the post-thawed sperm motility, viability, and membrane integrity in bulls. In addition, Ahmed et al. [73] revealed that tris-citric acid extender supplemented with 1.0% green tea extract improved semen quality parameters, antioxidant enzymes, total antioxidant capacity, lipid peroxidation, and DNA fragmentation in buffalo bull sperm. There have been several studies evaluating plants to treat different aspects of male infertility such as erectile dysfunction, sexual desire, and sperm abnormalities have been done [94–96]. However, when comparing the number of plant species evaluated scientifically to the number of plants used in male reproductive healthcare, there is much room for improvement.

5.4 *Cnidoscolus aconitifolius* (Chaya)

The plant has been used traditionally for both medicinal and non-medicinal purposes [83]. These types of medication may cause challenges; for example, in popular phytotherapy, it is very difficult to control the dose and quality of the product, which can lead to risks and damage to health [97]. However, the total percentage of morphologically abnormal sperm increased at 4th-week posttreatment in the group treated with 20% extract and slightly increased at 4th-week posttreatment in the group treated with 30% extract [83]. Moreover, *Cnidoscolus aconitifolius* (Chaya) on the semen extender decreases in live sperm, sperm concentration, sperm gross and individual motility, and normal sperm morphology, following DMN administration in rats [98]. Ajani and Oyeyemi [83] reported no significant effect on the sperm morphology as the number of morphologically abnormal sperm was within the normal percentage of 10% of *Cnidoscolus aconitifolius* (Chaya) on ram semen. However, little information is unknown on the effect of *Cnidoscolus aconitifolius* (Chaya) epididymal sperm, especially on the livestock animals, such as boar, cock, bull, ram, and buck.

5.5 *Cyclopia intermedia* (Honeybush)

Over the past few decades, there has been a growing focus on doing ethnopharmacological research on medicinal plants found in the South African fynbos region. Honeybush (Cyclopia spp.) has gained popularity as a tea among these plants, primarily because of its beneficial characteristics and lack of caffeine. Honeybush, scientifically known as Cyclopia intermedia, is a compact, ligneous shrub cultivated in the elevated terrains of the Langkloof district, situated between the Eastern and Western Cape provinces of South Africa. Unlike other types of herbal teas, honeybush is not extensively grown and the majority of the tea available for sale is obtained from naturally occurring plant populations. The plant has been found to possess antioxidant, antimutagenic, and antibacterial characteristics in several types of cells. However, its impact on reproductive function remains uncertain.

The plant's leaves, stems, and flowers are collected for the purpose of creating herbal tea infusion. This infusion is known by different names, such as Heuningtee, Bergtee, Boertee, Bossiestee, and Bush tea. In a study conducted by Ros-Santaella et al. [10], it was found that honeybush extract at a concentration of 12.5 µg/mL had positive effects on sperm motility and kinetic parameters. Additionally, it helped maintain the integrity of the sperm's plasma membrane and reduced lipid peroxidation in samples subjected to Fe^{2+}/ascorbate. Furthermore, the researchers demonstrated that honeybush extract has a safeguarding effect on pig sperm against the harmful consequences of oxidative stress. This is achieved by maintaining the sperm motility and protecting the integrity of their plasma membrane, while also reducing the levels of MDA. In addition, when semen is stored at a temperature of 17°C, the honeybush extract improves the quality of sperm, including its motility, acrosome integrity, and mitochondrial membrane, from 48 to 120 hours. The optimal concentration of honeybush extract for diluted boar semen, whether with or without oxidative stress, was 12.5 μg/mL. Honeybush extract serves as a cost-effective and viable source of antioxidants for the conservation of boar semen.

5.6 *Echinacea purpurea* (Asteraceae)

Echinacea purpurea (Asteraceae) is a perennial medicinal herb with important immunostimulatory and anti-inflammatory properties [99]. Merati and Farshad [19] showed a statistically significant improvement in the number of viable sperm upon the addition of 10 and 20 mg/l ginger extract or 20 mg/l echinacea extract to extenders of ram epididymal semen. Previous studies by Yamashiro et al. [100] and Merati and Farshad [19] reported that supplements containing 10 mg/L ginger and 20 mg/L echinacea extract significantly improved acrosome status compared to other extenders. Motamedi et al. [101] reported that *E. purpurea* extracts have protective effects on sperm characteristics at 24 hours after testicular ischemia/reperfusion (I/R) injuries in rats. In addition, the caftaric acid content of the Echinacea spp. protects sperm synthesis from xenotoxins, that is, synthetic pyrethroids, which increases oxidative damage to the testicles due to levels of malondialdehyde, catalase, superoxide dismutase, glutathione-S-transferase, and sperm abnormalities. In addition, oxidative damage can reduce testicular glutathione concentration, sperm count, and motility [102]. However, little information is available on the *Echinacea purpurea* on the epididymal sperm morphology and viability of livestock.

5.7 *Foeniculum vulgare* (fennel)

This plant has been used as an additive in boar [103] and ram [104] semen diluents during semen cryopreservation. In boar ejaculates, semen was frozen by using three concentrations of fennel (2.5, 5, and 10 g/100 mL) [105]. The plant extract had a positive effect on sperm motility and viability, especially at the highest concentration tested [72]. In ram, semen was cryopreserved by the use of three extract concentrations: 5, 10, and 15 mg/mL [106]. Fennel improved several sperm variables, such as viability, motility, kinetics, plasma membrane integrity, and mitochondrial activity, and reduced the MDA content [17]. The most suitable extract concentration for the cryopreservation of ram semen was 10 mg/mL with a concentration of 200×10^6 sperm/mL [10].

5.8 *Ilex paraguariensis* (yerba mate)

This plant extract has been used as an additive in the cryopreservation process of boar epididymal semen [107]. Tested concentrations ranged from 2.5 to 10 g/L. In all the sperm parameters assessed after thawing, the mate tea reduced lipid peroxidation (MDA levels) and protected DNA against oxidative damage. Better results were obtained with an extract concentration of 10 g/L in the semen extender [82]. The supplementation of a freezing extender with *Ilex paraguariensis* reduced sperm lipid membrane peroxidation and protected DNA against oxidative damage during cryopreservation at 120-minute post-thawing in epididymal boar sperm [82]. The use of *I. paraguariensis* as a dietary supplement in large and small ruminants reduced oxidative stress and improved their production and reproductive performance [108]. Little is known about the effects of *Ilex paraguariensis* on the epididymal sperm and the viability in livestock.

5.9 *L. leucocephala*

Leucaena plays an important role as a protein source for ruminants throughout Southeast Asia [109]. Although *L. leucocephala* has many positive nutritional benefits, it possesses the toxic non-protein-free amino acids [110]. Hmohn et al. [110] reported that feeding of toxic levels of *L. leucocephala* did not affect the semen quality (volume, motility, and live sperm percentage) of local bucks in Myanmar. Sperm viability, sperm count per ejaculation, and average curve speed of sperm in leucine-supplemented boars increased [111]. These studies suggest that Leu may play an important role in the development of testes and spermatogenesis. The incidence of some sperm abnormalities was also found to be higher in some epididymal regions and ejaculates of stressed boars. A high semen-collection frequency brings about an altered resorption and secretion pattern of the epididymal fluid, which results in defective sperm maturation and abnormal development of sperm motility [112].

5.10 *Moringa oleifera* leaf

Moringa oleifera leaf extracts (MOLE) are rich in many nutrients and active compounds, such as polyphenols, vitamins C and E, protein, calcium, amino acids, and several natural antioxidants. This MOLE extract has been reported to enhance the male reproductive potentials by lowering the amount of free radical sperm membrane [113]. Supplementation of MOLE into the dairy bull semen extender as a source of antioxidants has shown improvement in sperm quality [114] and antimicrobial agent [115].

5.11 *Phoenix dactylifera* (date palm)

Palm (*Phoenix dactylifera*) fruit is one of the plants that has long been emphasized in traditional medicine for the improvement of fertility potential [116]. It has been used as an additive during the chilling of buffalo and bull semen [117] and cryopreservation of buffalo and stallion semen [118]. *Phoenix dactylifera* extract was supplemented to semen extenders with or without egg yolk. After 7 days of buffalo semen chilling and semen thawing [117], significant improvements on sperm motility, viability, and membrane integrity were observed. The better results were obtained using a set of concentrations ranging from 10 to 50 mg/mL of *Phoenix dactylifera* [119]. However, some treatments have shown detrimental effects on sperm functionality [10]. Similarly, after thawing of stallion semen [120], date palm extract exerted a protective effect on sperm motility and acrosome and membrane integrity at a range of concentrations from 20 to 30 mg/mL of *Phoenix dactylifera*. Some concentrations of *Phoenix dactylifera* extract had negative effects on several sperm parameters. In addition, the addition of this plant extract to bull semen (both cooled at 5°C and cryopreserved) had a positive effect on sperm motility, viability, and plasma membrane integrity [121]. In addition, cryopreserved samples treated with this plant had a higher percentage of intact acrosomes and sperm with normal morphology than the control group [122]. To date, there is little information about the effect of *Phoenix dactylifera* on the viability of epididymal sperm and domestic animals.

5.12 *Urtica dioica* (Urticaceae)

Urtica dioica L. is a wild herbaceous perennial blooming plant that is commonly known as stinging nettle and has nutritional and medicinal properties [123]. Moreover, the aqueous extracts of *Urtica dioica* group exhibited greater live sperm percentage in comparison with the aqueous extracts of *Urtica dioica* palm group at 48 hours and 3-month post-cryopreservation in Holstein bulls [121]. Aqueous extracts of *Urtica dioica* palm are used as a traditional medicine to enhance male fertility by improving sperm count, motility, morphology, and DNA integrity, increasing testicular and epididymal weights [71]. However, increasing the dose of *U. dioica* significantly increased motility, count, normal morphology of sperm, seminiferous tubules diameter, and testosterone in all groups compared to control and testicular weight in 20 and 50 mg/kg doses in comparison with control group in mice [124]. However, more experiments are needed on the effect of *Urtica dioica* on epididymal sperm cells in domestic animals.

5.13 *Zingiber officinale* (ginger)

The rhizomes of this plant have been used as an additive for cryopreservation of ram epididymal semen [19]. There is evidence that ginger enhances semen quality *via* improving different sperm parameters, mainly count, viability, motility, morphology, and DNA integrity [125]. Gholami-Ahangaran et al. [125] observed a significant increase in viability and sperm motility, a decrease in structural abnormalities, and a greater integration of sperm cytoplasmic membranes in roosters receiving ginger powder. Ginger extract may be potentially useful in enhancing healthy sperm characteristics and the management of male infertility especially in those with low sperm morphology. Ginger has protective effects against oxidative stress in livestock. Most of the studies were done in mice and rabbits; however, little information is unknown on epididymal sperm, especially on the livestock.

6. Semen extenders used during cryopreservation of semen

In the last few years, the optimization of animal-origin-free-based semen extenders has been prioritized to cryopreserve ram semen [2, 126–128]. Sperm undergoes significant changes as a result of the cryopreservation process, which lowers the cells' ability for reproduction. The freezing extender's composition plays a crucial role in shielding sperm from harm [129]. According to Emamverdi et al. [128], one of the primary causes of decreased sperm quality during liquid storage and cryopreservation is the aberrant sperm metabolism that results in the generation of ROS. The semen needs to be diluted using the right conditions and extenders to avoid this kind of stress.

6.1 Soya lecithin-based semen extenders

Soybeans include lecithin, which is a type of phospholipid. This phospholipid can be used as a replacement for high-molecular-weight lipoprotein and phospholipids from egg yolk. It helps to avoid or reduce damage to the sperm plasma membrane that can happen during processes, including extension, chilling, and cryopreservation [36]. Soya lecithin is a naturally occurring combination of phosphatidylcholine and numerous fatty acids, including stearic, oleic, and palmitic acid, that can be used as an alternative to animal-derived ingredients found in semen extenders [130]. While there is no denying the hygienic benefits of soy lecithin-based semen extenders over egg yolk, there is ongoing disagreement over soy lecithin's effectiveness. Commercially available soy lecithin has been tested in several species in recent years, and results have shown that it is comparable to egg yolk [131, 132], better or worse [133].

In our previous studies, exogenous soybean lecithin has been evaluated for its efficiency against cold shock during the freeze-thawing process [2, 128, 129]. It has been demonstrated that lecithin has suitable protective properties for ram [127], goat [134], buffalo [135], and stallion sperm [136, 137] against cryo-damage during the freezing process. Although soybean lecithin reduces the difficulties of egg yolk in extenders, such as contamination, standardization, and agglutination [126], it provides the substrate for lipid peroxidation around sperm. As a generalization, soybean lecithin phosphatidylcholine tends to contain higher proportions of arachidonic and docosahexaenoic acids and more of the C18 unsaturated fatty acids [138]. The use of soy lecithin-based extenders has been assessed for the processing and freezing of bovine semen. However, the potential of extenders derived from soybean milk has not been widely investigated [36].

7. Methods used to extract plants for cryopreservation of livestock semen

Extraction is the separation of active portions of plant using selective solvents through standard procedures [139]. These techniques separate the soluble plant metabolites and leave behind the insoluble cellular marc. The products obtained from plants are relatively complex mixtures of metabolites, after removing the solvent in dry powder form, and are intended for oral or external use [140]. The classes of preparations known as decoctions, infusions, fluid extracts, tinctures, pilular (semisolid) extracts, or powdered extracts have been utilized [141].

The main goal of standardized extraction procedures for crude plant parts is to attain the therapeutically desired portions and to eliminate unwanted material by treatment with a selective solvent [142]. The extract that is obtained, after standardization, may be used in the form of tinctures or fluid extracts. These plant extracts contain a complex mixture of secondary metabolites, such as alkaloids, glycosides, terpenoids, flavonoids, and lignans [143]. In order to be used orally or externally, an extract may be further processed through various techniques of fractionation to isolate individual chemical entities, such as vincristine, vinblastine, hyoscyamine, hyoscine, pilocarpine, forskolin, and codeine [141].

7.1 Methods used for plant extractions

The methods used for plant extraction vary depending on the type of plant and the desired chemical composition needed. Some common methods for extracting plants include:

7.1.1 Steam distillation

This method involves passing steam through the plant material to extract active material or secondary metabolites. The steam is then condensed, separating the plant material from the water.

7.1.2 Solvent extraction

Solvents, such as hexane, ethanol, or supercritical carbon dioxide, can be used to extract oils from plant material. The solvent dissolves the oil, which is then separated from the solvent.

7.1.3 Cold pressing

This method involves pressing the plant material to extract active material without using heat. Cold-pressed materials are often used for culinary and cosmetic purposes.

7.1.4 Enfleurage

This traditional method involves placing plant material on a layer of fat to absorb the essential oils. The fat is then collected, and the oil is separated from the fat.

7.1.5 Supercritical fluid extraction

This method uses supercritical carbon dioxide as a solvent to extract plant material. The supercritical carbon dioxide has properties of both a liquid and a gas, making it an effective solvent for extracting active material.

8. The efficacy of plant extracts on the success of livestock semen cryopreservation

Plant extract supplementation has been proposed as a new strategy for controlling reproductive disorders. Recently, plant extracts have emerged as a popular and affordable natural source of supplements to preserve and enhance sperm function during semen preservation [10]. These plants are widely available, easily accessible, and have little to no negative effects. There is now enough evidence about the effects of plant extracts on fertility to recommend them as a treatment for livestock infertility. Most plant species are considered abundant sources of naturally occurring antioxidants with reduced cytotoxicity conventionally used to preserve sperm. These antioxidants can also work as ROS scavengers to reduce the harmful effects of oxidative stress on sperm function. Plant extracts can be used in several ways, including powdered form, freshly harvested, dried, diced, or combined with solvents, such as ethanol, water, or petroleum ether. Solvents are a key component of traditional medicine [144]. These plants and the substances they produce have been used as dietary supplements to treat infertility and enhance human, animal, and wildlife reproductive function [10, 145, 146]. They consist of mixtures of complex polymeric phenols, which are among the most researched phytochemicals due to their wide range of beneficial biological activities and health-prong consequences.

Antioxidants have been proven to have some degree of protective ability against the harmful effects of ROS when added to semen cryopreservation diluents [147–149]. According to Siecherle et al. [148], Baghshahi et al. [149], and Benhenia et al. [147], this supplementation improves sperm motility, viability, and post-thawing fertility. Reactive oxygen metabolites are eliminated by an antioxidant defense mechanism, which shields cells from structural harm. Effective antioxidant defense systems function in three ways: They disrupt free radical chain reactions, stop ROS from interacting with cell components, and eliminate the byproducts of ROS interactions with cellular macromolecules [150]. Similarly, research studies have focused on using plant extracts with antioxidant qualities to enhance the quality of cryopreserved semen. These extracts include well-known antioxidant phytochemicals [151]. In this context, tissues from most plants have been shown to have high antioxidant activity because they contain components, such as ascorbic acid, flavonoids, polyphenolics, and carotenes [152], which reduce levels of free radicals and the risk of cell death [153].

The semen cryopreservation process is primarily responsible for the structural damage to sperm. The cryopreservation of semen induces alterations in the resident electron transport chain and the mitochondrial membrane of sperm [154], leading to an overabundance of ROS, such as hydrogen peroxide (H_2O_2), nitric oxide (NO), or superoxide anion (O_2^-), which, in turn, affects sperm capacitation and the acrosome reaction [155]. In an attempt to lessen these effects, plant extracts have been added to the formulation of the semen diluents. Certain plant extracts may not have the ability to prevent these mechanical changes. Nonetheless, certain plant extracts may be a better fit for the cryopreservation procedure than for refrigerating sperm samples since they can serve as alternatives to cryoprotectants, such as egg yolk [85]. Although plant extracts are still in their infancy as preservatives for semen storage, their potential is great, and they can be lucrative and natural substitutes for the most widely used additives in sperm preservation today.

9. Future implications of using plant extract for livestock semen cryopreservation

Due to the variety of plant extracts, cryoprotectants, additives, and physical freezing methods available, determining the most optimal cryopreservation protocol is challenging. The current World Health Organization guidelines provide instructions for both, stating that any commercially tested and available plant extracts and CPAs may be used [156]. Therefore, there is an urgent need to generate comparative data to establish a standardized, optimized protocol for each species using different plant extracts. Plant extracts offer a promising alternative to the commonly used antioxidants in semen preservation. Many traditional antioxidants are, in fact, derived from plants (e.g., epicatechin gallate, quercetin, and aspalathin) as their inclusion in the semen extender is not profitable because of their high price [10].

Moreover, these compounds are typically used individually, whereas plant extracts, in their entirety, can offer a diverse array of secondary metabolites (e.g., phenols, terpenoids) with a powerful antioxidant effect and antimicrobial properties [157]. Recent advances in phytotherapy research have brought promising applications related to drug delivery. For instance, plant cells cultured *in vitro* release extracellular vesicles that can be easily harvested for pharmaceutical practice [158]. Considering that some plant extracellular vesicles have shown inhibitory effects against pathogenic microorganisms [158], they may be studied as possible substitutes for antibiotics in the semen preservation process. Recent studies have revealed that certain polyphenols can influence cellular metabolic pathways and convert various compounds into different forms. This process not only provides powerful antioxidant effects but also triggers a series of secondary protective effects [159, 160]. Certain polyphenols have been demonstrated to enhance sperm metabolism by improving the efficiency of mitochondrial respiration [159]. However, the source of the raw plant material, the extraction method, and the concentration of the plant extract are critical factors influencing its effects on sperm.

10. Conclusion

The plant extracts are usually added to the semen extender [161] and used during cryopreservation. This could be helpful to emerging farmers because it is regarded as an affordable natural source additive [162] to preserve and improve sperm functionality [161]. Many plant species (e.g., *Aloe vera*, Albizia harveyi, and Argania spinosa) are the influential sources of natural antioxidants, which have scavenging ability toward the development ROS for palliating the deleterious effects of oxidative stress on sperm function [33]. Moreover, the natural compound from plants proves to have antimicrobial properties and increase the activity of several antioxidant enzymes [10].

Plant, such as *Moringa oleifera*, is the assessable and cheap antioxidant source that can be planted in many regions around the world [163]. Many of its plant organs are harvested and used for medicinal purposes due to its potential uses, as a source of antioxidants and antibiotics [164]. The report by Falowo et al. [32] has indicated the higher phenolics and flavonoids in *Moringa oleifera* leaf extract (MOLE). The MOLE is an essential source of natural antioxidants, such as ascorbic acid, and phenolic compounds [165].

Acknowledgements

The Agricultural Research Council, Animal Production (Germplasm Conservation and Reproductive Biotechnologies) colleagues are acknowledged for their support.

Conflict of interest

The authors declare no conflict of interest.

Author details

Masindi Lottus Mphaphathi[1*], Maleke Dimpho Sebopela[1],
Mahlatsana Ramaesela Ledwaba[1], Mamonene Angelinah Thema[1], Rolly Maapola[1],
Nkhumeleni Cathbert Negota[2], Thabang Luther Mashilo[3]
and Thivhilaheli Richard Netshirovha[4]

1 Agricultural Research Council, Animal Production, Germplasm, Conservation and Reproduction Biotechnologies, Pretoria, South Africa

2 Faculty of Science, Engineering and Agriculture, Department of Animal Science, Reproduction, and Physiology, University of Venda, South Africa

3 Department of Agriculture, Land Reform, and Rural Development, Animal Production, Animal Improvement, Riviera, Pretoria, South Africa

4 Department of Agriculture and Animal Health, College of Agriculture and Environmental Sciences, University of South Africa, South Africa

*Address all correspondence to: masindim@arc.agric.za

References

[1] Mphaphathi M, Ledwaba M, Thema M. The current status of semen and oocytes cryopreservation. In: Biomedical Engineering. London, UK: IntechOpen; 2023. DOI: 10.5772/intechopen.107404

[2] Sharafi M, Borghei-Rad SM, Hezavehei M, Shahverdi A, Benson JD. Cryopreservation of semen in domestic animals: A review of current challenges, applications, and prospective strategies. Animals (Basel). 2022;**12**(23):3271. DOI: 10.3390/ani12233271

[3] Ledwaba MR, Mphaphathi ML, Thema MA, Pilane CM, Nedambale TL. Investigation of the efficacy of dithiothreitol and glutathione on *in vitro* fertilization of cryopreserved large white boar semen. Animals. 2022;**12**:1137. DOI: 10.3390/ani12091137

[4] Ozimic S, Ban-Frangez H, Stimpfel M. Sperm cryopreservation today: Approaches, efficiency, and pitfalls. Current Issues in Molecular Biology. 2023;**45**(6):4716-4734. DOI: 10.3390/cimb45060300

[5] Thema MA, Mphaphathi ML, Ledwaba MR, Nedambale TL. Sperm cryopreservation in windsnyer boars; principles, technique and updated outcomes. Animal Reproduction. 2023;**20**(3):e20220100

[6] Seshoka MM, Mphaphathi ML, Nedambale TL. Comparison of four different permitting and combination of two best cryoprotectants on freezing Nguni sperm evaluated with the aid of computer aided sperm analysis. Cryobiology. 2016;**72**:232-238

[7] Kulazsis R. Freezing of angora buck semen with skimmed milk-based extender containing various amounts of egg yolk. Journal of Lalahan Livestock Research Institute. 2010;**50**(2):81-89

[8] Mphaphathi ML, Seshoka MM, Luseba D, Sutherland B, Nedambale TL. The characterization and cryopreservation of Venda chicken semen. Asian Pacific Journal of Reproduction. 2016;**5**(2):132-139

[9] Pintus E, Ros-Santaella JL. Impact of oxidative stress on male reproduction in domestic and wild animals. Antioxidants (Basel). 2021;**10**(7):1154. DOI: 10.3390/antiox10071154

[10] Ros-Santaella JL, Pintus E. Plant extracts as alternative additives for sperm preservation. Antioxidants (Basel). 2021;**10**(5):772. DOI: 10.3390/antiox10050772

[11] Cojkic A, Hansson I, Johannisson A, Morrell JM. Effect of some plant-based substances on microbial content and sperm quality parameters of bull semen. International Journal of Molecular Sciences. 2023;**24**(4):3435. DOI: 10.3390/ijms24043435

[12] Al-Khayri JM, Rashmi R, Toppo V, Chole PB, Banadka A, Sudheer WN, et al. Plant secondary metabolites: The weapons for biotic stress management. Metabolites. 2023;**13**(6):716. DOI: 10.3390/metabo13060716

[13] Ullah A, Munir S, Badshah SL, Khan N, Ghani L, Poulson BG, et al. Important flavonoids and their role as a therapeutic agent. Molecules. 2020;**25**(22):5243. DOI: 10.3390/molecules25225243

[14] Noh S, Go A, Kim DB, Park M, Jeon HW, Kim B. Role of antioxidant natural products in management of infertility: A review of their

medicinal potential. Antioxidants (Basel). 2020;**9**(10):957. DOI: 10.3390/antiox9100957

[15] Morrell MJ, Malaluang P, Cojkic A. Alternatives to antibiotics in semen extenders used in artificial insemination. In: The Global Antimicrobial Resistance Epidemic - Innovative Approaches and Cutting-Edge Solutions. London, UK: IntechOpen; 2022. DOI: 10.5772/intechopen.104226

[16] Santos-Sánchez FN, Salas-Coronado R, Villanueva-Cañongo C, et al. Antioxidant compounds and their antioxidant mechanism. In: Antioxidants. London, UK: IntechOpen; 2019. DOI: 10.5772/intechopen.85270

[17] Malo C, Gil L, Cano R, Martínez F, Galé I. Antioxidant effect of rosemary (*Rosmarinus officinalis*) on boar epididymal spermatozoa during cryopreservation. Theriogenology. 2011;**75**(9):1735-1741. DOI: 10.1016/j.Theriogenology.2011.01.013. Epub 2011 Feb 18

[18] Zarfeshany A, Asgary S, Javanmard SH. Potent health effects of pomegranate. Advanced Biomedical Research. 2014;**3**:100. DOI: 10.4103/2277-9175.129371

[19] Merati Z, Farshad A. Ginger and echinacea extracts improve the quality and fertility potential of frozen-thawed ram epididymal spermatozoa. Cryobiology. 2020;**92**:138-145. DOI: 10.1016/j.Cryobiol.2019.12.003. Epub 2019 Dec 27

[20] Tvrda E, Greifova H, Mackovich A, Hashim F, Lukac N. Curcumin offers antioxidants protection to cryopreserved bovine semen. Czech Journal of Animal Science. 2017;**63**:247-255

[21] Abdelnour AA, Hassan MAE, Mohammed AK, Alhimaidi AR, Al-Gabri N, Al-Khaldi K, et al. The effect of adding different levels of curcumin and its nanoparticles to extender on post-thaw quality of cryopreserved rabbit sperm. Animals. 2020;**10**(9):1508

[22] Santonastaso M, Mottola F, Iovine C, Colacurci N, Rocco L. Protective effects of curcumin on the outcome of cryopreservation in human sperm. Reproductive Sciences. 2021;**28**(10):2895-2905. DOI: 10.1007/s43032-021-00572-9. Epub 2021 Apr 16

[23] El-Harairy M, Abdel-Khalek AE, Khalil WA, Khalifa EI, El-Khateeb AY, Abdulrhmn AM. Effect of aqueous extracts of *Moringa oleifera* leaves or atrium lappa roots on lipid peroxidation and membrane integrity of ram sperm preserved at cool temperature. Journal of Animal and Poultry Production. 2016;**7**(12):467-473

[24] Sobeh M, Hassan SA, El-Raey MA, Khalil WA, Hassan MAE, Wink M. Polyphenolics from *Albizia harveyi* exhibit antioxidants activities and counteract oxidative damage and ultra-structural changes of cryopreserved bull semen. Molecules. 2017;**27**:1993

[25] Shokry DM, Badr M, Orabi S, Khalifa H, Eldaim MA, El-Seedi H. *Moringa oleifera* leaves extract enhances fresh and cryopreserved semen characters of Barki rams. Italian Journal of Animal Science. 2020;**20**(1):1175-1186

[26] Elsheshtawy R, Elnattat W. Assessment of buffalo semen preservability using tris extender enriched with *Moringa oleifera* extract. Egyptian Journal of Veterinary Sciences. 2020;**51**(2):235-239. DOI: 10.21608/ejvs.2020.23502.1150

[27] Ros-santaella JL, Pintus E. Plant extracts as alternative additives for sperm preservation. Antioxidants. 2021;**10**:772

[28] Aitken RJ. Reactive oxygen species as mediators of sperm capacitation and pathological damage. Molecular Reproduction and Development. 2017;**84**(10):1039-1052

[29] Cunha ATM, Faria OAC, Guimarães ALS. Bovine sperm capacitation: Physiological changes and evaluations. JSM In vitro Fertilization. 2017;**2**(1):1011

[30] Fitri A, Toharmat T, Astuti DA, Tamura H. The potential use of secondary metabolites in *Moringa oleifera* as an antioxidant source. Media Peternakan. 2015;**38**(3):169-175

[31] Bustani GS, Baiee FH. Semen extenders: An evaluative overview of preservative mechanisms of semen and semen extenders. Veterinary World. 2021;**14**(5):1220

[32] Falowo AB, Muchenje V, Hugo CJ, Charimba G. In vitro antimicrobial activities of *Bidens pilosa* and *Moringa oleifera* leaf extracts and their effects on ground beef quality during cold storage. CyTA-Journal of Food. 2016;**14**(4):541-546

[33] Abd Eldaim MA, Shaban Abd Elrasoul A, Abd Elaziz SA. An aqueous extract from *Moringa oleifera* leaves ameliorates hepatotoxicity in alloxan-induced diabetic rats. Biochemistry and Cell Biology. 2017;**95**(4):524-530

[34] Abdull Razis AF, Ibrahim MD, Kntayya SB. Health benefits of *Moringa oleifera*. Asian Pacific Journal of Cancer Prevention. 2014;**15**(20):8571-8576

[35] Singh P, Agarwal S, Singh H, Singh S, Verma PK, Butt MS, et al. Effects of ascorbic acid as antioxidant semen additive in cryopreservation of cross-bred cattle bull semen. International Journal of Current Microbiology and Applied Sciences. 2020;**9**(7):3089-3099. DOI: 10.20546/ijcmas.2020.907.364

[36] Layek SS, Mohanty TK, Kumaresan A, Parks JE. Cryopreservation of bull semen: Evolution from egg yolk based to soybean based extenders. Animal Reproduction Science. 2016;**172**:1-9. DOI: 10.1016/j.anireprosci.2016.04.013

[37] Singh P, Agarwal S, Singh H, Verma PK, Pandey AK, Kumar S. Antioxidant effects of *Aloe vera* as semen additive in cryopreservation of cattle bull semen. International Journal of Current Microbiology and Applied Sciences. 2020;**9**(9):1625-1635

[38] Quispe C, Villalobos M, Bórquez J, Simirgiotis M. Chemical composition and antioxidant activity of *Aloe vera* from the pica oasis (Tarapacá, Chile) by UHPLC-Q/Orbitrap/MS/MS. Journal of Chemistry. 2018;**2018**:1-12

[39] Pinto SCC, Almeida DS, Alves MBR, Florez-Rodriguez SA, Abreu Junior GS, Britto-e-Alves N, et al. Does supplementation of vitamin C, reduced glutathione or their association in bovine frozen semen? Arquivo Brasileiro de Medicina Veterinária e Zootecnia. 2020;**72**(1):9-17

[40] Patel DK, Chaudhari CF, Modi LC, Chaudhari NF, Chaudhary MM, Tyagi KK, et al. Effect of different egg yolk levels in tris extender with and without *Aloe vera* and onion crude extracts on surti buck semen at refrigerated temperature. Indian Journal of Veterinary Sciences & Biotechnology. 2022;**18**(4):44-49

[41] Lemma A. Effect of cryopreservation on sperm quality and fertility. Artificial Insemination in Farm Animals. 2011;**12**:191-216

[42] Yeste M. Sperm cryopreservation update: Cryodamage, markers, and factors affecting the sperm freezability in pigs. Theriogenology. 2016;**85**(1):47-64

[43] Long JA. Avian semen cryopreservation: What are the biological challenges? Poultry Science. 2006;**85**(2):232-236

[44] Szurek EA, Eroglu A. Comparison and avoidance of toxicity of penetrating cryoprotectants. PLoS One. 2011;**6**(11):e27604

[45] Bhattacharya S. Cryopretectants and their usage in cryopreservation process. In: Cryopreservation Biotechnology in Biomedical and Biological Sciences. IntechOpen; 2018. DOI: 10.5772/intechopen.80477

[46] Bhattacharya MS. A review on cryoprotectant and its modern implication in cryonics. Asian Journal of Pharmaceutics. 2016;**10**(3):1-6

[47] Best BP. Cryoprotectant toxicity: Facts, issues, and questions. Rejuvenation Research. 2015;**18**(5):422-436

[48] Fahy GM, Lilley TH, Linsdell H, Douglas MS, Meryman HT. Cryoprotectant toxicity and cryoprotectant toxicity reduction: In search of molecular mechanisms. Cryobiology. 1990;**27**(3):247-268

[49] Fahy GM. Cryoprotectant toxicity neutralization. Cryobiology. 2010;**60**(3):45-53

[50] Liu S, Li F. Cryopreservation of single-sperm: Where are we today? Reproductive Biology and Endocrinology. 2020;**18**:1-2

[51] Raju R, Bryant SJ, Wilkinson BL, Bryant G. The need for novel cryoprotectants and cryopreservation protocols: Insights into the importance of biophysical investigation and cell permeability. Biochimica et Biophysica Acta (BBA)-General Subjects. 2021;**1865**(1):129749

[52] Mphaphathi M, Ledwaba M, Thema M. The current status of semen and oocytes cryopreservation. In: Cryopreservation-Applications and Challenges. London, UK: IntechOpen; 2023

[53] Ugur MR, Saber Abdelrahman A, Evans HC, Gilmore AA, Hitit M, Arifiantini RI, et al. Advances in cryopreservation of bull sperm. Frontiers in Veterinary Science. 2019;**6**:268

[54] Lv C, Wu G, Hong Q, Quan G. Spermatozoa cryopreservation: State of art and future in small ruminants. Biopreservation and Biobanking. 2019;**17**(2):171-182

[55] Silva AM, Pereira AF, Comizzoli P, Silva AR. Cryopreservation and culture of testicular tissues: An essential tool for biodiversity preservation. Biopreservation and Biobanking. 2020;**18**(3):235-243

[56] Peris-Frau P, Soler AJ, Iniesta-Cuerda M, Martín-Maestro A, Sánchez-Ajofrín I, Medina-Chávez DA, et al. Sperm cryodamage in ruminants: Understanding the molecular changes induced by the cryopreservation process to optimize sperm quality. International Journal of Molecular Sciences. 2020;**21**(8):2781

[57] Guthrie HD, Welch GR. Determination of intracellular reactive oxygen species and high mitochondrial membrane potential in percoll-treated viable boar sperm using fluorescence-activated flow cytometry. Journal of Animal Science. 2006;**84**(8):2089-2100

[58] Harstine BR, Utt MD, DeJarnette JM. Integrating a semen quality control program and sire fertility at a large artificial insemination organization. Animal. 2018;**12**(s1):s63-s74

[59] Hu E, Liao TW, Tiersch TR. A quality assurance initiative for

commercial-scale production in high-throughput cryopreservation of blue catfish sperm. Cryobiology. 2013;**67**(2):214-224

[60] Aljaser FS. Cryopreservation methods and frontiers in the art of freezing life in animal models. In: Animal Reproduction. London, UK: IntechOpen; 2022. p. 13

[61] Xiang J, Mlambo R, Shaw I, Seid Y, Shah H, He Y, et al. Cryopreservation of bioflavonoid-rich plant sources and bioflavonoid-microcapsules: Emerging technologies for preserving bioactivity and enhancing nutraceutical applications. Frontiers in Nutrition. 2023;**10**:1232129

[62] Tkachev AV, Tkacheva OL, Gazzavi-Rogozina LV, Zubova TV, Smolovskaya OV, et al. Modern technology of poultry semen cryopreservation. In: Proceedings of IV International Scientific and Practical Conference Modern S&T Equipments and Problems in Agriculture. Kemerovo (Russia). 2020. pp. 235-244. DOI: 10.32743/kuz.mepa.2020.235-244

[63] Lohmiller JJ, Swing SP, Hanson MM. Reproduction and breeding. In: The Laboratory Rat. Burlington: Academic Press and American College of Laboratory Animal Medicine; 2020. pp. 157-179. DOI: 10.1016/B978-012074903-4/50009-1

[64] Shokry DM, Abd Eldaim MA, Badr MR, Khalifa HK, Orabi SH, Hassan AM, et al. Enhancement impact of *Moringa oleifera* leaves extract–base extender on cryopreservation and fertilization of Barki ram sperms: Comparative study with vitamin E and selenium combination. Italian Journal of Animal Science. 2021;**20**(1):1175-1186

[65] Ozioma EO, Chinwe OA. Herbal Medicines in African Traditional Medicine. Herbal Medicine. London: Intechopen; 2019;**10**:191-214. DOI: 10.5772/intechopen.80348

[66] Sadogh A, Gorji N, Moeini R. Herbal foodstuffs in Avicenna's recommended diet to improve sperm quality and increase male fertility; an evidence-based approach. Journal of Complementary and Integrative Medicine. 2022;**19**(1):47-70

[67] Santos CS, Silva AR. Current and alternative trends in antibacterial agents used in mammalian semen technology. Animal Reproduction. 2020;**17**:e20190111

[68] Tavares WR, Barreto MDC, Seca AM. Aqueous and ethanolic plant extracts as bio-insecticides—Establishing a bridge between raw scientific data and practical reality. Plants. 2021;**10**(5):920

[69] Dumanović J, Nepovimova E, Natić M, Kuča K, Jaćević V. The significance of reactive oxygen species and antioxidant defense system in plants: A concise overview. Frontiers in Plant Science. 2021;**11**:552969

[70] Khalil WA, Ismail AA, Sindi RA, Abdel-Khalek AKE, Yousif AI, Swelum AA, et al. Ethanolic purslane (*Portulaca oleracea*) leaf extract improves the quality of cryopreserved goat semen. Andrologia. 2023;**2023**(1):7771525

[71] El-Sheshtawy RI, El-Nattat WS, Ali AH, Sabra HA. The effect of phoenix dactylifera pollen grains tris-infusion on semen preservability of local bull breeds. Global Veterinaria. 2014;**13**(5):728-732

[72] Dias TR, Alves MG, Tomás GD, Socorro S, Silva BM, Oliveira PF. White tea as a promising antioxidant medium additive for sperm storage at room temperature: A comparative study with green tea. Journal of Agricultural and Food Chemistry. 2014;**62**:608-617. DOI: 10.1021/jf4049462

[73] Ahmed H, Jahan S, Khan A, Khan L, Khan BT, Ullah H, et al. Supplementation of green tea extract (GTE) in extender improves structural and functional characteristics, total antioxidant capacity and in vivo fertility of buffalo (*Bubalus bubalis*) bull spermatozoa. Theriogenology. 2020;**145**:190-197. DOI: 10.1016/j.theriogenology.2019.10.024

[74] Hmohn ZZW, Kyaw H, Phyu SH, Hnin CLY, Po SP, Sint TT, et al. Effect of toxic levels of *Leucaena leucocephala* on semen quality of goats in Myanmar. International Journal of Veterinary Sciences and Animal Husbandry. 2017;**2**(6):9-12

[75] Suleman S, Kanwal MA, Malik F, Ali R, Siddique S, Kanwal N, et al. Jambul (*Syzygium cumini*) pulp extract enhances viability, motility, and *in vitro* fertilizability of cryopreserved bovine semen. Biopreservation and Biobanking. 2020;**19**:53-59

[76] Tvrda E, Michalko J, Lius J, Vukovic NL, Ivanis E, Michal D, et al. Characterization of the Omija (*Schisandra chinensis*) extract and its effects on the bovine sperm vitality and oxidative profile during *in vitro* storage. Evidence-based Complementary and Alternative Medicine. 2020;**2020**:7123780, 15 pages. DOI: 10.1155/2020/7123780

[77] Abdramanov A, Massanyi P, Sarsembayeva N, Usenbayev A, Alimov J, Tvrdá. The *in vitro* effect of elderberry (*Sambucus nigra*) extract on the activity and oxidative profile of bovine spermatozoa. Journal of Microbiology, Biotechnology and Food Sciences. 2017;**6**(6):1319-1322. DOI: 10.15414/jmbfs.2017.6.6.1319-1322

[78] Monton A, Gil L, Malo C, Olaciregui M, Gonzalez N, de Blas I. Sage (*Salvia officinalis*) and fennel (*Foeniculum vulgare*) improve cryopreserved boar epididymal semen quality study. CryoLetters. 2015;**36**:83-90

[79] Shen T, Jiang ZL, Liu H, Li QW. Effect of salvia miltiorrhiza polysaccharides on boar spermatozoa during freezing-thawing. Animal Reproduction Science. 2015;**159**:25-30

[80] Mehdipour M, Kia HD, Nazari M, Najafi A. Effect of lecithin nanoliposome or soybean lecithin supplemented by pomegranate extract on post-thaw flow cytometric, microscopic and oxidative parameters in ram semen. Cryobiology. 2017;**78**:34-40. DOI: 10.1016/j.cryobiol.2017.07.005

[81] Wurlina W, Hariadi M, Safitri E, Susilowati S, Meles DK. The effect of crude guava leaf tannins on motility, viability, and intact plasma membrane of stored spermatozoa of Etawa crossbred goats. Veterinary World. 2020;**13**(3):530-537

[82] Luño V, Gil L, Olaciregui M, Gonzalez N, Jerez RA, de Blas I. Rosmarinic acid improves function and *in vitro* fertilising ability of boar sperm after cryopreservation. Cryobiology. 2014;**69**:157-116

[83] Ajani OS, Oyeyemi MO. Spermatozoa characteristics and morphology of West African dwarf rams treated with aqueous extract of *Cnidoscolus aconitifolius* (Chaya). International Journal of Applied. 2014;**9**(1):33-40

[84] Olugbenga OM, Olukole SG, Adeoye AT, Adejoke AD. Semen characteristics and sperm morphological studies of the West African dwarf buck treated with *Aloe vera* gel extract. Iranian Journal of Reproductive Medicine. 2011;**9**(2):83

[85] Souza ALP, Lima GL, Peixoto GCX, Silva AM, Oliveira MF, Silva AR.

Use of *Aloe vera*–based extender for chilling and freezing collared peccary (*Pecari tajacu*) semen. Theriogenology. 2016;**85**:1432-1438. DOI: 10.1016/j.theriogenology.2016.01.007

[86] Nass SJ, Miller DJ, Winer MA, Ax RL. Male accessory sex glands produce heparin-binding proteins that bind to cauda epididymal spermatozoa and are testosterone dependent. Molecular Reproduction and Development. 1990;**25**:237-246

[87] Fakhrildin MB, Sodani IJ. Effect of *Aloe vera* extracts on *in vitro* human sperm parameters for asthenozoospermic patients. Thi-Qar Journal of Science. 2014;**5**:1991-8690

[88] Morton JF. Rooibos tea, *Aspalathus linearis*, a caffeineless, low-tannin beverage. Economic Botany. 1983;**37**:164-173

[89] Nakano M, Nakashima H, Itoh Y. Anti-human immunodeficiency virus activity of oligosaccharides from rooibos tea (*Aspalathus linearis*) extracts *in vitro*. Leukemia. 1997;**11**:08876924

[90] Marnewick JL, Van der Westhuizen FH, Joubert E, Swanevelder S, Swart P, Gelderblom WC. Chemoprotective properties of rooibos (*Aspalathus linearis*), honeybush (*Cyclopia intermedia*) herbal and green and black (*Camellia sinensis*) teas against cancer promotion induced by fumonisin B1 in rat liver. Food and Chemical Toxicology. 2009;**47**(1):220-229

[91] McKay DL, Blumberg JB. A review of the bioactivity of South African herbal teas: Rooibos (*Aspalathus linearis*) and honeybush (*Cyclopia intermedia*). Phytotherapy Research: An International Journal Devoted to Pharmacological and Toxicological Evaluation of Natural Product Derivatives. 2007;**21**(1):1-6

[92] Schulz H, Joubert E, Schütze W. Quantification of quality parameters for reliable evaluation of green rooibos (*Aspalathus linearis*). European Food Research and Technology. 2003;**216**:539-543

[93] Khan H, Khan M, Qureshi MS, Ahmad S, Gohar A, Ullah H, et al. Effect of green tea extract (*Camellia sinensis*) on fertility indicators of post-thawed bull spermatozoa. Pakistan Journal of Zoology. 2017;**49**(4):1243-1249. DOI: 10.17582/journal.pjz/2017.49.4.1243.1249

[94] Rakuambo NC, Meyer JJM, Hussein AA. Xanthone with activity against erectile dysfunction isolated from *Securidaca longepedunculata*. Fitoterapia. 2004;**75**:497-499

[95] Meyer JJM, Rakuambo NC. Novel xanthones from *Securidaca longepedunculata* with activity against erectile dysfunction. Journal of Ethnopharmacology. 2008;**119**:599-603

[96] Yakubu MT, Afolayan AJ. Effect of aqueous extract of *Bulbine natalensis* baker stem on the sexual behaviour of male rats. International Journal of Andrology. 2009;**32**:629-636

[97] Bautista-Robles V, Guerrero-Reyes G, Sánchez-Torres GI, Parada-Luna FD, Barrios-Gutiérrez JJ, Vázquez-Cerero D, et al. *Cnidoscolus aconitifolius*: Therapeutic use and phytochemical properties. Literature review. Revista de la Facultad de Medicina. 2020;**68**(3):446-452

[98] Somade OT, Ugbaja RN, Idowu MA, Akinloye OA. *Cnidoscolus aconitifolius* leaf extract and ascorbate confer amelioration and protection against dimethyl nitrosamine-induced renal toxicity and testicular abnormalities in rats. Toxicology Reports. 2021;**8**:1098-1108

[99] Manayi A, Vazirian M, Saeidnia. *Echinacea purpurea*: Pharmacology, phytochemistry and analysis methods. Pharmacognosy Reviews. 2015;**9**(17):63-72

[100] Yamashiro Y. Gut microbiota in health and disease. Annals of Nutrition & Metabolism. 2018;**71**:242-246. DOI: 10.1159/000481627

[101] Motamedi S, Asghari A, Aahandideh A, Abedi G, Mortazavi P. Effects of Echinacea purpurea extract on sperm characteristics and hematology following testicular ischemia-reperfusion injury in rat. Crescent Journal of Medical and Biological Sciences. 2018;**5**(2): 119-122

[102] Abdallah FB, Fetoui H, Zribi N, Fakhfakh F, Keskes L. Protective role of caffeic acid on lambda cyhalothrin-induced changes in sperm characteristics and testicular oxidative damage in rats. Toxicology and Industrial Health. 2012;**28**(7):639-647

[103] Vinitha UG, Sathasivam R, Muthuraman MS, Park SU. Intensification of supercritical fluid in the extraction of flavonoids: A comprehensive review. Physiological and Molecular Plant Pathology. 2022;**118**:101815. DOI: 10.1016/j.pmpp.2022.101815

[104] Mazza G, Kay CD. Bioactivity, absorption and metabolism of anthocyanins. In: Daayf F, Lattanzio V, editors. Recent Advances in Polyphenols Research. West Sussex, UK: Blackwell Publishing Ltd.; 2008

[105] Jovičić M, Chmelíková E, Sedmíková M. Cryopreservation of boar semen. Czech Journal of Animal Science. 2020;**65**(4):115-123. DOI: 10.17221/47/2020-CJAS

[106] Abdellatif MA, Gabr SA, El-Sherbieny MA, Yousif AI. Effect of aqueous and methanolic green tea extracts on motility, normal forms, and kinetic parameters of spermatozoa in rahmani ram semen equilibrated at 5°C for 4 hours. Journal of Animal and Poultry Production. 2022;**13**(5):63-70

[107] Padayachee A, Netzel G, Netzel M, Day L, Zabaras D, Mikkelsen D, et al. Binding of polyphenols to plant cell wall analogues—Part 1: Anthocyanins. Food Chemistry. 2012;**134**:155-161. DOI: 10.1016/j.foodchem.2012.02.082

[108] Mazur M, Zwyrzykowska-Wodzińska A, Wojtas E, Zachwieja A, Salejda AM. Effect of yerba mate (*Ilex paraguariensis* A. St.-Hil.) supplementation on oxidative stress in ruminants. Chilean Journal of Agricultural Research. 2019;**79**):316-322. DOI: 10.4067/S0718-58392019000200316

[109] Shelton HM, Brewbaker JL. *Leucaena leucocephala* – the most widely used forage tree legume. In: Gutteridge. 1994. Available from: http://fao.org.com

[110] Hmohn ZZW, Kyaw H, Phyu HS, Hnin CLY, Sint SPT, Bo M, et al. Effect of toxic levels of *Leucaena leucocephala* on semen quality of goats in Myanmar. International Journal of Veterinary Sciences and Animal Husbandry. 2017;**2**(6):09-12

[111] Lin Y, Li J, Wang K, Fang Z, Che L, Xu S, et al. Effects of dietary L-leucine supplementation on testicular development and semen quality in boars. Frontiers in Veterinary Science. 2022;**9**:904653. DOI: 10.3389/fvets.2022.904653

[112] Pruneda A, Pinart E, Briz MD, Sancho S, Garcia-Gil N, Badia E, et al. Effects of a high semen-collection frequency on the quality of sperm from ejaculates and from six epididymal regions in boars. Theriogenology. 2005;**63**:2219-2232

[113] El-Seadawy IE, Kotp MS, El-Maaty AMA, Fadl AM, El-Sherbiny HR, Abdelnaby EA. The impact of varying doses of moringa leaf methanolic extract supplementation in the cryopreservation media on sperm quality, oxidants, and antioxidant capacity of frozen-thawed ram sperm. Tropical Animal Health and Production. 2022;**54**(6):344

[114] El-Nagar HA. Effect of some semen extenders as a natural source of antioxidants on quality of frozen Friesian semen. Journal of Animal and Poultry Production. 2017;**8**(9):391-397

[115] Hammad ME, Wafa WM, Gabr AA, Elkishky AY. Different types and levels of *Moringa oleifera* leaf extract as a source of antibiotics in Friesian bull semen extender. Journal of Animal and Poultry Production. 2019;**10**(3):67-71

[116] Zare M, Haghpanah T, Asadi Shekari M, Eftekhar-Vaghefi SH. The prophylactic effect of date palm (*Phoenix dactylifera* L.) fruit extract on testicular toxicity induced by formaldehyde: An experimental study. International Journal of Reproductive BioMedicine. 2020;**18**:275-286. DOI: 10.18502/ijrm.v13i4.6890

[117] El-Sheshtawy RI, El-Nattat WS, Samy IAS, Shahba MI, Al-Sedawy I. Chilled and post-thawed semen characteristics of buffalo semen diluted in tris extender enriched with date palm pollen grains (TPG). Asian Pacific Journal of Reproduction. 2016;**5**(3):252-255

[118] Kumar P, Kumar D, Jerome A. Buffalo semen cryopreservation: An update. In: Current Concepts in Bovine Reproduction. Singapore: Springer Nature Singapore; 2022. pp. 243-262. DOI: 10.1007/978-981-19-0116-4_13

[119] Ghafoor K, Sarker MZ, Al-Juhaimi FY, Babiker EE, Alkaltham MS, Almubarak AK. Extraction and evaluation of bioactive compounds from date (*Phoenix dactylifera*) seed using supercritical and subcritical CO_2 techniques. Food. 2022;**11**(12):1806

[120] El-Sisy GA, El-Badry DA, El-Sheshtawy RI, El-Nattat WS. Effects of *Phoenix dactylifera* pollen grains extract supplementation on post-thaw quality of Arabian stallion semen. Bulgarian Journal of Veterinary Medicine. 2018;**21**:40-49. DOI: 10.15547/bjvm.1044

[121] Mohamed OA, Abdulkareem TA. Some post-cryopreserved semen characteristics of holstein bulls as influenced by adding aquoeus extract of *Urtica dioica* and date palm pollen powder to tris extender. Plant Archives. 2020;**20**:461-467

[122] Larson JL, Miller DJ. Simple histochemical stain for acrosomes on sperm from several species. Molecular Reproduction and Development. 1999;**52**(4):445-449

[123] Bhusal KK, Magar SK, Thapa R, Lamsal A, Bhandari S, Maharjan R, et al. Nutritional and pharmacological importance of stinging nettle (*Urtica dioica* L.): A review. Heliyon. 2022;**8**(6):e09717. DOI: 10.1016/j.heliyon.2022.e09717

[124] Jalili C, Mohammad SR, Naseri A. Protective effect of *Urtica dioica* L against nicotineinduced damage on sperm parameters, testosterone and testis tissue in mice. Iranian Journal of Reproductive Medicine. 2014;**12**(6):401-408

[125] Gholami-Ahangaran M, Karimi-Dehkordi M, Javar AA, Maziar Salehi H, Ostadpoor M. A systematic review on the effect of ginger (*Zingiber officinale*) on improvement of biological

and fertility indices of sperm in laboratory animals, poultry and humans. Veterinary Medicine and Science. 2021;**7**(5):1959-1969

[126] Aires VA, Hinsch KD, Mueller-Schloesser F, Bogner K, Mueller-Schloesser S, Hinsch E. *In vitro* and *in vivo* comparison of egg yolk-based and soybean lecithin-based extenders for cryopreservation of bovine semen. Theriogenology. 2003;**60**(2):269-279. DOI: 10.1016/S0093-691X(02)01369-9

[127] de Paz P, Esteso MC, Alvarez M, Mata M, Chamorro CA, Anel L. Development of extender based on soybean lecithin for its application in liquid ram semen. Theriogenology. 2010;**74**(4):663-671. DOI: 10.1016/j.theriogenology.2010.03.022

[128] Emamverdi M, Zhandi M, Zare Shahneh A, Sharafi M, Akbari-Sharif A. Optimization of ram semen cryopreservation using a chemically defined soybean lecithin-based extender. Reproduction in Domestic Animals. 2013;**48**(6):899-904. DOI: 10.1111/rda.12183

[129] Pillet E, Labbe C, Batellier F, Duchamp G, Beaumal V, Anton M, et al. Liposomes as an alternative to egg yolk in stallion freezing extender. Theriogenology. 2012;**77**(2):268-279. DOI: 10.1016/j.theriogenology.2011.08.001

[130] Crespilho AM, Sá Filho MF, Dell'Aqua JA Jr, Nichi M, Monteiro GA, Avanzi BR, et al. Comparison of *in vitro* and *in vivo* fertilizing potential of bovine semen frozen in egg yolk or new lecithin based extenders. Livestock Science. 2012;**149**(1-2):1-6. DOI: 10.1016/j.livsci.2012.05.011

[131] Forouzanfar M, Sharafi M, Hosseini SM, Ostadhosseini S, Hajian M, Hosseini L, et al. *In vitro* comparison of egg yolk–based and soybean lecithin–based extenders for cryopreservation of ram semen. Theriogenology. 2010;**73**(4):480-487. DOI: 10.1016/j.theriogenology.2009.10.005

[132] Yildiz C, Bozkurt Y, Yavas I. An evaluation of soybean lecithin as an alternative to avian egg yolk in the cryopreservation of fish sperm. Cryobiology. 2013;**67**(1):91-94. DOI: 10.1016/j.cryobiol.2013.05.008

[133] Celeghini ECC, de Arruda RP, de Andrade AFC, Nascimento J, Raphael CF, Rodrigues PHM. Effects that bovine sperm cryopreservation using two different extenders has on sperm membranes and chromatin. Animal Reproduction Science. 2008;**104**(2-4):119-131. DOI: 10.1016/j.anireprosci.2007.02.001

[134] Salmani H, Nabi MM, Vaseghi-Dodaran H, Rahman MB, Mohammadi-Sangcheshmeh A, Shakeri M, et al. Effect of glutathione in soybean lecithin-based semen extender on goat semen quality after freeze-thawing. Small Ruminant Research. 2013;**112**(1-3):123-127. DOI: 10.1016/j.smallrumres.2012.12.015

[135] Akhter S, Ansari MS, Andrabi SMH, Rakha BA, Ullah N, Khalid M. Soya-lecithin in extender improves the freezability and fertility of buffalo (*Bubalus bubalis*) bull spermatozoa. Reproduction in Domestic Animals. 2012;**47**(5):815-819. DOI: 10.1111/j.1439-0531.2011.01973.x

[136] Papa FO, Felício GB, Melo-Oña CM, Alvarenga MA, De Vita B, Trinque C, et al. Replacing egg yolk with soybean lecithin in the cryopreservation of stallion semen. Animal Reproduction Science. 2011;**129**(1-2):73-77. DOI: 10.1016/j.anireprosci.2011.10.006

[137] Nouri H, Towhidi A, Zhandi M, Sadeghi R. The effects of centrifuged

egg yolk used with INRA plus soybean lecithin extender on semen quality to freeze miniature Caspian horse semen. Journal of Equine Veterinary Science. 2013;**33**(12):1050-1053. DOI: 10.1016/j.jevs.2013.03.184

[138] Scholfield CR. Composition of soybean lecithin. Journal of the American Oil Chemists' Society. 1981;**58**(10):889-892. DOI: 10.1007/BF02659652

[139] Abubakar AR, Haque M. Preparation of medicinal plants: Basic extraction and fractionation procedures for experimental purposes. Journal of Pharmacy & Bioallied Sciences. 2020;**12**(1):1-10. DOI: 10.4103/jpbs.JPBS_175_19. Epub 2020 Jan 29

[140] Heinrich M, Jalil B, Abdel-Tawab M, Echeverria J, Kulić Ž, McGaw LJ, et al. Best practice in the chemical characterisation of extracts used in pharmacological and toxicological research-the ConPhyMP-guidelines. Frontiers in Pharmacology. 2022;**13**:953205. DOI: 10.3389/fphar.2022.953205

[141] Handa SS, Khanuja SP, Longo G, Rakesh DD. Extraction technologies for medicinal and aromatic plants. Earth, Environmental and Marine Sciences and Technologies; 2008. pp. 260. Available from: http://www.di.ens.fr/users/longo/download.html

[142] Stéphane F, Kezetas Jean Jules B, El-Saber Batiha G, Ali I, Ndjakou Bruno L. Extraction of bioactive compounds from medicinal plants and herbs. In: Natural Medicinal Plants. London, UK: IntechOpen; 2022. DOI: 10.5772/intechopen.98602

[143] Zhang L, Xu Q, Li L, Lin L, Yu J, Zhu J, et al. Antioxidant and enzyme-inhibitory activity of extracts from *Erigeron annuus* flower. Industrial Crops and Products. 2020;**148**:112283

[144] Mukherjee PK. Quality Control of Herbal Drugs: An Approach to Evaluation of Botanicals. Business Horizons. 1st ed. New Delhi: Pharmaceutical Publishers; 2002. pp. 380-422

[145] Zhao J, Jin Y, Du M, Liu W, Ren Y, Zhang C, et al. The effect of dietary grape pomace supplementation on epididymal sperm quality and testicular antioxidant ability in ram lambs. Theriogenology. 2017;**97**:50-56. DOI: 10.1016/j.theriogenology.2017.04.010

[146] Pini T, Makloski R, Maruniak K, Schoolcraft WB, Katz-Jaffe MG. Mitigating the effects of oxidative sperm DNA damage. Antioxidants. 2020;**9**:589. DOI: 10.3390/antiox9070589

[147] Benhenia K, Lamara A, Fatmi S, Iguer-Ouada M. Effect of cyclodextrins, cholesterol and vitamin E and their complexation on cryopreserved epididymal ram semen. Small Ruminant Research. 2016;**141**:29-35

[148] Sicherle CC, Maia MS, Bicudo SD, Rodello L, Azevedo HC. Lipid peroxidation and generation of hydrogen peroxide in frozen-thawed ram semen supplemented with catalase or trolox. Small Ruminant Research. 2011;**95**(2-3):144-149

[149] Baghshahi H, Riasi A, Mahdavi AH, Shirazi A. Antioxidant effects of clove bud (*Syzygium aromaticum*) extract used with different extenders on ram spermatozoa during cryopreservation. Cryobiology. 2014;**69**(3):482-487

[150] Fraczek M, Kurpisz M. The redox system in human semen and peroxidative damage of spermatozoa. Postepy Higieny i Medycyny Doswiadczalnej (Online). 2005;**59**:523-534

[151] Allai L, Druart X, Öztürk M, Ben Moula A, Nasser B, El Amiri B.

Protective effects of *Opuntia ficus*-indica extract on ram sperm quality, lipid peroxidation and DNA fragmentation during liquid storage. Animal Reproduction Science. 2016;**175**:1-9. DOI: 10.1016/j.anireprosci.2016.09.013

[152] Jayawardana BC, Liyanage R, Lalantha N, Iddamalgoda S, Weththasinghe P. Antioxidant and antimicrobial activity of drumstick (*Moringa oleifera*) leaves in herbal chicken sausages. LWT – Food Science and Technology. 2015;**64**:1204-1208. DOI: 10.1016/j.lwt.2015.07.028

[153] Wang H, Lu P, Li Z, Yuan C, Liu H, Zhao J, et al. Oligomeric proanthocyanidins and bamboo leaf flavonoids improve the quality of bull semen cryopreservation. Molecules. 2022;**27**(3):1144. DOI: 10.3390/molecules27031144

[154] Leahy T, Gadella BM. Sperm surface changes and physiological consequences induced by sperm handling and storage. Reproduction. 2011;**142**:759. DOI: 10.1530/REP-11-0310

[155] Aitken RJ. Free radicals, lipid peroxidation and sperm function. Reproduction, Fertility and Development. 1995;7(4):659-668. DOI: 10.1071/RD9950659

[156] Khursheed A, Rather MA, Jain V, Rasool S, Nazir R, Malik NA, et al. Plant based natural products as potential ecofriendly and safer biopesticides: A comprehensive overview of their advantages over conventional pesticides, limitations and regulatory aspects. Microbial Pathogenesis. 2022;**173**:105854

[157] Rahman SU, Huang Y, Zhu L, Feng S, Khan IM, Wu J, et al. Therapeutic role of green tea polyphenols in improving fertility: A review. Nutrients. 2018;**10**(7):834. Available from: https://www.mdpi.com/2072-6643/10/7/834

[158] Woith E, Guerriero G, Hausman JF, Renaut J, Leclercq CC, Weise C, et al. Plant extracellular vesicles and nanovesicles: Focus on secondary metabolites, proteins and lipids with perspectives on their potential and sources. International Journal of Molecular Sciences. 2021;**22**(7):3719. Available from: https://www.mdpi.com/1422-0067/22/7/3719

[159] Ferramosca A, Lorenzetti S, Di Giacomo M, Lunetti P, Murrieri F, Capobianco L, et al. Modulation of human sperm mitochondrial respiration efficiency by plant polyphenols. Antioxidants. 2021;**10**(2):217. Available from: https://www.mdpi.com/2076-3921/10/2/217

[160] Olson KR, Briggs A, Devireddy M, Iovino NA, Skora NC, Whelan J, et al. Green tea polyphenolic antioxidants oxidize hydrogen sulfide to thiosulfate and polysulfides: A possible new mechanism underpinning their biological action. Redox Biology. 2020;**37**:101731. Available from: https://www.sciencedirect.com/science/article/pii/S2213231720309368

[161] Raheja N, Choudhary S, Grewal S, Sharma N, Kumar N. A review on semen extenders and additives used in cattle and buffalo bull semen preservation. Journal of Entomology and Zoology Studies. 2018;**6**(3):239-245. Available from: https://www.academia.edu/download/56583678/A_review_on_semen_extenders_and_additives.pdf

[162] Vongsak B, Sithisarn P, Gritsanapan W. Simultaneous HPLC quantitative analysis of active compounds in leaves of *Moringa oleifera* lam. Journal of Chromatographic Science. 2014;**52**(7):641-645. Available from: https://academic.oup.com/chromsci/article-abstract/52/7/641/521924

[163] Sokunbi O, Ajani O, Lawanson A, Amao E. Antibiotic potential of *Moringa leaf* (*Moringa oleifera* lam.) crude extract in bull semen extender. European Journal of Medicinal Plants. 2015;**9**(2):1-8. Available from: http://go7publish.com/id/eprint/2393/

[164] Hodas F, Zorzenon MR, Milani PG. *Moringa oleifera* potential as a functional food and a natural food additive: A biochemical approach. Anais da Academia Brasileira de Ciencias. 2021;**22**(93):e20210571. Available from: https://www.scielo.br/j/aabc/a/fbjxDWcDJGLsRjPNtVCj5cf/

[165] Mostek A, Dietrich MA, Słowińska M, Ciereszko A. Cryopreservation of bull semen is associated with carbonylation of sperm proteins. Theriogenology. 2017;**92**:95-102. Available from: https://www.sciencedirect.com/science/article/pii/S0093691X17300110

Section 5

Nutrition and Environment

Chapter 6

Effects of Natural Feed Additives on Reducing the Carbon Footprint in Broiler Farms

Roshan Riaz, Mükremin Ölmez, Özlem Karadağoğlu and Tarkan Şahin

Abstract

The world's population is continuously growing, resulting in an increasing need for meat, particularly poultry. As the poultry industry expands to fulfill this demand, it faces the dual challenge of increasing production while reducing its environmental impact. Even though broiler production has comparatively lower environmental impacts than other livestock species, it remains a significant source of greenhouse gas emissions. Emissions and nutrient losses occur at various stages of chicken production, emphasizing the need for effective management strategies to minimize environmental impact. Natural feed additives are viable solutions to this issue. These additives enhance nutrient utilization and improve gut health, resulting in better broiler performance and reduced emissions. By optimizing digestive efficiency and minimizing waste by-products, natural feed additives offer a promising approach to lowering the environmental footprint of poultry production. The chapter compiles recent research findings, showing the effectiveness of these additives in broiler production in mitigating the environmental impacts. By integrating these innovative solutions, the poultry industry can sustainably meet the growing demand for meat, thereby contributing to global food security while reducing its environmental impact.

Keywords: broiler, carbon footprints, environment, life cycle assessment, natural feed additives

1. Introduction

The global population is projected to continue growing over the next 50–60 years, peaking at around 10.3 billion people in the mid-2080s, up from 8.2 billion in 2024 [1]. This population surge will inevitably lead to increased food demand, particularly for meat [2]. This growth in meat consumption is driven by demographic factors such as urbanization and population growth and by nutritional preferences reflecting transitions towards diets richer in animal protein. There has been a significant shift from red meat to white meat, particularly poultry [3, 4]. To satisfy this increasing demand, poultry farming has been consistently expanding, which has led to environmental concerns related to poultry production and underscored the importance of sustainable animal diets. In 2023, global broiler meat production reached approximately 103.5 million metric tons and is projected to rise to about 104.2 million metric tons by 2024 [5]. The majority of this increase is expected in China, Brazil, and the European Union, while decreases are anticipated in Pakistan, Turkey, South Africa, and the Islamic Republic of Iran. In China, poultry meat production is forecasted to grow by more than 2% due to increased investments in breeding programs, which have boosted the supply of domestically produced breeding stock [6].

Although the broiler industry has a comparatively lower carbon footprint (CF) compared to other livestock, it still significantly contributes to greenhouse gas emissions (GHG) [7]. The effects of poultry farming on the climate have spurred extensive research aimed at enhancing scientific understanding in this field. However, there is still a critical need for further investigation into the influence of poultry production on GHG emissions and the development of effective mitigation practices to reduce this environmental burden. Despite the substantial contribution of food systems to GHG emissions, opportunities for reduction have received limited attention because these emissions are seen as an inevitable aspect of meeting global nutritional needs. However, consumer demand is rising for products that have a minimal environmental impact.

Natural feed additives in animal diets can enhance productivity and performance by improving digestibility and maintaining beneficial gut microflora [8], thereby benefiting the environment [9]. Several studies have been conducted on the effects of natural feed additives on the reduction of emissions from broilers. Natural feed additives, including phytogenic compounds, prebiotics, probiotics, organic acids, and enzymes, represent a promising research area for animal nutrition, health, and environmental impacts.

This chapter aims to integrate the latest research on natural feed additives in broiler nutrition and manure management to mitigate the carbon footprints of broiler houses. By exploring these natural alternatives, the chapter seeks to provide insights into effective strategies for reducing the environmental impact of poultry farming while meeting the increasing global demand for meat.

2. Methodology

This study provides a comprehensive understanding of the environmental benefits of natural feed additives in reducing the carbon footprint of broiler farms. The study employed a thorough up-to-date literature review, focusing on studies that examined the effects of phytogenic compounds, prebiotics, probiotics, organic acids, and enzymes on emissions from broiler production. The studies were selected based on their relevance, research quality, and publication in reputable journals. An analytical framework was used to assess the impact of these additives on key environmental indicators, including GHG emissions, as well as on broiler production performance and nutrient utilization. This chapter provides valuable insights into the potential benefits and limitations of natural feed additives as a sustainable strategy for reducing the environmental impact of poultry farming.

3. Green house gas emissions from livestock sector

Principal sources of CF in livestock production primarily originate from several key sources. Nitrous oxide (N_2O) emissions arise from fertilizer use, cultivation, and manure handling. Methane (CH_4) emissions are predominantly generated through enteric fermentation and manure management processes. Additionally, carbon dioxide (CO_2) emissions are associated with on-farm fuel combustion, operations involving heavy machinery, heating, electricity consumption, fertilizer application, and land-use changes for feed production [10].

The GHGs of livestock are categorized into direct and indirect emissions. Globally, direct GHG from the livestock, including CH_4 from enteric fermentation and both CH_4 and N_2O from manure, amount to approximately 3.7 giga-tonnes (Gt) of CO_2 equivalent (CO_2eq), representing about 60% of total livestock emissions. Indirect emissions, which account for 40% (2.6 Gt CO_2eq), are derived from the fertilizers manufacturing and pesticides for feed production, field applications, blending and pelleting of concentrated feed, synthetic fertilizer application, crop residue decomposition, rice cultivation, manure deposition, and the processing and transportation of feed, live animals, and livestock products [11].

3.1 Emissions from different livestock species

The production of animal protein is a significant contributor to global GHG emissions. In 2015, emissions from livestock production amounted to 6.2 Gt CO_2eq, which constitutes 12% of total anthropogenic emissions estimated at 50–52 Gt CO_2eq. Cattle are the leading contributors among livestock species, generating about 3.8 Gt CO_2eq annually and representing roughly 62% of total livestock-related emissions. Pigs, chickens, buffaloes, and small ruminants contribute 14, 9, 8, and 7% of livestock emissions, respectively (**Figure 1**). Meat production shares a major proportion of emissions (67%), followed by milk (30%) and eggs (3%). Compared to ruminants, monogastric animals are relatively minor emitters of GHGs [11].

3.2 Emission sources and emission percentages

Methane accounts for 54% of all livestock emissions, while CO_2 and N_2O contribute 31 and 15%, respectively. The sources and specific GHGs vary significantly between different geographic locations, animal species, and production systems. In ruminant systems, enteric CH_4 constitutes a substantial proportion of total emissions,

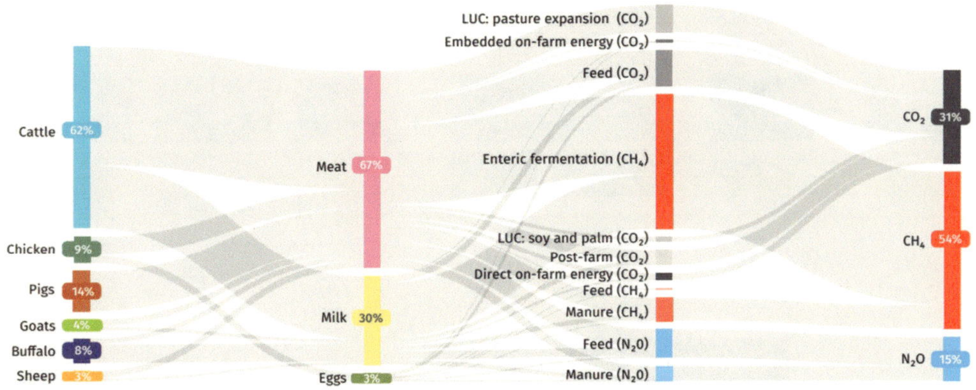

Figure 1.
Species, product type, sources of emissions, and GHG contribution (percentage) to total livestock sector GHG emissions [11].

whereas in mono-gastric systems, feed production, land-use change, and manure management are the primary contributors. Enteric fermentation contributes approximately 46% of total emissions, followed by manure storage and processing, which accounts for around 12.8% of emissions. Production of feed is responsible for 12.64% of the sector's emissions. This comprehensive analysis highlights the significant impact of livestock production on global GHG emissions and underscores the need for targeted mitigation strategies to reduce emissions from this sector [11].

3.3 Emissions intensity: Developed vs. developing world

The developed world, while contributing to high overall emissions, demonstrates much lower emissions intensities compared to the developing world. This is largely due to advancements in livestock nutrition, genetics, health, and management practices in developing countries, which help to reduce CH_4 and carbon dioxide CO_2 emissions intensities, also aided by reduced land-use needs. In contrast, many developing regions face high emissions levels from livestock, largely because of high emissions intensities. These elevated intensities arise from lower productivity and the large number of animals, resulting in inefficient resource use and higher GHG emissions per unit of animal product [11–13].

4. Broiler production and carbon footprints

The broiler industry, although having a comparatively lower carbon footprint than other animal production systems, still makes a considerable contribution to GHG emissions. The various processes involved in broiler raising ultimately result in emissions to the environment [14]. The primary sources include on-farm management practices and associated inputs to the farm [12]. Therefore, the efficiency with which broilers convert feed into meat and associated inputs are critical factor in determining the intensity of environmental impacts associated with broiler production. **Figure 2** illustrates the range of greenhouse gas emissions associated with some segments of the chicken supply chain.

The main sources of CF in broiler production are manure, energy consumption, and feed. Feed production involves a series of practices that, in some way, significantly influence the CF of broiler production. Additionally, energy used for

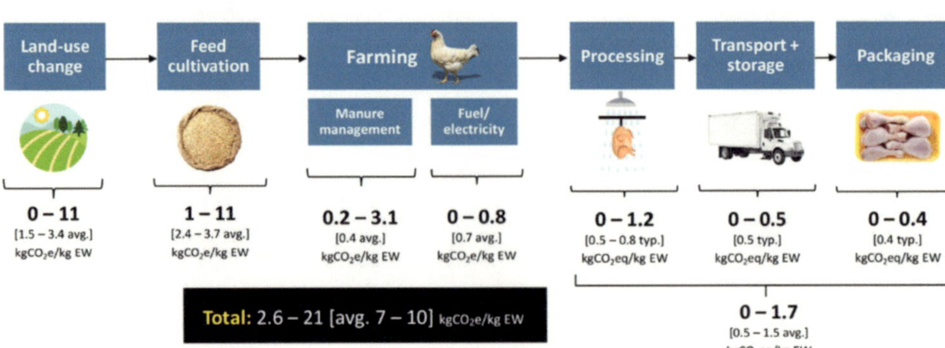

Figure 2.
Greenhouse gas emission ranges from some segments of supply chain of chicken [15].

husbandry practices at the farm further contributes to emissions. The CH_3, a potent GHG, is also released during manure management processes. The main GHGs linked to on-farm activities are carbon dioxide, methane, and nitrous oxide, with nitrous oxide emissions in modern industrial broiler farms primarily originating from bedding materials [10, 16, 17].

4.1 Nutritional strategies in mitigating GHG emissions at broiler farm

A number of strategies should be implemented to reduce the environmental impacts of emissions from poultry production. Mitigation strategies include a series of practices from raw material production to the utilization of the final product. It is difficult to describe the mitigation practices at each step. However, in broiler production, the strategies can be summarized as practices related to feed utilization, on-farm production, and management practices, manure management (including its use for energy production or as a fertilizer), and energy management through cost-effective and renewable solutions [17]. On-farm major contributors to the emissions from the broiler house include feed and manure management practices. Several studies have been conducted on nutritional manipulation to mitigate the impacts of broiler production. Replacement of soybean with alternative protein sources [18, 19] palm oil replacement with cottonseed oil [20], low protein diet formulation [21], balanced diet with low protein [22], and insect meal as alternative protein source [23] are some example of the nutritional strategies to reduce the CF of the broilers. However, natural feed additives have also been proven to be a strategy to mitigate CF of broilers through both supplementation in diets and manure management.

5. Natural feed additives as a promising approach in reducing carbon foot prints at broiler farms

A detailed illustration of the impact of natural feed additives on the broiler farm is presented in **Figure 3**.

5.1 Phytogenic feed additives

Phytogenic feed additives are feed additives derived from herbs, spices, and other plants. These additives are available in various forms, including solid, dried, and ground forms, as well as extracts or essential oils. Phytogenic feed additives encompass a wide range of substances, primarily extracts from plant materials such as flowers, buds, seeds, leaves, twigs, bark, herbs, wood, fruits, and roots [26]. Phytogenic compounds possess several beneficial effects on the broilers, such as promoting growth performance, intestinal functions, immunomodulation, and improved health [27–29].

Despite the ongoing research, the exact mechanism of the effects of the phytogenic feed additives on the mitigation of emissions from broilers need to be elucidated. However, the mode of action of the phytogenic feed additives can be described on the basis of the active compounds present in the source plant or their extract. Within phytogenic feed additives, the concentration of active compounds can vary significantly based on several factors. These include the specific plant part used (such as seeds, leaves, roots, or bark), the season in which the plant is harvested, and the geographical origin of the plant [26]. The mode of action involves increased growth and production performance through an antimicrobial effect on the gut microflora,

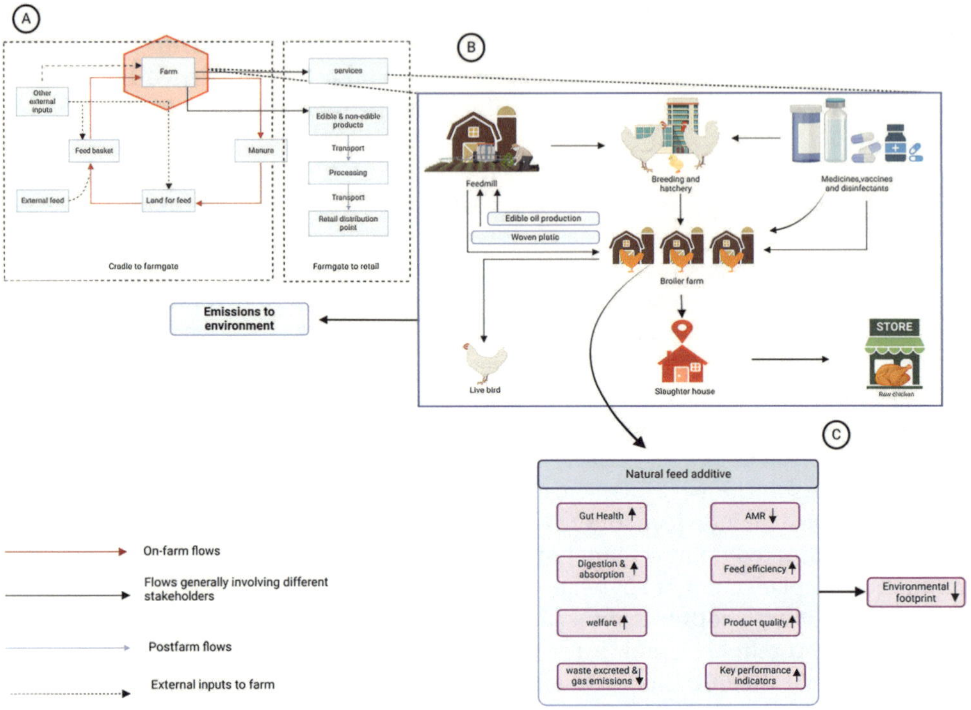

Figure 3.
Summary of the effects of natural feed mitigating additives on the mitigation of environmental footprints from the broiler farm and some associated inputs to the broiler farm [24, 25].

reduced toxic metabolite levels in the intestine, and relief of the stress to the birds. Improved feed efficiency promotes a more complete oxidative breakdown of organic matter in the intestine, leading to a decrease in odor and lower concentrations of harmful gases in the excreta [30]. Enhanced nutrient assimilation can decrease the amount of organic matter that contributes to the production of noxious gases such as NH_3 and H_2S [31]. Ultimately, this will result in reduced carbon footprints in broiler production. However, the results on the effect of phytogenic feed additives are inconsistent. Cross et al. [32] reported that broiler chickens fed a diet containing thyme herb exhibited similar body weight gain (BWG) to those on a control diet. A study on the *Camellia L.* plant showed a negative impact on body weight and feed intake in broiler chickens [33].

Several studies have been conducted on the effects of phytogenic feed additives on the reduction of emissions from broilers. A study on the combination of 900 mg/kg of ginseng and 900 mg/kg of artichoke extract reported that incorporating a 0.05% concentration of this herbal mixture into the diet significantly enhanced both the average body weight gain and feed conversion ratio over the entire experimental period. Furthermore, this herbal mixture group exhibited a substantial reduction in NH_3 and H_2S emissions in the excreta [30]. Similarly, in broiler birds fed with fermented ginger, there was a significant reduction in fecal NH_3 and CO_2 emissions [34]. Li et al. [35] also reported that when phytoncide was supplemented at 0.5 and 1 g per kg of basal diet, it significantly reduced emissions of NH_3 and hydrogen sulfide H_2S gases. Yucca may also be utilized in broiler diets to mitigate emissions. A study demonstrated that incorporating 0.12 g/kg of Yucca into the broiler diet effectively reduced ammonia concentrations in poultry houses [36]. The fruit *Punica granatum L.* has long been used in traditional medicine due to its numerous phytobiotic characteristics, which possess antimicrobial and immunomodulatory properties. Supplementing broiler diets with 1% *Punica granatum L.* significantly reduced the emissions of ammonia and methanethiol from broiler excreta [37]. Bostami et al. [38] reported that supplementation with fermented pomegranate byproducts significantly reduced fecal ammonia emissions in all fermented pomegranate byproducts supplemented groups. Citrus extract has demonstrated beneficial effects on growth performance in broiler chickens [39]. Citrus extract additives also help mitigate the emissions from the broiler. A study reported that the inclusion of citrus extract 25 kg bag in feed at a rate of 250 g per ton of feed resulted in a reduction of 6 tons of CO_2 equivalent (CO_2 eq) emissions over the life cycle of poultry production [40]. It is important to note that not all the phytogenic feed additives results in decreased emissions from the broiler birds. A feeding trial was conducted to assess the impact of dietary inclusion of silymarin seed at supplementation levels of 0, 0.02, 0.04, and 0.06%. The results indicated that the addition of silymarin to the broiler diet did not lead to significant differences in the excretion levels of CO_2, H_2S, and NH_3. However, results showed an increase in BWG and nutrient digestibility [39]. This shows that silymarin ultimately reduces the carbon footprints of the broiler production per kg of meat produced.

In an experiment investigating alternatives to antibiotic growth promoters, three different types of essential oil were evaluated: cinnamon oil, clove oil, and ajwain oil. The results showed that nitrogen excretion per kilogram of body weight was reduced in the cinnamon oil group fed diet supplemented with CNO at 0.3 g/kg diet compared to the control, clove oil (0.6 g/kg diet), and ajwain oil (0.4 g/kg diet) groups [41]. Similarly, Park and Kim [42] reported that essential oil supplementation decreased the ammonia emission from the broiler chickens. Chemicals and bioactive compounds capable of altering gut microflora might be responsible for enhanced nutrient retention [42]. In addition, essential oil disrupts the cell wall of harmful bacteria and improves the gut health and environment, which results in improved nutrient digestibility, growth performance, and reduced emissions [27, 42].

5.2 Probiotics and prebiotics

Prebiotics and probiotics have gained significant attention as supplements in poultry nutrition, particularly for broilers. Prebiotics, which are indigestible components of feed, positively impact the host by selectively enhancing the growth and activity of beneficial bacteria in the intestinal tract [43]. Probiotics are living microorganisms that, when administered in adequate amounts, provide beneficial effects on the host [44]. Numerous probiotic strains are employed in poultry nutrition, including various strains of *Lactobacillus*, *Bifidobacterium*, *Bacillus*, and *Enterococcus*. Probiotics exert their effects through multiple mechanisms, including the enhancement of intestinal barrier function, the modulation of the host's immune response, competition with pathogenic bacteria for attachment sites and nutrients, and the production of antimicrobial compounds such as bacteriocins and organic acids [45]. The incorporation of prebiotics and probiotics into broiler diets has a vital role in broiler production performance and health. These positive outcomes stem from the alteration of gut microbiota, fortification of gut barrier function, and inhibition of harmful bacteria, resulting in the sustainable and productive rearing of broilers [46]. However, some studies reported no effect of the use of single or multiple-strain probiotics or symbiotics [47, 48]. The variation in the results could be due to the strains of probiotic, prebiotic, and raising conditions of the birds.

Prebiotics and probiotics hold significant importance in reducing the carbon footprint of broiler production through several mechanisms. By improving gut health and nutrient absorption, these additives enhance feed conversion rates, leading to reduced undigested protein excretion and, consequently, decreased ammonia emissions from broiler houses [49]. Improved nutrient digestibility [50], driven by probiotics such as *Bacillus subtilis* and *Bacillus licheniformis*, means that broilers require less feed to achieve the same weight gain, decreasing overall carbon emissions associated with feed production and transport [51]. Furthermore, prebiotics and probiotics decrease the production of odorous gases like hydrogen sulfide and methane [52] by modulating microbial activity in the gut and feces, contributing to a lower greenhouse gas footprint. Improved feed efficiency, facilitated by a healthy gut environment, further minimizes the need for feed, thus reducing the carbon footprint from feed-related activities [53].

Probiotics, particularly *Bacillus subtilis* and *Bacillus licheniformis* strains, have exhibited considerable potential in diminishing carbon footprints in broiler production. These strains have been found to improve growth performance, nutrient absorption, and gut microbial balance, which directly correlates with reduced emissions of harmful gases, such as NH_3, H_2S, CO_2, and acetic acid. *Bacillus subtilis* supplementation led to a significant increase in weight gain (6%) and improvement in feed conversion ratio (FCR) (4%); emission reductions included a 12% decrease in NH_3 and an 8% decrease in H_2S emissions. Similarly, a 7% increase in BWG and a 5% improvement in (FCR) were seen when broilers were supplemented with *Bacillus licheniformis*. The reduction in emissions included a 10% decrease in CO_2 and a 7% decrease in acetic acid emissions [46].

Probiotics have been demonstrated to modify the gut microbiota, enhancing beneficial bacteria such as *Lactobacillus* while decreasing detrimental bacteria like *Salmonella*. The investigation demonstrated that the combined effect of tri-strain probiotics (*Bacillus subtilis*, *Clostridium butyricum*, and *Lactobacillus acidophilus*) in broiler rations led to a significant decrease in NH_3 by 25% emissions from excreta.

A study on the effects of dietary metallo-protease and *Bacillus velezensis* CE 100 in broiler chickens resulted in a 15% reduction in litter moisture content, which led to a 10% decrease in the release of dimethyl sulfide ($(CH_3)_2S$) in and other volatile compounds metallo-protease group and *Bacillus velezensis* CE 100 group achieved a 12% reduction in litter moisture, contributing to an 8% decrease in greenhouse gas emissions. Combined, these additives resulted in a 20% overall reduction in greenhouse gas emissions, highlighting their potential in lowering the carbon footprint in broiler production [54]. In another study, the inclusion of a symbiotic additive containing *Bacillus subtilis*, inulin, and *Saccharomyces cerevisiae* led to significant decreases in urinary nitrogen in the excreta, reducing ammonia emissions from the manure by 8%. Specifically, *Bacillus subtilis* contributed to a 5% reduction in ammonia emissions due to enhanced gut health and nutrient absorption. Inulin, a prebiotic, further reduced ammonia emissions by 2% through its role in promoting beneficial bacterial growth and improving gut efficiency. *Saccharomyces cerevisiae* added 1% reduction by enhancing overall feed digestibility and reducing waste. This combined reduction in ammonia emissions signifies a substantial positive impact on lowering the carbon footprint associated with broiler production [55].

Yeast, particularly *Saccharomyces cerevisiae*, significantly reduces the carbon footprint in broilers. By enhancing nutrient utilization, yeast improves digestibility and absorption, leading to less undigested feed and, consequently, lower methane and ammonia emissions from manure. Demonstrated that yeast culture supplementation in broilers indicated 10% reduction in ammonia emissions in broiler houses supplemented with higher doses of yeast, suggesting a strong correlation between yeast dosage and emission reduction [56].

5.3 Organic acids

Incorporating organic acids as feed additives in poultry nutrition offers a promising strategy for reducing emissions while enhancing poultry health and productivity. A study showed that the inclusion of a blend of organic acids and medium-chain fatty acids in a corn-soybean meal-based diet for broilers significantly enhanced body weight gain, feed conversion ratio, and dry matter digestibility. Additionally, it linearly increased *Lactobacillus* concentrations while reducing ammonia emissions, though no impact was observed on hydrogen sulfide emissions [57].

5.4 Enzymes

Enzyme supplementation in the diets of broilers has demonstrated considerable potential in reducing carbon footprints connected with poultry production. Research has demonstrated that enzymes can enhance nutrient utilization, leading to a decrease in nitrogen excretion and subsequent reductions in NH_3 emissions [58]. Javaid et al. [59] reported that a commercial enzyme mix containing phytase, xylanase, amylase, and protease decreased nitrogen excretion in broiler excreta by 14.8%. Although there were no significant differences in growth performance and FCR between enzyme-treated birds and non-enzyme-treated birds, the reduction in nitrogen excretion suggests improved nutrient utilization. Similarly, another study demonstrated that protease supplementation can reduce CP content in the excreta up to 16.6% without substantial effect on BWG and FCR [60]. Nguyen et al. [61] study indicated that a cocktail of protease and xylanase not only improved body weight gain and dry matter retention but also significantly decreased NH_3 and H_2S emissions. These results suggest that enzymes can play a crucial role in making poultry production more sustainable by mitigating the environmental impacts associated with feed conversion and waste emissions. Different studies on enzyme supplementation in broiler diets provides persuasive evidence of the beneficial effects of enzymes on emissions. Protease, when supplemented at different levels, showed a consistent trend of reducing NH_3 emissions, particularly at higher doses. Park and Kim [42] observed that a 0.02% protease supplementation resulted in lower NH_3 emissions without affecting H_2S or total mercaptans. Overall, these studies indicate that enzyme supplementation not only enhances feed efficiency and growth performance but also plays a significant role in reducing harmful emissions, thereby contributing to a lower environmental impact of broiler production systems.

Ammonia gas is primarily produced due to the inefficient utilization of feed protein, leading to nitrogen excretion as uric acid. A study demonstrated that supplementing broiler diets with protease reduced NH_3 emissions, although it had no effect on reducing H_2S levels [42]. Furthermore, contrary to the advantages of exogenous enzymes in minimizing ammonia emissions in broiler chicken houses, certain have produced diverging results. Lourenco et al. [62] uncovered that the utilization of protease failed to impact ammonia emissions. This absence of impact on ammonia emissions corresponds to their observation that protease did not enhance protein digestibility in broiler chickens. Moreover, Akter et al. [63] documented that incorporating phytase into feed could actually lead to an increase in ammonia concentrations in broiler litter. This inconsistency in results was attributed to experimental conditions.

Life cycle assessment studies have shown that enzyme supplementation in poultry diets can bring about significant environmental benefits. In particular, the use of protease has been found to significantly reduce environmental impacts in both feed production and broiler management. Leinonen et al. [64] reported that protease supplementation resulted in a marked decrease in ammonia emissions, which is vital for mitigating air pollution in the poultry industry. Similarly, Oxenboll et al. [65] research demonstrated that protease use can reduce water and air pollution by lowering nitrous compound emissions, thus addressing eutrophication and acidification. Hickmann et al. [66]'s study found that β-mannanase supplementation, although variable in its effect on acidification, generally reduced the need for soybean oil in feed, lowering environmental impacts by up to 5.6 and 1.1%. An investigation into amylase use in Brazilian chicken feed revealed a potential 6% reduction in greenhouse gas emissions [67]. These findings collectively highlight the role of enzyme supplementation in promoting environmental sustainability in poultry production.

5.5 Natural additives for litter treatment

Emissions from broiler houses, particularly NH_3, show significant environmental and health concerns. Mitigating these emissions is crucial for sustainable poultry production. Recent studies have demonstrated the effectiveness of several natural feed additives, including alum, zeolite, clinoptilolite, biochar, urease inhibitors, and probiotics, in mitigating NH_3 emissions from broiler houses. Each of these additives contributes uniquely to emission reduction and litter quality improvement.

Alum is particularly effective, reducing daily average in-house NH_3 concentration by 42% and overall NH_3 emission rate by 47%. It also increases litter nitrogen content and reduces the carbon-to-nitrogen (C/N) ratio without significantly affecting N_2O and CH_4 emissions, although it does reduce CO_2 emissions significantly [68].

Zeolite has shown effectiveness in studies, reducing moisture, pH, and ammonia levels while increasing nitrogen retention. Zeolite supplementation (200–600 g/kg of litter) was observed to reduce moisture content and ammonia volatilization [69]. Additionally, adding 5–100 g/kg zeolite to litter consistently reduced moisture content, pH, and ammonia volatilization across multiple flocks [70].

Clinoptilolite, another zeolite variant, also contributes to emission reductions, decreasing NH_3 and N_2O emissions by 28 and 34%, respectively, without affecting CO_2 losses. In addition to mineral additives, biochar and urease inhibitors have proven effective [71]. Pereira et al. [72] study showed that these additives reduce NH_3 emissions by 40–60% and global warming potential emissions by 31%. Biochar was particularly effective, achieving a 58% reduction, while urease inhibitors required reapplication to maintain a 41% reduction. Some studies highlighted the effectiveness of clay and biochar amendments in reducing emissions broiler manure. Chen et al. [73] research showed that adding clay (0–10% by dry weight) to chicken manure and wheat straw significantly reduced N_2O (25.3–63.4%), CH_4 (26.01–50.24%), and ammonia (8.5–70.5%) emissions, although CO_2 emissions increased at 10% clay. Kalus et al. [74] study found that beech-woo biochar (2 and 4%) and a biochar-glycerin-aluminosilicates mix (3 and 6%) as dietary additives reduced ammonia emissions by up to 17%, though odor emissions remained unaffected.

Probiotics offer a biological approach to emission reduction. Mahardhika et al. [75] research highlighted that probiotics in drinking water or sprayed on litter significantly reduced ammonia concentrations in broiler cages. Another study indicated a 35% reduction in ammonia emissions with probiotic-treated litter [76].

6. Conclusion

In conclusion, the environmental impact of chicken farming, although less than that of other livestock, still contributes significantly to emissions of ammonia, nitrous oxide, nitrogen oxides, and methane. These uncontrolled emissions pose significant challenges and could negatively impact future food sustainability. Available strategies for reducing emissions are numerous, but no single solution is comprehensive or long-lasting. However, nutritional strategies, particularly the use of natural feed additives, have shown promising results. These additives enhance nutrient absorption, gut health, and immune function, leading to improved performance in broilers. Moreover, they significantly reduce the emission of harmful gases per kilogram of meat produced, making them the most effective current method. Thus, it is crucial to incorporate nutritional strategies that include natural feed additives to achieve a more sustainable and environmentally friendly poultry farming system.

Conflict of interest

The authors declare no conflict of interest.

Author details

Roshan Riaz, Mükremin Ölmez*, Özlem Karadağoğlu and Tarkan Şahin
Department of Animal Nutrition and Nutritional Diseases, Faculty of Veterinary Medicine, Kafkas University, Kars, Turkey

*Address all correspondence to: mukremin.olmez@hotmail.com

References

[1] United Nations Department of Economic and Social Affairs, Population Division. World Population Prospects 2024: Summary of Results (UN DESA/POP/2024/TR/NO. 9). New York, NY: United Nations Publications; 2024

[2] González N, Marquès M, Nadal M, Domingo JL. Meat consumption: Which are the current global risks? A review of recent (2010-2020) evidences. Food Research International. 2020;**137**:109341

[3] Arellano M, Bond S. Some tests of specification for panel data: Monte Carlo evidence and an application to employment equations. The Review of Economic Studies. 1991;**58**(2):277-297. DOI: 10.2307/2297968

[4] Mathijs E. Exploring future patterns of meat consumption. Meat Science. 2015;**109**:112-116. DOI: 10.1016/j.meatsci.2015.05.007

[5] Shahbandeh M. Chicken Meat Production Worldwide from 2012 to 2024. New York, NY: Statista Inc.; 2024

[6] FAO. Meat Market Review: Emerging Trends and Outlook. Italy: FAO Rome; 2023

[7] Leinonen I, Kyriazakis I. How can we improve the environmental sustainability of poultry production? Proceedings of the Nutrition Society. 2016;**75**(3):265-273. DOI: 10.1017/S0029665116000094

[8] Demir E, Sarica Ş, Özcan M, Sui MM. The use of natural feed additives as alternatives for an antibiotic growth promoter in broiler diets. British Poultry Science. 2003;**44**(S1):44-45. DOI: 10.1080/713655288

[9] Placha I, Gai F, Pogány SM. Natural feed additives in animal nutrition—Their potential as functional feed. Frontiers Media SA. 2022;**9**:1062724

[10] Cederberg C, Hedenus F, Wirsenius S, Sonesson U. Trends in greenhouse gas emissions from consumption and production of animal food products–implications for long-term climate targets. Animal. 2013;7(2):330-340. DOI: 10.1017/S1751731112001498

[11] FAO. Pathways towards Lower Emissions–A Global Assessment of the Greenhouse Gas Emissions and Mitigation Options from Livestock Agrifood Systems. Rome: FAO. 77p; 2023. DOI: 104060/cc9029en

[12] Herrero M, Henderson B, Havlík P, Thornton PK, Conant RT, Smith P, et al. Greenhouse gas mitigation potentials in the livestock sector. Nature Climate Change. 2016;**6**(5):452-461. DOI: 10.1038/nclimate2925

[13] Tahir MN, Riaz R, Bilal M, Nouman HM. Current standing and future challenges of dairying in Pakistan: A status update. In: Milk Production, Processing and Marketing. London: Intechopen; 2019. pp. 1-24. DOI: 10.5772/intechopen.83494

[14] De Vries M, de Boer IJ. Comparing environmental impacts for livestock products: A review of life cycle assessments. Livestock Science. 2010;**128**(1-3):1-11. DOI: 10.1016/j.livsci.2009.11.007

[15] Daesoo Kim RP, Harris N, Kim C, Thoma G. Measuring and Mitigating GHGs: Chicken. Gland, Switzerland: World Wildlife Fund (WWF); 2022

[16] Bist RB, Subedi S, Chai L, Yang X. Ammonia emissions, impacts,

and mitigation strategies for poultry production: A critical review. Journal of Environmental Management. 2023;**328**:116919. DOI: 10.1016/j.jenvman.2022.116919

[17] Zisis F, Giamouri E, Mitsiopoulou C, Christodoulou C, Kamilaris C, Mavrommatis A, et al. An overview of poultry greenhouse gas emissions in the Mediterranean area. Sustainability. 2023;**15**(3):1941. DOI: 10.3390/su15031941

[18] Fatica A, Fantuz F, Wu M, Tavaniello S, Maiorano G, Salimei E. Soybean vs. pea bean in the diet of medium-growing broiler chickens raised under semi-intensive conditions of inner Mediterranean areas: Growth performance and environmental impact. Animals. 2022;**12**(5):649. DOI: 10.3390/ani12050649

[19] Yuan C, Song H, Xy Z, Yj J, At Z, Azzam MM, et al. Effect of expanded cottonseed meal on laying performance, egg quality, concentrations of free gossypol in tissue, serum and egg of laying hens. Animal Science Journal. 2014;**85**(5):549-554. DOI: 10.1111/asj.12169

[20] Abín R, Laca A, Laca A, Díaz M. Environmental assesment of intensive egg production: A Spanish case study. Journal of Cleaner Production. 2018;**179**:160-168. DOI: 10.1016/j.jclepro.2018.01.067

[21] Van Emous R, Winkel A, Aarnink A. Effects of dietary crude protein levels on ammonia emission, litter and manure composition, N losses, and water intake in broiler breeders. Poultry Science. 2019;**98**(12):6618-6625. DOI: 10.3382/ps/pez508

[22] Mosnier E, Van der Werf H, Boissy J, Dourmad J-Y. Evaluation of the environmental implications of the incorporation of feed-use amino acids in the manufacturing of pig and broiler feeds using life cycle assessment. Animal. 2011;**5**(12):1972-1983. DOI: 10.1017/S1751731111001078

[23] Bovera F, Loponte R, Marono S, Piccolo G, Parisi G, Iaconisi V, et al. Use of Tenebrio molitor larvae meal as protein source in broiler diet: Effect on growth performance, nutrient digestibility, and carcass and meat traits. Journal of Animal Science. 2016;**94**(2):639-647. DOI: 10.2527/jas.2015-9201

[24] JAPFA. Life Cycle Assessment: JAPFA. 2019. Available from: https://www.japfacomfeed.co.id/life-cycle-assessment

[25] Vellinga TV, Gerber P. Greenhouse Gas Emissions from the Global Livestock Sector-Methodology for a Life Cycle Assessment. Wageningen, Hollanda: Wageningen University & Research Center; 2010

[26] Wang J, Deng L, Chen M, Che Y, Li L, Zhu L, et al. Phytogenic feed additives as natural antibiotic alternatives in animal health and production: A review of the literature of the last decade. Animal Nutrition. 2024;**17**:244-264. DOI: 10.1016/j.aninu.2024.01.012

[27] Yitbarek MB. Phytogenics as feed additives in poultry production: A review. International Journal of Extensive Research. 2015;**3**:49-60

[28] Karadağoğlu Ö, Şahin T, Ölmez M, Yakan A, Özsoy B. Changes in serum biochemical and lipid profile, and fatty acid composition of breast meat of broiler chickens fed supplemental grape seed extract. Turkish Journal of Veterinary and Animal Sciences. 2020;**44**(2):182-190. DOI: 10.3906/vet-1906-37

[29] Ölmez M, Şahin T, Karadağoğlu Ö, Yörük MA, Kara K, Dalğa S. Growth performance, carcass characteristics, and fatty acid composition of breast and thigh meat of broiler chickens fed gradually increasing levels of supplemental blueberry extract. Tropical Animal Health and Production. 2021;**53**:1-8. DOI: 10.1007/s11250-020-02542-w

[30] Hossain MM, Cho SB, Kang D-K, Nguyen QT, Kim IH. Comparative effects of dietary herbal mixture or guanidinoacetic acid supplementation on growth performance, cecal microbiota, blood profile, excreta gas emission, and meat quality in Hanhyup-3-ho chicken. Poultry Science. 2024;**103**(4):103553. DOI: 10.1016/j.psj.2024.103553

[31] Amad AA, Wendler K, Zentek J. Effects of a phytogenic feed additive on growth performance, selected blood criteria and jejunal morphology in broiler chickens. Emirates Journal of Food and Agriculture (EJFA). 2013;**25**(7):549-554. DOI: 10.9755/ejfa.v25i7.12364

[32] Cross D, McDevitt R, Hillman K, Acamovic T. The effect of herbs and their associated essential oils on performance, dietary digestibility and gut microflora in chickens from 7 to 28 days of age. British Poultry Science. 2007;**48**(4):496-506. DOI: 10.1080/00071660701463221

[33] Khalaji S, Zaghari M, Hatami K, Hedari-Dastjerdi S, Lotfi L, Nazarian H. Black cumin seeds, Artemisia leaves (*Artemisia sieberi*), and *Camellia L.* plant extract as phytogenic products in broiler diets and their effects on performance, blood constituents, immunity, and cecal microbial population. Poultry Science. 2011;**90**(11):2500-2510. DOI: 10.3382/ps.2011-01393

[34] Hassan MR, Choe HS, Ryu KS. A comparison of feeding multi-probiotics and fermented ginseng byproducts on performance, intestinal microflora and immunity of broiler chicks. Korean Journal of Poultry Science. 2012;**39**(4):253-260. DOI: 10.5536/KJPS.2012.39.4.253

[35] Li H, Zhao P, Lei Y, Hossain M, Kim I. Phytoncide, phytogenic feed additive as an alternative to conventional antibiotics, improved growth performance and decreased excreta gas emission without adverse effect on meat quality in broiler chickens. Livestock Science. 2015;**181**:1-6. DOI: 10.1016/j.livsci.2015.10.001

[36] Cabuk M, Alcicek A, Bozkurt M, Akkan S. Effect of *Yucca schidigera* and natural zeolite on broiler performance. International Journal of Poultry Science. 2004;**3**(10):651-654

[37] Ahmed ST, Yang C-J. Effects of dietary *Punica granatum* L. by-products on performance, immunity, intestinal and fecal microbiology, and odorous gas emissions from excreta in broilers. The Journal of Poultry Science. 2017;**54**(2):157-166. DOI: 10.2141/jpsa.0160116

[38] Bostami A, Ahmed S, Islam M, Mun H, Ko S, Kim S, et al. Growth performance, fecal noxious gas emission and economic efficacy in broilers fed fermented pomegranate byproducts as residue of fruit industry. International Journal of Advanced Research. 2015;**3**(3):102-114

[39] Shanmugam S, Park JH, Cho S, Kim IH. Silymarin seed extract supplementation enhances the growth performance, meat quality, and nutrients digestibility, and reduces gas emission in broilers. Animal Bioscience. 2022;**35**(8):1215. DOI: 10.5713/ab.21.0539

[40] Bui H, Cisse SH, Ceccaldi M, Perrin A, Benarbia MEA, Chicoteau P.

Mitigating the environmental impacts from pig and broiler chicken productions: Case study on a citrus extract feed additive. Animals. 2023;**13**(23):3702. DOI: 10.3390/ani13233702

[41] Chowdhury S, Mandal GP, Patra AK. Different essential oils in diets of chickens: 1. Growth performance, nutrient utilisation, nitrogen excretion, carcass traits and chemical composition of meat. Animal Feed Science and Technology. 2018;**236**:86-97. DOI: 10.1016/j.anifeedsci.2017.12.002

[42] Park J, Kim I. Effects of a protease and essential oils on growth performance, blood cell profiles, nutrient retention, ileal microbiota, excreta gas emission, and breast meat quality in broiler chicks. Poultry Science. 2018;**97**(8):2854-2860. DOI: 10.3382/ps/pey151

[43] Salem HM, Saad AM, Soliman SM, Selim S, Mosa WF, Ahmed AE, et al. Ameliorative avian gut environment and bird productivity through the application of safe antibiotics alternatives: A comprehensive review. Poultry Science. 2023;**102**(9):102840. DOI: 10.1016/j.psj.2023.102840

[44] Riaz R, Karadağoğlu Ö. In: Ölmez M, editor. Veteriner Bilimlerinde Probiyotiklerin Terapötik, Proflaktik ve Fonksiyonel Kullanımı. 1st ed. Gaziantep: Özgür Yayınları; 2023. pp. 1-8

[45] Yun W, Lee D, Choi Y, Kim I, Cho J. Effects of supplementation of probiotics and prebiotics on growth performance, nutrient digestibility, organ weight, fecal microbiota, blood profile, and excreta noxious gas emissions in broilers. Journal of Applied Poultry Research. 2017;**26**(4):584-592. DOI: 10.3382/japr/pfx033

[46] Biswas S, Kim MH, Baek DH, Kim IH. Probiotic mixture (*Bacillus subtilis* and *Bacillus licheniformis*) a potential in-feed additive to improve broiler production efficiency, nutrient digestibility, caecal microflora, meat quality and to diminish hazardous odour emission. Journal of Animal Physiology and Animal Nutrition. 2023;**107**(4):1065-1072. DOI: 10.1111/jpn.13784

[47] Şahin T, Aksu Elmalı D, Kaya İ, Sarı M, Kaya Ö. The effect of single and combined use of probiotic and humatein quail (*Coturnix coturnix* Japonica) diet on fatttening performance and carcass parameters. Kafkas Universitesi Veteriner Fakültesi Dergisi. 2011;**17**(1):1-5

[48] Şahin T, Kaya I, Unal Y, Elmali DA. Dietary supplementation of probiotic and prebiotic combination (combiotics) on performance, carcass quality and blood parameters in growing quails. Journal of Animal and Veterinary Advances. 2008;7(11):1370-1373

[49] Aalaei M, Khatibjoo A, Zaghari M, Taherpou K, Akbari-Gharaei M, Soltani M. Effect of single-and multi-strain probiotics on broiler breeder performance, immunity and intestinal toll-like receptors expression. Journal of Applied Animal Research. 2019;**47**(1):236-242. DOI: 10.1080/09712119.2019.1618311

[50] Toghyani M, Toghyani M, Tabeidian SA. Effect of probiotic and prebiotic as antibiotic growth promoter substitutions on productive and carcass traits of broiler chicks. In: International Conference on Food Engineering and Biotechnology. Singapore: ACSIT Press; 2011

[51] Tayeri V, Seidavi A, Asadpour L, Phillips CJ. A comparison of the effects of antibiotics, probiotics, synbiotics and prebiotics on the performance and carcass characteristics of broilers. Veterinary Research Communications. 2018;**42**:195-207. DOI: 10.1007/s11259-018-9724-2

[52] Taherpour K, Moravej H, Taheri HR, Shivazad M. Effect of dietary inclusion of probiotic, prebiotic and butyric acid glycerides on resistance against coccidiosis in broiler chickens. The Journal of Poultry Science. 2012;**49**(1):57-61. DOI: 10.2141/jpsa.011031

[53] Khaksefidi A, Rahimi S. Effect of probiotic inclusion in the diet of broiler chickens on performance, feed efficiency and carcass quality. Asian-Australasian Journal of Animal Sciences. 2005;**18**(8):1153-1156. DOI: 10.5713/ajas.2005.1153

[54] Park CJ, Sun SS. Effect of dietary metallo-protease and bacillus velezensis CE 100 supplementations on growth performance, footpad dermatitis and manure odor in broiler chickens. Animal Bioscience. 2022;**35**(10):1628

[55] Helal F, El-Badawi A, El-Naggar S, Shourrap M, Aboelazab O, Abu HS. Probiotics role of Saccharomyces cerevisiae and *Bacillus subtilis* in improving the health status of rabbits' gastrointestinal tract. Bulletin of the National Research Centre. 2021;**45**(1):66. DOI: 10.1186/s42269-021-00522-0

[56] Hoque M-R, Jung H-I, Kim I-H. Effect of yeast culture (*Saccharomyces cerevisiae*) supplementation on growth performance, excreta microbes, noxious gas, nutrient utilization, and meat quality of broiler chicken. The Journal of Poultry Science. 2021;**58**(4):216-221. DOI: 10.2141/jpsa.0190144

[57] Nguyen DH, Kim IH. Protected organic acids improved growth performance, nutrient digestibility, and decreased gas emission in broilers. Animals. 2020;**10**(3):416. DOI: 10.3390/ani10030416

[58] Sugiharto S, Agusetyaningsih I, Widiastuti E, Wahyuni H, Yudiarti T, Sartono T. Dietary supplementation of enzymes: An approach to mitigate ammonia emission during broiler production. Iranian Journal of Applied Animal Science. 2023;**13**(4):615-625

[59] Javaid A, Younas F, Ullah I, Yasinzai M. Impact of an indigenously produced multi-enzyme complex from *Bacillus subtilis* KT004404 on growth and blood parameters in broiler chicken. PLoS One. 2022;**17**(7):e0271445. DOI: 10.1371/journal.pone.0271445

[60] Nazemzadeh S, Heshmat G, Ansari H. Effect of supplemented ProAct (CT) protease enzyme on performance and the amount of protein excreted in feces of broiler chickens. Journal of Livestock Science. 2017;**8**:115-121

[61] Nguyen D, Tran H, Yun H, Kim I. Influence of a cocktail of protease and xylanase in different energy densities of corn-and soybean-meal-based diet on growth performance, nutrient digestibility, carcass quality, and gas emission in broilers. Canadian Journal of Animal Science. 2017;**98**(2):271-278. DOI: 10.1139/cjas-2017-0091

[62] Lourenco JM, Nunn SC, Lee EJ, Dove CR, Callaway TR, Azain MJ. Effect of supplemental protease on growth performance and excreta microbiome of broiler chicks. Microorganisms. 2020;**8**(4):475. DOI: 10.3390/microorganisms8040475

[63] Akter M, Graham H, Iji PA. Response of broiler chickens to diets containing different levels of sodium with or without microbial phytase supplementation. Journal of Animal Science and Technology. 2019;**61**(2):87. DOI: 10.5187/jast.2019.61.2.87

[64] Leinonen I, Williams AG. Effects of dietary protease on nitrogen emissions from broiler production: A holistic

comparison using life cycle assessment. Journal of the Science of Food and Agriculture. 2015;**95**(15):3041-3046. DOI: 10.1002/jsfa.7202

[65] Oxenboll K, Pontoppidan K, Fru-Nji F. Use of a protease in poultry feed offers promising environmental benefits. International Journal of Poultry Science. 2011;**10**(11):842-848

[66] Hickmann FM, Andretta I, Létourneau-Montminy M-P, Remus A, Galli GM, Vittori J, et al. β-Mannanase supplementation as an eco-friendly feed strategy to reduce the environmental impacts of pig and poultry feeding programs. Frontiers in Veterinary Science. 2021;**8**:732253. DOI: 10.3389/fvets.2021.732253

[67] Gernaey B, Sorbara JOB, Nielsen PH. Environmental assessment of amylase used as digestibility improvement factor for intensive chicken production in Brazil. Sustainability. 2018;**10**(8):2735. DOI: 10.3390/su10082735

[68] Eugene B, Moore PA Jr, Li H, Miles D, Trabue S, Burns R, et al. Effect of alum additions to poultry litter on in-house ammonia and greenhouse gas concentrations and emissions. Journal of Environmental Quality. 2015;**44**(5):1530-1540. DOI: 10.2134/jeq2014.09.0404

[69] Ezenwosu C, Udeh V, Onyimonyi A. Litter treatment using zeolite as a management protocol to enhance broiler performance and control ammonia emission in broiler production in the humid tropics. Asian Journal of Research and Review in Agriculture. 2022;**4**(1):58-72

[70] Schneider A, Almeida DD, Yuri F, Zimmermann O, Gerber M, Gewehr C. Natural zeolites in diet or litter of broilers. British Poultry Science. 2016;**57**(2):257-263. DOI: 10.1080/00071668.2016.1150962

[71] Pereira JL, Martins F, Bonifácio G, Garcia C, Teixeira J, Trindade H. Biochar as an alternative litter additive to mitigate gaseous emissions from broiler housing and subsequent storage. Agronomy. 2024;**14**(7):1595. DOI: 10.3390/agronomy14071595

[72] Pereira JL, Ferreira S, Pinheiro V, Trindade H. Ammonia and greenhouse gas emissions following the application of clinoptilolite on the litter of a breeding hen house. Environmental Science and Pollution Research. 2019;**26**:8352-8357. DOI: 10.1007/s11356-019-04429-2

[73] Chen H, Awasthi MK, Liu T, Zhao J, Ren X, Wang M, et al. Influence of clay as additive on greenhouse gases emission and maturity evaluation during chicken manure composting. Bioresource Technology. 2018;**266**:82-88. DOI: 10.1016/j.biortech.2018.06.073

[74] Kalus K, Konkol D, Korczyński M, Koziel JA, Opaliński S. Effect of biochar diet supplementation on chicken broilers performance, NH3 and odor emissions and meat consumer acceptance. Animals. 2020;**10**(9):1539. DOI: 10.3390/ani10091539

[75] Mahardhika B, Mutia R, Ridla M, editors. Efforts to reduce ammonia gas in broiler chicken litter with the use of probiotics. In: IOP Conference Series: Earth and Environmental Science. Bristol: IOP Publishing; 2019

[76] Pezzuolo A, Sartori C, Vigato E, Guercini S. Effect of Litter Treatment with Probiotic Bacteria on Ammonia Reduction in Commercial Broiler Farm. Jelgava: Latvia University of Life Sciences and Technologies Faculty of Engineering; 2019. pp. 1631-1635

Section 6

Fattening Methodology and Beef Production

Chapter 7

Beef Cattle

Ioan Hutu and Gary William Onan

Abstract

The type of cattle feeding system significantly influences the final slaughter weight, degree of carcass fatness vs. muscle, and beef quality. Three basic systems, based on the age at which cattle are finished and marketed, include fattening calves, young cattle and bulls, and adult cattle (mainly cull cows). The level of daily energy intake, or plane of nutrition, affects the intensity of the fattening phase, with higher nutrient density diets yielding higher daily gains and increased gain efficiency. Longer finishing periods result in greater fat deposition and higher eating quality. Feeding strategies are classified as intensive, semi-intensive, and extensive. Housing options for finishing cattle include confinement, pasture, or a combined system. The choice of system depends on local resources and cattle genetics, with significant breed differences in maturity age, feed efficiency, weight gain rate, and market weight.

Keywords: beef cattle, feeding and fattening systems, housing, breeding season, calving season

1. Introduction

In the EU, beef comprises 10% of meat production. Worldwide, beef accounts for 13% of total meat production. In Europe, and even more so in Eastern Europe, beef is important socially and economically as a means for sustaining rural populations. During the most recent 10 years, beef production has seen a decrease in the EU due in part to epizootic disease pressures and reductions in reproductive rate. Two primary beef production systems are found in Europe: (i) those based on grazing, most often found in NW Europe (and USA); (ii) those primarily utilizing cereal grains for raising cattle, particularly found in Central, Eastern, and Mediterranean Europe [1].

2. Methodology

The methods outline the approaches and procedures used to investigate the influence of cattle feeding systems on slaughter weight, carcass composition, and beef quality. This study utilized a combination of field observations, controlled experiments, and statistical analyses to ensure comprehensive and accurate results. The study was designed to describe and compare three different cattle feeding systems:

intensive, semi-intensive, and extensive in order to determine how these feeding systems impact the final slaughter weight, degree of carcass fatness vs. muscle, and the eating quality of beef. The case selection included diverse sources regarding beef farms and cattle from different housing systems, breeds, and age groups to provide a comprehensive comparison across different feeding strategies and cattle types. In order to evaluate how different nutritional plans and feeding environments influence growth rates, carcass composition, and meat quality, the feeding systems were categorized as intensive, semi-intensive, and extensive. Data on cattle growth, carcass composition, and meat quality were used for comparing the outcomes of different feeding systems. The chapter aims to inform cattle farmers and the beef industry. By understanding the effects of different feeding systems, stakeholders can make informed decisions to optimize beef production for better quality and efficiency.

The chapter was designed to provide comprehensive insights into the effects of various cattle feeding systems on beef production. This methodological approach ensures the reliability and applicability of the findings, contributing valuable knowledge to the field of animal husbandry and meat science.

3. Beef feeding systems

The type of cattle feeding system used, to a large degree, determines the final slaughter weight, degree of carcass fatness vs. muscle, and the eating quality of the beef. Three basic systems defined by the age at which cattle are finished and marketed may be identified: (i) fattening of calves; (ii) fattening of young cattle and young bulls; and (iii) fattening of adult cattle (mostly cull cows, and most of those are marketed as is, without additional feeding).

The plane of nutrition (level of daily energy intake) determines the intensity of the finishing phase (fattening phase) of cattle feeding. Diets with higher nutrient density produce higher average daily gains and increase feed efficiency (kg gain/kg feed dry matter). Finishing periods of longer duration also result in carcasses with greater fat deposition and higher eating quality. Therefore, feeding strategies may be classified as: (i) *intensive*, (ii) *semi-intensive* and *extensive* [2].

Housing of finishing cattle may also differ and include: (i) *confinement* (enclosed or open barns or sheds), (ii) *pasture*, (iii) a *combined system* where cattle are initially on pasture and then finished in confinement. Which system is used depends on local crop and feed resources as well as the genetic base of the cattle utilized. There are substantial differences among cattle breeds regarding the age at which they become physiologically mature (which coincides with market age in many market areas), their feed utilization efficiency, rate at which they gain weight, and final market weight attained [1, 2].

3.1 Fattening systems and slaughtering age

Slaughter age of cattle depends on the type of production system used [3]. Slaughter age is younger in intensive systems and older in extensive ones [1, 2]:

- *Intensive production*—market calves for veal, young bulls, or steers, which are slaughtered before12 months old;

- *Semi-intensive production*—market heifers and bulls or steers between 12 and 24 months old;

- *Extensive production*—market animals aged over 30 months; extensive systems also include cow-calf herds, which market weaned calves for further semi-intensive feeding.

- *Cull breeding animals, cows and bulls*, are also marketed for beef.

Slaughter age has a significant impact on the eating quality and yield of meat from the various classes of cattle.

3.2 Relationship of feeding system and cattle growth

The cattle growth curve shape is influenced by feeding systems. With intensive systems, growth is more rapid, so the slope of the curve is steeper and will flatten off at a higher weight when compared to extensive systems. Animals of different sex conditions also exhibit differences in rate and extent of growth, with intact bulls growing faster and to heavier weights than heifers or steers (castrates).

Milk-fed veal calves (white veal) are raised in intensive year-round systems where calves receive a milk-only diet with the use of milk replacer powder. These calves can achieve an average daily gain of 1.1 kg [4] and are typically marketed at a weight of 140–160 kg at around 90–120 days of age. These calves produce 86–100 kg carcasses distinguished by very light pinkish colored meat.

Veal (rose veal) is produced by using greater amounts of dry feeds in the calves' diets and obtaining a weight of 300–480 kg. With a 60% dressing yield (carcass weight divided by live weight), these animals will produce 180–290 kg of carcasses. Veal may be obtained from animals as old as 1 year under current EU regulatory definitions [5].

Any animals older than 1 year of age at the time of marketing produce beef. This may come from steers (castrated at any age 1 week or older), young heifers or bulls, and from culled breeding stock (cows or bulls). Typically, these animals reach weights of around 600 kg or more and come from semi-intensive or extensive feeding systems [1].

4. Intensive systems: veal production

Intensive feeding systems are characterized by rapid gain (>900 g/day) with animals slaughtered at younger ages (≤18 months) with high dressing yields (55% for dairy breeds and over 60% for others—often as high as 65–67%). The meat from these animals typically has very high eating quality. Intensive systems are found both in smaller farms and in very large farms with high concentrations of animals [1].

4.1 Details for feeding white veal

Production of white veal (milk-fed veal or Class V veal under EU classification) is one of the most intensive of all feeding systems. It utilizes young calves from a variety of breeds, either dairy, beef, or crosses, such as Simmental, Schwyz, Dappled Romanian (Romanian Simmental), Charolais, or surplus dairy calves—all dairy bull calves and any heifer calves not needed as replacements for the milking herd [6]. Calves are obtained and put on their feeding regime as soon as the colostrum feeding period is finished (about 2–5 days of age). These calves will be fed for 50–150 days, depending on the desired weight at marketing, which will range from 100 to 200 kg. For each 1 kg gain, 9–11 kg milk will be consumed (the energy density of milk is 3100–3950 kcal/kg). Average daily gain for these calves is about 1200 g, and dressing yield is typically over 60%.

4.1.1 Production steps for milk-fed veal

Feeding milk-fed veal (white veal) calves encompasses several steps [2].

Ideally, *calves purchased (acquisition of the calves)* for white veal production will come from farms that practice good calf management. This includes excellent disease control and assurance that calves have consumed a minimum of 35 kg of colostrum and have been taught to nurse from nipple bottles or to drink from buckets. Calves should be at least 10 days old and weigh 40–45 kg at time of purchase.

Calves should be *transported* from their birth farm to the buyer's farm in clean, disinfected trailers, considering *stress reduction during transport and handling*. Steps should be taken to minimize temperature stress by controlling drafts and trying to maintain trailer temperature in the general range of 20°C. Trailers should allow 0.7–1.0 m^2 for each calves. When the duration of transport exceeds 6 hours, prophylactic treatment (antibiotics and vitamins A, D3, and E) or anti-stress medication (Combelen or Romtiazin) should be administered. All calf movement and handling should be conducted in a manner that capitalizes on natural animal behaviors in order to minimize stress. Upon arrival at the feeding farm, it is suggested that each calf receive 6–8 L of 10% glucose solution at 36–37°C and split out into 2 or 3 separate feedings. On the second day, half of the glucose solution will be replaced by milk or milk replacer. Prophylactic antibiotics and vitamin treatments should be re-administered as well. On the third day, the normal feeding regimen can be fully instituted.

A milk-based diet will be utilized during the entire *feeding period*. It may utilize whole fresh milk or various types of milk replacers. Milk replacers are typically formulated from non-fat dry milk (NFDM or skimmed milk powder) powder that has been fortified with a fat supplement and minerals and vitamins. Calves are fed 8–10% of their body weight of reconstituted milk replacer per day (4–15 L daily).

The high milk diets generate [2] the following results: reticulo-rumen and omasum development is retarded; the calves largely remain monogastric throughout their lives; dressing yield is high due to limited gastric system development; rapid growth due to diets with high digestibility and high nutrient density; light pink meat—low *Fe* content of the diet results in lower myoglobin development in the animals' muscle tissues; and under current EU regulations, calves must be offered dry feed *ad libitum*, which is typically a mix of chopped straw and grains even having a negative effect on emission of greenhouse gasses and land occupation [7].

There are several principles [1, 2] for proper feeding: slowly increase the amount of milk replacer fed each day during the first 2 weeks as the calf's digestive system adapts; decrease the volume of reconstituted milk replacer by increasing the concentration of powder from 130 g up to 220 g powder/liter of liquid; and increasing the concentration of the milk replacer powder will necessitate offering water to the calves, typically twice per day.

Calves are housed in climate-controlled buildings in most situations. "All-in-all-out" management is practiced. This means that when a group of calves is ready for market, they are all sold, the building is thoroughly cleaned and disinfected, and then the new replacement calves are brought in. This practice breaks the disease cycle between groups of calves. Basic building environmental precepts are: temperature maintained between 16 and 20°C year-round; relative humidity, below 75%; constant uniform ventilation with no drafts; and low light levels in order to encourage reduced activity and therefore more efficient gain.

Calves may be housed within *individual cages, individual stands,* or in *group pens* within the structure. The front of the stall contains bucket holders for milk replacer, water, and dry feed. The advantage of individual housing is that observation of the health status of each calf is easier and feed intake can be precisely monitored. If group pens are used, 4–6 animals typically share one pen. Pens are 8–12 m^2 in total area. It should be noted that most modern veal operations utilize group pens, as these are considered to offer improved overall animal well-being, particularly the opportunity for the calves to maintain natural social interactions.

4.2 Feeding of rose veal (red veal)

Under current EU regulations, any calf that is over 8 months of age falls into Class Z veal, or what has traditionally been known as rose veal (red veal). These calves are by definition older (8–12 months of age), have not been fed a milk diet their entire life, and will have greater color intensity (more red) in their carcass musculature—see Veal Classification [8].

Calves are initially raised in the same manner as for white veal. The acquisition, transport, health protocols, housing, and initial feeding regimen are essentially the same. The differentiation occurs at about 2 months of age. In rose veal production systems, calves are typically weaned off milk replacer and placed on 100% dry feeds at that age. Some producers may feed lesser amounts of milk replacer to an older age (3–4 months), but that is not common in the larger commercial enterprises. So in essence, then, the feeding strategy may be broken out into two phases: *phase I* is the first 2 months when calves receive a largely milk diet and are introduced to dry feeds, and *phase II* consists of the remainder of the feeding period during which the calves receive dry feeds only.

The immediate post-weaning ration is a blend of grains and dry hay with about 18% crude protein content (dry matter [DM] basis). As calves grow, the ration is adjusted to include other feedstuffs, such as corn (maize), silage, and chopped straw. These diets are often blended with a total mixed ration (TMR) blender. A typical ration might include 4 parts grains and protein supplement: 2 parts corn silage: 1 part chopped straw (the straw helps maintain proper rumen physiology by stimulating rumination). During the growth period, these calves will gain about 1.5–1.6 kg/day. Calves will be marketed at 8–12 months of age (although in some instances as young as 6 months, depending on growth rate) at weights ranging from 300 to 480 kg. Dressing yields for these calves around are at least 55% and may be as high as 65%.

4.3 Housing during phase I

During the first 2 months, calves for rose veal production are housed in a manner essentially identical to those raised for milk-fed veal. Again, this is typically a confinement structure where temperature can be controlled to some degree in an effort to maintain a constant 16–18°C temperature throughout the year. Relative humidity should be 70–75% with 0.2–0.3 m/s airflow. Calves may be housed in *individual stalls* or *small group pens*.

The *individual stalls* would be constructed the same as described above for milk-fed veal. Once calves are weaned, they are moved to group pens.

Housing in group pens is typically designed for 15–20 calves/pen and is often arranged in a double row. Space allowance is 1.2–1.4 m^2 per calf. Pen fronts are equipped with a manger for feeding concentrates and hay. The manger should allow for 30–35 cm per feeding space. Manger access may be through stanchions that allow for locking of the animals for health protocols, etc. The pens also contain an automatic water fountain and a mineral feeding tube for *ad lib* salt. If the pens have a slatted floor for manure removal, slats should have 3.0 cm spacing.

4.4 Housing during phase II

Upon weaning, calves may be moved to a different barn or shed, which does not require as sophisticated climate control. Natural ventilation controlled with doors, windows, and partial coverings on those, such as tarps, provides adequate ability to control temperature and air movement for these animals. In winter, temperature can be maintained in the 10–15°C range except during extreme cold. Calves are housed in group pens with 20–30 calves per pen. Space requirement is 1.7 m^2 per animal up to 350 kg body weight, and 2.2 m^2 per animal if taken to heavier weights prior to market. Per animal space allowance is controlled by altering the number of animals per pen. Pens are equipped with a manger along one side with 20 cm/feeding space and with an automatic water fountain and mineral feeder. The floor is again a slatted type with 5 cm spacing between slats. Under-floor removal of manure may be via a water flush system or a mechanical scraper.

4.5 Feeding strategy

Feed rations for rose veal calves are broken out into two general phases: milk feeding period (phase I) and the post-weaning period (phase II). The post-weaning period actually encompasses gradual dietary changes as the calves grow and their digestive capacity increases, and the nutrient density of their rations can therefore decrease [1].

4.5.1 Phase I: Milk-fed period

This period lasts for about 60 days, although this is variable depending between farms and may vary from 42 to 90 days. It is typical to feed veal calves higher levels of milk than would be used for female dairy replacement calves. Veal calves are often fed up to 6 L of reconstituted milk replacer per day, split between two feedings. Calf starter is offered free choice (*ad libitum*) beginning at 7–10 days of age. Calf starters are composed of grains, such as rolled corn (maize), rolled oats, or wheat milling by-products. In addition, protein supplements, such as soybean, sunflower, or canola (rape) oil meals along with mineral and vitamin supplements are included. Calf starters contain around 3500 kcal metabolizable energy (ME)/kg dry matter (DM) and between 20 and 25% crude protein (CP). Calves will be consuming about 1 kg of calf starter/day by weaning time and have a body weight of about 90 kg.

4.5.2 Phase I to phase II transition

This period lasts for 4–6 weeks. Calves begin this period consuming the same calf starter indicted above at a rate of about 1 kg/day. During this period, calf starter consumption will increase up to about 2.5 kg/day as their digestive system develops (rumen development proceeds at a rapid pace at this time) and its capacity increases. Once calves are consuming 2.5 kg starter/day, forages can be introduced into their diet. Initially, the forage source is high quality dry hay (alfalfa or grass). By the end of the transition period, calves will weigh about 130 kg and be consuming up to 4 kg DM/day. Gain efficiency during the transition will be around 0.3 kg gain/kg DMI (dry matter intake).

4.5.3 Phase II: Growing and finishing

Once calves have transitioned, it is common to introduce silages into their ration—either alfalfa, grass, or corn silage. Corn silage is often used as it has a higher energy density than other silages, which promotes more rapid gain. It is also common to offer the feed as a total mixed ration (TMR) for the remainder of the time until market. Using the TMR approach, all feed ingredients are blended fresh each day with a mobile mechanical mixer that creates a uniform blend, which, in essence, means the animal consumes a completely balanced ration with each bite it takes from the TMR mix. A typical TMR ration will include roughly 30% silage, 40% grains, 20% oil meal, and 10% chopped straw, along with mineral and vitamin supplements (DM basis). Energy density will be around 3100 kcal ME/kg DM, with protein content gradually decreasing from 18% CP at the beginning of Phase II down to about 14% CP as marketing time approaches. Average daily gain (ADG) can be expected to be 1.5 kg/day. As indicated earlier, market age and weight depend on the merchandizing specifications of the slaughter plant and may range from 300 to 450 kg of weight at 8–12 months of age. There is an important niche in rose veal production that utilizes pasture as a substantial portion of nutrients during Phase II. With this approach, pasture provides the forage portion of the diet, but considerable concentrate is still fed. Animals are housed outdoors in a pasture that may be managed using intensive rotational grazing practices.

4.6 Veal production for the dairy industry

Veal production is a critical adjunct for optimizing income for dairy producers. In essence, veal production adds value to "waste products" or by-products of the dairy industry. Those so-called by-products include dairy bull calves, which have no inherent value for the dairy producers themselves, along with utilization of waste or co-products from dairy manufacturing, including non-fat dry milk powder (NFDM) from butter manufacture and dried whey from cheese manufacture. Both NFDM and dried whey provide the primary ingredients for milk replacer powders that are used to feed veal calves. Therefore, three important by-products of dairy production have substantial value added by the veal industry. It should be noted that in addition to veal production, dairy breed bull calves can also be utilized for beef production by growing them to heavier weights over a longer time period. In the United States, the vast majority of dairy bull calves are castrated and finished as beef, reaching market at 1.5–2.5 years of age. The proportion fed for veal meat in the USA is much lower than in Europe. A later section of this chapter will outline the essential practices for dairy-beef production.

5. Extensive systems: beef cow-calf herds

The base or fundamental sector for the majority of beef produced in Western Europe, the United Kingdom, Australia, and North and South America is the beef cow-calf herd. These herds consist of cows from breeds of cattle, which have been selected specifically for meat production over the last 200 or more years. Typical breeds would include Angus, Hereford, Simmental (beef type), Charolais, Limousin, and many others. These beef cow herds most often obtain the bulk of their feed from pastures (or stored forages during the cold seasons in higher latitudes). The pastures range from intensively managed rotational grazing operations in locations such as Western Europe (in plain areas of Romania the grazing season is starting in-between April 15 and May 1, depending on the grass' height) or the Midwestern USA, to native semi-arid grasslands with limited carrying capacity (carrying capacity is the number of cows a unit of land can support for 1 year), such as those found in the Southwestern USA and Australia. These herds are maintained in a completely outdoor ambient environment. The only housing provided may be an open-front shed for cows during calving time in colder climates [1, 9].

5.1 Annual management cycle of a beef herd

Because of the basic reproductive cycle of cattle, a cow can experience only one birthing event per year. The gestation period for cattle is about 285 days, and following parturition (birth of the calf), there is a lactational inhibition of return to estrus (return to fertility) with a duration of 40–90 days in most beef breeds. Thus, the goal is for each cow to give birth once per year. Cattle are not a prolific species, so each cow, therefore produces one calf per year. The sale of that calf constitutes the entire income from that cow for that year [1, 9].

5.1.1 Calving season

It is common for most herds to calve in the spring of the year, as that is when pasture forage is most abundant and will easily supply the nutrient needs for a lactating beef cow. In temperate climates, most calves are born in the months of January, February, and March (**Figure 1**), and near mountain areas later, in April or May.

For those calving early in this period, in cold climates, open-front sheds are used to provide some shelter for the cow and her newborn calf. These sheds often have individual pens—one for each cow-calf pair. These pens are typically about 2.5 × 4 m in size. Cows are placed in the shed a day or two prior to the expected calving date and

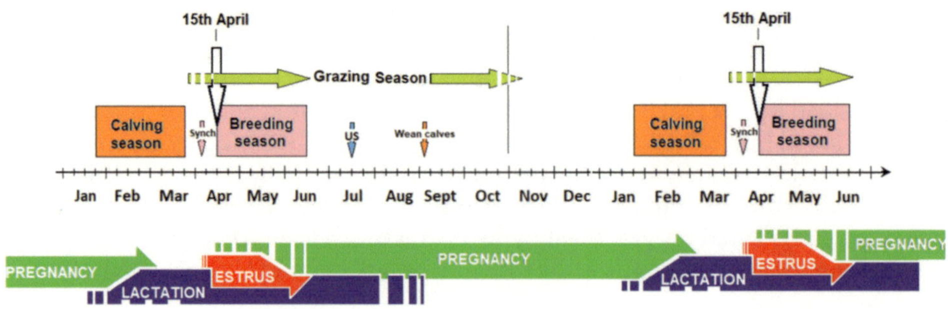

Figure 1.
Management cycle of beef cow-calf herd in plains area of Ester Europe. In the context of grazing season starting in middle of April and ending in late October, the 56-day calving season (January 24–March 10) and a 56-day breeding season (April 15–June 10) can be easily applied in Romania. The Ovsinch or Select Synch will start on April 6; the pregnancy control by ultrasound machine (US) can be performed on July 10, and the claves can be weaned at the end of August [9]. Source: Developed by the authors.

moved back out once the calf has established itself—usually by 2 days of age. At birth, the herdsperson will weigh the calf and apply some kind of identification, often an eartag. It is also critical that the calf be observed and helped if necessary to make sure it nurses and obtains an adequate amount of colostrum within the first 4–6 hours of life. When calving occurs later in the season, or in warmer regions, shelter for the cow and calf is not required, and most calves are born in the pasture. Even here, however, the herdsperson will obtain a birth weight within 12 hours, and identify the calves, and confirm that they are nursing adequately. Bull calves may be left intact, but in some countries (UK, USA, and Australia) they are castrated. Castrated bull calves (steers) are easier to handle and tend to fatten to a greater extent, with higher levels of intramuscular fat (marbling), thus producing beef with higher eating quality than that obtained from intact bulls. The best management practice is that bull calves are castrated before they attain an age of 1 month. Earlier castration is substantially less stressful for the animal and results in equal or better growth performance throughout the first 7 months than if calves are castrated later (such as at weaning time).

5.1.2 Breeding season

Starting around 50–60 days post-calving, breeding season begins (April–June, see **Figure 1**). In much of the world, breeding is still accomplished by using natural service with a bull. In such a system, bulls are placed in the pasture with the cow herd (about 1 mature bull per 40 cows) and allowed to remain there for the next 60 days. More progressive operations utilize artificial insemination (AI) in the first 3–6 weeks of breeding season. This requires more work, either in the form of daily estrus detection or administration of estrus synchronizing products for cows calved with 2 months before the treatment. AI, however, has the following benefits: (i) access to improved or elite genetics, (ii) opportunity to use multiple bulls whose genetics best complement those of each cow, (iii) precise knowledge of breeding dates and, therefore, better prediction of calving dates, (iv) no need to buy expensive bulls and bear the cost of maintaining them for 10 months of each and (v) year in which they are not being used.

Some herds use AI exclusively. Many AI herds will utilize a "clean-up" bull which is introduced to the cow herd after two estrous cycles (40–50 days into the breeding season, sometimes the last 4–8 weeks). The clean-up bull will typically impregnate any cows who did not conceive due to the AI insemination(s) they may have received. When handling cows for AI, it is also a good time to handle the calves and administer their first round of vaccinations and other health protocols as well.

5.1.3 Pregnancy determination

Good managers will follow the breeding season with pregnancy determination in order to clearly identify any open cows (cows that did not conceive). Pregnancy determination may be by manual rectal palpation of the uterus or by use of ultrasound scans. Depending on the method used, the timing may be 25–60 days post-breeding. Most managers will do a confirmatory pregnancy determination on all cows at 60–90 days post-breeding. Usually, 1 month after the end of the breeding season, the pregnancy control by ultrasound scans is performed. Occasionally, a cow that is pregnant early in gestation may lose the embryo and be open later. Open cows are non-productive, and are therefore culled and shipped to slaughter once their current calf is weaned.

5.1.4 Weaning

The next major event in the annual cow herd cycle is weaning the calves. Weaning age may vary, but a common age is around 6 or 7 months (beginning of August–end of September, in plane areas of Romania, see **Figure 1**). In the USA, the "standard" weaning age to which production records are adjusted is 205 days. Most European bull studs calculate an estimated breeding value for 200-day growth, which again, would be a typical weaning age for herds specifically focused on beef production. Weaning may be abrupt, where calves and cows are separated and one or the other group is moved to a different location. Abrupt weaning places calves under considerable stress. They tend to spend the next 2 or 3 days walking constantly in search of their mothers. While doing this, they do not spend much time eating and thus do not gain weight, and in fact, they may lose a small amount of weight immediately post-weaning. A weaning practice that is gaining more adherents is called fenceline weaning [10]. This practice has the calves and cows separated but maintained on adjacent pastures for at least a week. It is critical that the pastures be separated with a very high-quality fence (3 or 4-strand high-tension electric fence with a charge maintained at a minimum of 2500 V) to prevent break through by the calves. This system lowers stress on the calves as it allows them to maintain sight and smell contact with their mothers while they adjust to the absence of nursing and full transition to plant-based feeds. The calves' pasture should have abundant, excellent-quality vegetative grass. If grass is limited, calves can be offered limited quantities of dry grains to supplement their nutrition. After one to 3 weeks, cows and/or calves may be moved to separate locations.

5.1.5 Dry cow period

Once calves are weaned, they will be placed on grower rations or maintained on pastures (see later sections). Cows dry off at weaning and are typically in the second trimester of gestation, so their nutrient requirements are low. They can be maintained for the next 3–4 months on low-quality forages or crop residues. During the last 50–60 days prior to expected calving, cows will require increased energy and protein intake as the nutrient demands for the rapidly growing fetus and impending lactation become paramount.

5.2 Cow herd health management

The keys to maintaining high health status in a beef cow-calf herd are as follows: (i) establish a proactive vaccination program; (ii) provide proper nutrition, including vitamin and mineral balance; (iii) maintain a clean environment for cattle by using paved lots where cattle congregate and cleaning them regularly; and (iv) limit the introduction of outside stock (the extreme example would be maintaining a closed herd by exclusive use of AI).

5.2.1 Vaccination of cows

A robust vaccination program is one of the most effective methods for maintaining high health status in a cattle herd. All cattle in the herd should be vaccinated annually for all endemic diseases for which vaccines are available. Many cattle vaccines require two doses before an effective immune response develops in the animals. The initial dose and rappel should be separated by about 3 weeks' time, and about 3 weeks is required for full antibody response to develop in the animals following administration of the second dose (There are exceptions to this timing for particular products, but in general this is a good guideline for planning administration of vaccinations). Cows should be at peak immunity when the highest probability of disease challenge occurs. For many diseases, this is at breeding time. Therefore, cows and naïve heifers should be vaccinated prior to breeding. Naïve animals will require two doses, so the initial dose must be administered 6 weeks or more prior to the anticipated breeding date. Mature cows who have experienced previous vaccinations will only require a single booster dose, which must be administered at least 3 weeks prior to breeding.

Common vaccines administered to cow herds pre-breeding in many areas of the world include infectious bovine rhino-tracheitis (IBR), bovine viral diarrhea (BVD), parainfluenza III (PI3), bovine respiratory syncytial virus (BRSV), *Campylobacter fetus* (vibrio), and five strains of leptospirosis. There are vaccine products available that include all these antigens within one formulation. Pre-calving cows are often vaccinated with products that transfer immunity through the colostrum to the calf for prevention of calfhood diseases such as diarrhea (scours). Again, in general, the same timing constraints apply, as cows should achieve maximum immunity at calving so that antibody levels are high in their colostrum. Vaccines (products are available with all these antigens within one formulation) for the following may be administered at this time: Rotavirus, Coronavirus, *Escherichia coli*, and *Clostridium perfringens.*

5.2.2 Vaccination protocol for calves

At 1–3 months of age, calves should be administered vaccines for the following: IBR (intranasal formulation), 7-way Clostridial, and *Moraxella bovis* ("pinkeye" or infectious conjunctivitis).

At 2–3 weeks pre-weaning, calves should be administered a combination IBR-BVD-PI3-BRSV vaccine along with a Clostridial booster (depending on product used earlier). In all instances, read label directions for the vaccines being utilized. Different products of different types (modified live vs. killed, etc.) may require alterations in timing. Further, some modified live versions may hazard a risk of abortion if used in pregnant animals. In addition, administration of the correct dosage is critical, and again, the label is the definitive guide for determining correct dose. Refer to **Tables 1** and **2** for an example of a vaccination schedule for a beef cow herd.

5.2.3 Control of parasites

An effective parasite control program is important for overall health of cattle as well (**Tables 1** and **2**). The most common internal parasites in European cattle include *Ostertagia ostergi* (a gastrointestinal nematode), *Cooperia oncophora* (a gastrointestinal nematode), and *Fasciola hepatica* (liver fluke). Prophylactic treatments with common parasiticides such as avermectins and albendazoles are common for control of nematodes. For beef cow herds, a de-worming prior to breeding in mid-summer is often adequate. Calves may need more frequent treatment as they are much more susceptible to parasitism. Depending on the parasite load in the herd, calves may need more frequent treatments beginning at one to 2 months of age. In all cases, integrated pest management (IPM) best management practices should be utilized in order to prevent long-term buildup of resistance by the worm population to common anthelmintics. Key among IPM practices is routine monitoring of parasite load in the

Vaccine type	Current product used	Administered to what groups and how many times	Time of administration
IBR/BVD/PI3/BRSV/5-way Lepto	Cattle Master Gold FP5	Cows and Bred Heifers 1 time	3 weeks prior to calving (or breeding depending on operator preference)
E coli, Rotavirus & Coronavirus	Scour Guard 4(K)/C	Cows – 1 time Bred Heifers – 2 times	Heifers: 1st shot 6 weeks prior to calving. Cows & Heifers: 3 weeks prior to calving
Clostridium & Hemophilus somnus	Ultrabac-7/Somubac	Cows and Bred Heifers 1 time	3 weeks prior to calving
Vitamin & Mineral	Multimin	Cows and Bred Heifers 1 time	3 weeks prior to calving
Dewormer	Ivermectin	Cows and Bred Heifers 1 time	3 weeks prior to calving

Source: Developed by the authors.

Table 1.
Pre-calving vaccination protocol.

Vaccine type	Current product used	Administered to what groups and how many times	Time of administration
IBR/BVD/PI3/BRSV/5-way Lepto	Cattle Master Gold FP5	Calves 2 times (initial and booster)	September–October
Pasteurella	1 Shot	Calves one time	August–September
Clostridium & Hemophilus somnus	Ultrabac-7/Somubac	Calves 2 times (initial and booster)	August–September
Dewormer	Ivermectin	Calves when weaned	September–October

Source: Developed by the authors.

Table 2.
Weaning vaccination protocol.

herd through fecal egg count (FEC) determinations. Liver fluke is best controlled by preventing grazing in areas where snail populations are likely to be high (wet areas with stagnant water) as snails are a necessary intermediate host. There are some parasiticides that will control liver fluke (albendazole), but they are few. Macrocylic lactones (among them the avermectin class) are convenient dewormers as they also control ectoparasites (such as lice and mites) and are therefore often the anthelmintic of choice for cattle use.

5.3 Cow herd nutrition

Beef cows are among some of the least demanding of common livestock from a nutritional standpoint. Their nutritional requirements are highest during the first 3 months of lactation. But even at this time, they are substantially lower than those of dairy cattle. Late gestation (last 60 days) and late lactation (last 4 months) also have somewhat higher (and about equal to one another) nutrient requirements, but again, these are substantially lower than for dairy cattle. In essence, beef cows' nutritional requirements can easily be met, even at their highest, with good quality pasture as the only feed beyond some key mineral supplementation (salt and trace or microminerals).

5.3.1 Macronutrients

During early lactation, a ration consisting of only high-quality vegetative grass will supply enough energy to maintain body condition—nearly twice the protein requirement and approximately twice the Ca and P requirements for a typical high producing Angus cow, based on Beef Cow Ration Balancer [11]. Since early lactation is the period in the cow's life where her nutrient requirements are highest, it follows that a pasture diet will meet her requirements during the remainder of the annual cycle. In fact, post-weaning, her requirements are so low that very modest to poor quality forages or crop residues will be adequate. Greater attention to nutrition will be needed during the last 2 months prior to calving. Cows in colder climates may still be on stored feeds at this time, and particular attention must be paid to providing adequate energy, protein, and certain vitamins and minerals. For cows in warmer climates, or those calving later in the spring, new pasture may be the feed source during late gestation, and as such will be more than adequate to meet the cattle's nutrient requirements.

5.3.2 Micronutrient

The only other feed required while on pasture is a common cattle trace mineral salt (TMS), which is salt fortified with the most common microminerals (Cu, Fe, Mn, I, Zn, Co, and perhaps Se). In many areas of the world, one or more of the common microminerals are deficient in the native soils, which results in forages, which are deficient for livestock requirements. Since these minerals are quite cheap and easily supplemented by blending with salt, it is good nutritional insurance to offer trace mineral salt-free choice (*ad libitum*) to cattle at all times. Trace mineral salt composition is formulated such that as cattle consume their salt requirement (which they can self-regulate), the consumption of the trace minerals is controlled at proper levels for most situations. While all trace minerals are important for optimum health and performance, one in particular should be highlighted as it is occasionally overlooked, and that is selenium. Adequate intake of this element is critical for normal reproduction and immunity, but intake above requirements is also toxic. The requirement for cattle is low, about 0.3 mg/kg of diet, but may be toxic at levels as low as 5.0 mg/kg. Soil levels vary greatly across Europe, so local conditions must be taken into account when supplementing Se to cattle. If it is needed, use a TMS with Se; if not needed, use a TMS that has no added Se. It should be noted that vitamin supplementation for beef cows is typically not required while they are consuming pasture. Fresh forage has more than adequate levels of vitamins A and E to meet the demands of lactating beef cows, and since they are housed outdoors, they receive plenty of sunshine to engender more than adequate endogenous synthesis of vitamin D. If fed dried, preserved green forage (hay) during winter, vitamin levels may be adequate early on, but ADE diminishes in stored hay over time, so by late gestation, it is prudent to supplement these vitamins to beef cows as well. Vitamin ADE feed additives are relatively inexpensive and again offer good nutritional insurance during the cold seasons of the year.

For the *calves*, the combination of good quality pasture and milk from their dams assures that nutritional requirements of calves are met during the nursing period of their lives. A later section of this chapter will discuss post-weaning nutrition of calves.

5.3.3 Body condition

A key guide for general nutrition of beef cows is their body condition. It is a good indicator of energy intake status. Even though the energy needs of beef cows can be easily met with good quality pasture, situations often arise where pasture conditions are not good enough, or where stored feeds of poor quality must be utilized. This is particularly important when cattle are raised in semi-arid regions (such as the steppes of Eastern Europe or the high plains of the USA where pasture production is often not adequate), or drought conditions arise, or cattle are grazed on forb species of very poor quality. Pastures and other forages may require supplementation with grains or tubers, or other higher energy feedstuffs to maintain adequate energy intake under these conditions. Therefore, the body condition of the cows should be closely monitored to inform the need for energy supplementation. A common system for body condition scoring (BCS) of beef cattle uses a 1 to 9 scale. Cows that are at BCS 4 or lower at calving, will have reduced conception rates on the next breeding. Those that are BCS 5 or above will have typical conception rates (>90%). Cattle with BCS 8 or 9 will also experience reduced conception as well as increased dystocia and incidence of metabolic diseases. The goal therefore is to have each individual cow in the herd achieve a BCS of 5 or 6 by calving. Energy intake should be adjusted with the use of grains, if necessary in order to achieve that target.

5.4 Cow herd housing

In general, beef cow herds are maintained entirely outdoors. The only time where shelters may be used is for very early calving schedules in cold climates or highly inclement weather. Even at these times, the shelter is merely an open-front shed (open to the south) with no insulation or mechanical ventilation. Cows and their calves are usually kept in such sheds for only a few days just prior to and immediately following calving. It is a complete waste of financial capital to construct elaborate housing facilities for beef cow herds, as these animals have evolved to thrive in temperate climates, and even calves are extremely hardy once they are born, have nursed, and dried off.

5.5 Replacement heifers

At weaning time, any heifer calves that are selected for cow herd replacements are identified and often grown separately from steers or heifers for slaughter. Heifer replacements will be maintained on diets similar to those used in the growing phase of slaughter cattle, but often with higher levels of certain vitamins and minerals (such as Ca, P, and Co, for example). In addition, the energy content may be slightly lower and protein slightly higher. Heifers need to gain only about 0.6–1.0 kg/day to achieve 2/3 of mature weight by the age of 15 months. If they have conceived by 15 months, they will enter the herd and bear their first calf by 2 years of age. Achieving 2/3 of mature weight at breeding will allow for optimum conception rate and attainment of adequate maturity at the time of first calving. Since this is a slower growth rate than ideal for market animals, heifers should be fed lower energy rations—a lower grain-to-forage ratio. If heifers get too fat, excess udder fat develops and their milk production tissue development is hindered, which results in lifetime loss of milk production. They should be maintained at a BCS of 5 during their growth and development period.

6. Semi-intensive systems: grow-finish for beef production

Once calves are weaned, they may be managed in a variety of ways while they are being grown and finished to produce beef. In general, there are two phases that may be accomplished in differing ways. Those phases include:

Growing phase: Calves are maintained on diets of moderate energy density with the objective of maintaining an adequate growth rate with little fattening.

Finishing phase: Calves are placed on an energy intensive diet with the goal of attaining final market weight at a fatness level that optimizes carcass value.

6.1 Growing phase

Immediately upon completion of weaning, calves must be slowly transitioned to growing diets. There are two basic approaches for growth diets:

Feedlots: Placing calves in open-front sheds (**Figure 2**, center) with paved outdoor lots (some operations may use complete confinement facilities with insulation, mechanical ventilation, and slatted floors for manure removal—but this is not often economical) and providing harvested stored feeds.

Pasture back grounding or stocker systems: Maintaining calves on pastures for 2–6 months prior to movement to a feedlot for finishing.

6.1.1 Feedlot growing

Calves are transported to feedlot facilities (**Figure 2**, left) and slowly transitioned to grower diets. An important point is that the dietary transition be slow—covering a period of 2–4 weeks. If calf herd ownership is retained by the cow herd operator, the transition may be faster since calves are typically not transported any great distance, and the dietary history of the calves is known. Most cow herd owners, however, sell their calves after weaning. These sales may be through various marketing mechanisms and often result in calves being transported considerable distances prior to reaching the feeding facilities. In these instances, the transition to the grower diet should be slower. The initial feed offering upon arrival in the feedlot is usually *ad lib* dry hay for about 2 days, at which time a small amount of cereal grain (0.5 kg/head/day) along

Figure 2.
Growing calves in an open-front shed (left), an outside drive-by bunk feeder (center) and on a wheat pasture in the USA midsouth region (right). Source: Owned by Ioan Hutu and Gary Onan.

with some grass or maize silage may be introduced. During the following 3 weeks, the amount of grain and silage is gradually increased (about 0.5 kg DM of each per day) until calves are completely meeting their appetite with that ration. These rations typically take the form of a TMR and are delivered to calves in fenceline feed bunks. It is important that there be enough bunk space so that all calves can eat simultaneously during this transition phase (45 cm/head).

6.1.2 Grower rations

A typical grower ration is designed to obtain an average daily gain (ADG) of about 1.4 kg/day. This is usually achieved using a blend of forages (often silages) and grains with about 55–60% of dry matter (DM) from forage and 35–40% from grains, with the remainder consisting of protein supplements and mineral and vitamin supplements. The energy content of such a ration will be roughly 1.8 Mcal/kg NEm (net energy for maintenance) and 1.2 Mcal/kg NEg (net energy for gain). Crude protein (CP) content is often formulated at around 12–14%. The silages may include maize, alfalfa, or alfalfa/grass mix. The grains may include maize (either dry or ensiled (high-moisture corn—HMC), barley, or up to 1/3 wheat if market prices make it an economical feed. Grains could also include distiller's grains and solubles (DGS), either wet (directly from the distillery) or dried, from ethanol plants producing products for either fuel or human consumption. Spent grains from breweries are also commonly used. Protein sources may be any of the common oil meals, such as soybean, sunflower, canola (rape), and linseed. Calves may be maintained on the grower rations for 3–4 months or until they reach about 80% of expected market weight, at which time they are transitioned to a finishing (fattening) diet.

6.1.3 Health protocols

If calves have been properly vaccinated prior to and at weaning time, no additional vaccine administration is required. Some feedlots may, however, give vaccines for *Mannheimia haemolytica* and/or *Histophilus somni* depending on previous bovine respiratory disease history. In addition, a prophylactic injection of an antibiotic such as Draxxen (Tulathromycin) or Micotil (Tilmicosin) may be given upon arrival for the prevention of bovine respiratory disease.

6.1.4 Pasture growing

The practice of growing calves on pasture is referred to as backgrounding (implies preparation for eventual movement to a feedlot). Again, these calves may be retained by the cow herd operator but are more often sold to another farmer. If calves must be transported any considerable distance, they should be closely monitored upon arrival by placing in a pasture near the farmstead and given prophylactic treatments as indicated above for feedlot calves. Diet transition is more simple, however, as these calves will be on pasture of some type, which is often so close to what they are accustomed to consuming that no real transition is required. Pastures may be of any type, grass, legume/grass mixes, or vegetative grains. In the USA it is common to graze these calves on winter wheat that was planted 30–60 days earlier (**Figure 2**, right). Calves graze the wheat during late fall and into early winter, and again in the early spring. These wheat pastures are harvested for grain the next growing season, but cattle can graze them until the first hollow stems appear in the wheat plants. Stocking rates in the fall will be about one animal per ha. Calves will gain about 1 kg/day when back grounded on good quality pastures of either forage grasses or vegetative grains.

Since these calves gain slower than those grown in feedlots, they will be older when they enter the finishing phase, but the feed cost per kg of gain for pasture calves is often substantially less than that for feedlot calves, so this approach is very economically viable even if gains are slower [12, 13].

6.2 Finishing (fattening) phase

Finishing or fattening of cattle is achieved by feeding the stock rations with high energy density in order to promote rapid fat accumulation prior to marketing. The fattening enhances the eating quality of the beef from these animals. The degree to which they are fattened depends on cultural preferences and market forces (such as feed cost and value of fattened animals). Cultural preferences play a major role in determining the value of finishing for market. British, Canadian, and USA consumers typically prefer beef with higher levels of fatness, in particular, high levels of intramuscular fat (IMF) or marbling. Higher IMF has been documented to improve the juiciness and flavor of the beef product and perhaps contribute slightly to tenderness as well. Continental consumers typically prefer leaner beef and are not as concerned about high marbling levels in the product. Perhaps the most extreme market for fattened beef is the Kobe style beef sold in Japan, which has so much marbling in the muscle that there is whiter (fat) than red (lean) coloration in the uncooked product. Most cattle are finished in feedlot facilities on rations with very high cereal grain content. They may be, however, be successfully finished on pasture as well, though this is a small, but important niche for beef finishing. Pasture finishing is perhaps more common in Continental Europe than in Britain or North America. If calves have been grown in feedlots, they can be quickly and easily transitioned to finishing diets and will reach market weights at ages of 12–18 months (assuming the cattle possess good growth genetics). If the cattle come from pasture backgrounding systems, they will be older and take longer to transition and will reach market weights at ages of 18–24 months. The age difference between the two types of approaches serves an important function in that it spreads the marketing time more evenly throughout the year, even though calves are born somewhat seasonally. That stabilizes market quantities and prices, which is important for cattle farmers' and beef packers' financial planning. It also provides a uniformly priced and readily available year-round supply of beef for consumers.

6.2.1 Feed rations for finishing

Finishing rations are designed to generate gains of 1.5–2.0 kg/day in fattening cattle. This is achieved by using rations that are predominantly cereal grains. Such rations are typically comprised of about 80% DM from grains; 15–18% DM from forages (often silages); and the remainder mineral and vitamin supplements. Vegetal protein supplementation is often not required as the CP requirement is only 11–12% and can easily be met with the addition of urea to the ration. Urea can be used as a nitrogen source by rumen microbes to synthesize amino acids, which provide the animal with its protein needs. Energy content of finishing rations should be around 2.2 Mcal/kg NEm and 1.35 Mcal/kg NEg. This high energy density promotes rapid growth accompanied by rapid fattening. Grains, silages, and industrial co-products similar to those used for grower rations are suitable. Once cattle achieve the desired market weight, often around 500–600 kg depending on breed and type of cattle, they can be sold for beef. Dressing yield for these cattle is between 60 and 64%. Gender influences the weight at which cattle attain a degree of finish suitable for market also (Berg and Butterfield). In a general sense for cattle of the same genetics, heifers will attain a market-ready degree of fatness at lighter weights (about 45–50 kg lighter than steers), steers will be intermediate, and intact bulls heaviest (again 45–50 kg heavier than steers).

6.2.2 Health concerns for finishing cattle

The incidence of infectious disease is very low in finishing cattle if vaccinated earlier in life, as indicated in previous sections of this chapter. The primary diseases of concern in finishing cattle are metabolic in nature. Two important ones are bloat and laminitis. Both of these become more likely on high energy density diets and are therefore more common in finishing cattle. Bloat is the accumulation of gas foam in the rumen which creates tremendously high internal pressures that eventually lead to failure of the heart. Bloat comes about when the normal rumen gasses form a foam that the cattle are unable to eructate (belch) from their rumen. Highly digestible starch diets occasionally promote bloat. Its incidence can be reduced by adding buffers such as sodium bicarbonate or magnesium oxide, or by adding surfactants such as poloxalene to the TMR blend. Laminitis is a separation of the soft laminae in the hooves of cattle. It leads to extreme pain and the inability of cattle to walk well. It also often causes deformation of the third phalangeal joint in the hoof leading to permanent skeletal dysfunction. Laminitis is caused by buildup of endotoxins in the blood which originate from the gut. These endotoxins are components of gut bacteria that have died as a result of an acidic gut environment. Since starches are rapidly converted to volatile fatty acids in the rumen, cattle on high starch diets may become acidotic, which may precipitate incidence of laminitis. Prevention or mitigation is accomplished by increasing dietary energy intake slowly whenever changes are made, so that rumen microbes can adjust. Feeding of rumen buffers as indicated above for bloat is also an advantageous practice.

6.3 Pasture finishing

Cattle may be successfully finished on pasture without the use of any cereal grains or silages made from cereal crops. Pasture-finished beef is a popular niche product and commands a premium price in various developed countries. Pasture finishing would be considered an extensive management system. Onan and Hutu [12] outline an effective method for pasture finishing of cattle. In general, the three requisite practices are as follows:

1. *Wean heavy calves in the fall* (200-day weights in excess of 270 kg)—this requires a cow herd with high genetic propensity for milk production and growth; it also requires excellent pasture management using intensive rotational grazing principles [14].

2. *Encourage rapid growth during the winter* months by feeding excellent quality legume or grass silages. The limiting factor for rapid growth is energy intake. Exceptionally high-quality grass or legume silages will have energy values of about

1.3 Mcal/kg NEm and 0.8 Mcal/kg NEg. These energy values are high enough to maintain an average daily gain of 1.0–1.2 kg/day. The protein content of such feeds is more than adequate for growing cattle. Onan and Hutu [12] grew calves over winter with a ration consisting only of ad lib alfalfa silage and *ad lib* trace mineral salt, which produced an overall average daily gain of 1.2 kg/day. These calves should, therefore, weigh around 450 kg at the time of turning out to pasture in the spring.

3. *Encourage rapid gains on pasture* during the crop growing season of the calves' second year. This is accomplished by implementing intensive rotational grazing practices combined with a practice known as "top grazing." The finishing cattle are allowed only 1 day of access to each grazing paddock and then moved to a new paddock, even though extensive forage remains in the first paddock. This short grazing period allows the cattle to consume only the best quality portions of the plants (the new growth at the top of the plants), and therefore they consume only the highest quality forage available. This should result in gains in excess of 1.0 kg/day, with cattle reaching market weight and level of fatness by autumn (by 18–20 months of age). Remaining forage in each paddock can be grazed off by stock with lower nutrient demand, such as a cow herd or replacement heifer calves. The problem most often encountered with pasture finishing systems is an inability to maintain the minimum rate of gain required for the cattle to reach market weights by the end of the grazing season during their second year of life, that is, by 18–20 months of age. If this is not achieved, those cattle often do not reach market weight until they are 30 or more months of age. This results in a substantial decrease in eating quality, particularly in regard to tenderness. The frequent occurrence of pasture-finished beef being marketed at old ages or being underfinished at the market has led many consumers to think that pasture beef is undesirable from the standpoint of eating quality. If properly grown, as outlined earlier in this section, pasture beef will have eating quality equal to that of feedlot-finished beef.

6.4 Finishing dairy steers

An important sector of the beef industry is that which finishes steers derived from dairy cattle herds. Most often these will be Holsteins, but Brown Swiss, Schwyz and dairy-type Simmental steers work well also. Dairy steers exhibit some unique characteristics relative to beef cattle breeds. They are as follows: (i) marketed at heavier weights in some areas—as high as 800 kg; (ii) carry less prominent muscle, but nonethe-less produce carcasses with a total retail cutout nearly as high as beef breeds; (iii) less fat at market weight. The subcutaneous and intermuscular fat depots comprise a lower percentage of carcass weight than is the case for most beef breeds; (iv) more marbled—the amount of intramuscular fat is greater than for most beef breeds; (v) lower yielding for dress—about 58–59%; this is largely due to their larger gastrointestinal tract; and (vi) able to tolerate a high energy density diet for a longer period of time without developing metabolic disease problems (such as acidosis).

Because of these differences, dairy steers are managed somewhat differently than grow-finish beef cattle. Young dairy calves will originally be raised using management nearly identical to that used for rose veal. Instead of being sent to slaughter at 300–450 kg, they are moved into finishing lots and put on high-energy density rations to promote rapid growth and fattening. These calves are thus on high energy density diets their entire lives. Most beef breed cattle would not be able to tolerate that duration of high grain feeding and would likely develop metabolic problems. An alternative management process is to raise the calves similar to that for rose veal production but then move them to pasture backgrounding at lighter weights (around 200–300 kg). After a period of pasture growth and upon attainment of about 450–550 kg weight, these steers are then moved into feedlots for finishing. These are the type of cattle that are taken to the very heavy final weights of 700–800 kg. Once in the finishing phase, dairy steers rations are often very similar to those used for beef breed cattle, although they may contain a slightly higher percentage of cereal grains than what is typically used for beef breeds. There is one dairy steer feeding system that utilizes only cracked corn (maize) and a pelleted supplement fed *ad lib* through a large self-feeder (a "steer stuffer") designed for cattle. This is a proprietary method—see Tend-R-Leen, used in the Midwest region of the USA [15].

Finishing dairy steers for beef is an important sector for adding value to a "waste product" of the dairy industry, namely bull calves. If these bull calves are not utilized for veal production, finishing them (usually as steers) for beef production is the other viable alternative. In the USA, the vast majority of dairy steers are utilized in this manner, as the veal industry is much smaller than in Europe.

7. Mitigation strategies of greenhouse gas in beef cattle sector

Today, the beef cattle sector receives quite a lot of criticism due to its contributor to greenhouse gas (GHG) emissions and environmental impact, which are major drivers of climate change. Sources of GHG emissions in the beef industry can be: enteric fermentation by methane production (CH_4), manure production, feed production and consumption of fossil fuels for farms, transportation, and processing, which results in CO_2 emissions. The recommended mitigation strategies for GHG emissions in the beef sector should include:

- Improved feed efficiency through diet optimization [7], inhibitors for enteric methane, such as 3-nitrooxypropanol [16]—of feeding and genetic and genomic selection [17] for higher feed efficiency and lower methane emissions.
- Manure management by anaerobic digesters: (using methane for biogas) and improving the storage for reducing the methane and nitrous oxide emissions.
- Rotational grazing and integrating trees into pasture lands to enhance carbon storage increase sustainable land management.
- Increasing energy efficiency by using renewable energy and implementing energy-saving technologies and practices in beef production and processing.

Proper grazing management has been estimated to increase soil C storage on US rangelands from 0.1 to 0.3 Mg C ha^{-1} $year^{-1}$ and new grasslands have been shown to store as much as 0.6 Mg C ha^{-1} $year^{-1}$ [18]. Considering the lowest carbon sequestration is found in systems with beef from dairy bull calves slaughtered between 9 and 19 months, which accounts for 0.2–0.9 kg CO_2/kg carcass, while the highest carbon sequestration is observed in steer production and beef breed systems, contributing between 2.3 and 4.8 kg CO_2/kg carcass, reducing the feed intake [19], age [16] for slaughtering should be encouraged.

8. Conclusion

- The fattening system type has a direct impact on cattle growth performance, meat quality, and economic efficiency: Intensive systems provide rapid growth rates and high slaughter yields, making them ideal for white veal production, while semi-intensive and extensive systems result in slower growth but offer benefits in sustainability and nutritional quality.

- Tailoring feeding and maintenance strategies to the cattle's development stage is essential for achieving optimal performance and producing high-quality beef.

- Effective beef cattle management requires an integrated approach, with a focus on parasite control, balanced nutrition, and consistent monitoring of body condition.

- Sustainable beef production depends on responsible resource use, reducing greenhouse gas emissions, and maintaining a balance between biological efficiency and market demands.

Conflict of interest

The authors declare no conflict of interest.

Author details

Ioan Hutu[1]* and Gary William Onan[2]

1 University of Life Science "Regele Mihai I", Timisoara, TM, Romania

2 Wisconsin University, River Falls, WI, USA

*Address all correspondence to: ioan.hutu@fmvt.ro

References

[1] Hutu I, Onan WG. Animal Science: Traditional and Modern Perspectives. Timisoara: Agroprint; 2024. 520 p

[2] Acatincăi S. Producţiile bovinelor. Timişoara: Editura Mirton; 2000. 459 p

[3] Soulat J, Picard B, Monteils V. Influence of cattle category and slaughter age on Charolais-breed carcase and meat traits. Italian Journal of Animal Science. 2023;**22**(1):263-275. DOI: 10.1080/1828051X.2023.2182720

[4] Sargeant JM, Blackwell TE, Martin SW, Tremblay RR. Production practices, calf health and mortality on six white veal farms in Ontario. Canadian Journal of Veterinary Research. 1994;**58**(3):189-195

[5] Dudouet C. La production des bovins allaitants. Paris: Editions France Agricole; 1999. 390 p

[6] Bolton SE, von Keyserlingk MAG. The dispensable surplus dairy calf: Is this issue a "wicked problem" and where do we go from here? Frontiers in Veterinary Science. 2021;**8**:660934. DOI: 10.3389/fvets.2021.660934

[7] Mollenhorst H, Berentsen PBM, Berends H, Gerrits WJJ, de Boer IJM. Economic and environmental effects of providing increased amounts of solid feed to veal calves. Journal of Dairy Science. 2016;**99**(3):2180-2189. DOI: 10.3168/jds.2014-9212

[8] Veal Classification [Internet]. Available from: https://agriculture.ec.europa.eu/document/download/2fd345d9-fef1-40fc-94e7-f4608682a38d_en?filename=methodology-carcase-remainders_en.pdf) [Accessed: August 02, 2024]

[9] Hutu I, Bratu D, Lungu B, Iuliu T, Irina S, Otava G, et al. The management plan for beef cow-calf production in west Romania plains. In: International Multidisciplinary Conference on Sustainable Development - Section Animal Breeding and Pathology Today. Timisoara: University of Life Science "Regele Mihai I"; 30-31 May 2024

[10] Fenceline Weaning [Internet]. Available from: https://www.canr.msu.edu/uploads/resources/pdfs/fence_line_weaning_a_marketing_tool_for_your_calves_(2929).pdf [Accessed: August 01, 2024]

[11] OSU Cowculator [Internet]. Available from: https://extension.okstate.edu/programs/cowculator.html [Accessed: June 05, 2024]

[12] Onan G, Hutu I. Grass-fed beef. Lucrări Ştiinţifice Medicină Veterinară. Timisoara, Romania: Agroprint. 2016;**49**(1):142-153

[13] Onan G, Hutu I, Radunz A. Residual feed intake as a measure of feed efficiency. Lucrări Ştiinţifice Medicină Veterinară. Timisoara, Romania: Agroprint. 2016;**49**(1):154-162

[14] Undersander D, Albert B, Cosgrove D, Johnson D, Peterson P. Pastures for Profit. Madison, WI, USA: University of Wisconsin Extension; 2002. Available from: https://learningstore.extension.wisc.edu/products/pastures-for-profit-a-guide-to-rotational-grazing-p96#:~:text=Make%20a%20profit%20from%20your,(40%20pages%3B%202014) [Accessed: August 04, 2024]

[15] Get to Know the Tend-R-Leen® Dairy-Beef Program. Available from:

https://formafeed.com/tend-r-leen-dairy-beef-program/ [Accessed: July 12, 2024]

[16] Samsonstuen S, Åby BA, Crosson P, Beauchemin KA, Aass L. Mitigation of greenhouse gas emissions from beef cattle production systems. Acta Agriculturae Scandinavica, Section A—Animal Science. 2020;**69**(4):220-232

[17] Pickering NK, Oddy VH, Basarab J, et al. Animal board invited review: Genetic possibilities to reduce enteric methane emissions from ruminants. Animal. 2015;**9**(9):1431-1440. DOI: 10.1017/S1751731115000968

[18] Schuman GE, Janzen HH, Herrick JE. Soil carbon dynamics and potential carbon sequestration by rangelands. Environmental Pollution. 2002;**116**(3):391-396

[19] Mogensen L, Kristensen T, Nielsen NI, Spleth P, Henriksson M, Swensson C, et al. Greenhouse gas emissions from beef production systems in Denmark and Sweden. Livestock Science. 2015;**174**:126-143

Chapter 8

Sustainable Production of Certified Beef Cattle Farming in the Pantanal

Urbano Gomes Pinto de Abreu, Eliane Gonçalves Gomes and Silvio Henrique Ribeiro Balduino

Abstract

The Pantanal is a Brazilian biome with a peculiar beef cattle production system. In November 2018, the Government of Mato Grosso do Sul, Brazil, approved legislation that made the tax incentive for organic and sustainable meat production official. Despite the initial difficulty in organizing the production chain and, in parallel, going through 2 years of pandemic, in 2022 the production system was consolidated. In order to evaluate the progress of the sustainable and organic meat production system in the Pantanal, data from nine slaughterhouses accredited by the Pantanal Sustainable Livestock Program were subjected to unstructured modeling using the K-means method. Two groups were identified (C1 and C2). The first (C1) had a smaller number of animals per batch, and most batches had females. Group C2 had a larger number of animals, mainly males. C1 and C2 received, respectively, 53 and 47% of the total tax exemption (BRL 8,495,587.53). The development of a sustainable livestock system with quality and certification for bovine at slaughter proved to be an important option for the development of livestock farming in the Pantanal.

Keywords: beef cattle, production systems, livestock development, unstructured modeling, wetland

1. Introduction

The term sustainable agriculture is used to describe everything from organic farming to conventional agriculture that maximizes economic returns with minimum environmental impact. However, the concept of sustainability is based on a holistic philosophy with a series of principles and values, which also involves specific practices [1]. Currently, a large portion of the world's population is concerned about the use of chemical compounds, growth promoters, and chemical additives in animal production. In addition, it is important to consider aspects related to animal welfare and environmental conservation, among others, which are negatively associated with more intensive livestock production systems, leading some consumers to increasingly choose to consume organic products [2].

Data from the Brazilian organic food market since the 1980s were analyzed [3]. These authors found that a social ideology and initiatives opposed to conventional agriculture were the basis of the organic production systems. However, as the patterns of supply and demand changed, particularly in the last 25 years, the expansion of the organic products market has gone beyond ideological motivation. The Brazilian agribusiness sector has been adapting to this reality. The main reasons for this expansion are changes in the awareness of Brazilian consumers. A substantial number, especially in large cities, are concerned about nutritional quality and aspects of the production systems of the food they eat.

According to Ref. [4], in the Brazilian case, the dissemination of sustainable practices in the beef chain is desirable to provide coordination for the organization of production, processing, and distribution, which generates opportunities for the development of sustainable livestock farming. In 2004, WWF-Brazil and the Brazilian Association of Organic Producers (ABPO) launched the Sustainable Livestock in the Pantanal project. The certified organic livestock farming in this region was grounded in three pillars, namely environmental conservation, animal welfare, and decent working conditions. These are the premises of all organic and sustainable meat protocols [5].

The competitiveness of the organic production chain is currently achieved through product differentiation strategies. In the beef chain in Rio Grande do Sul, a Brazilian Southern state, there was a need to establish partnerships between the agents in the chain to obtain a differentiated product [6]. This is in line with the process underway in Mato Grosso do Sul, a state in the Center-West region of Brazil, where ABPO, WWF-Brazil, different private companies, and the state government are establishing strategic partnerships to develop the production chain.

In 2018, the government of the state of Mato Grosso do Sul approved legislation that formalizes the tax incentive to produce organic and sustainable meat. It created the Subprogram to Support the Production of Sustainable Meat in the Pantanal, within the scope of the Mato Grosso do Sul Livestock Advancement Program (PROAPE), through a joint resolution of the State Secretariat for Finance and the State Secretariat for the Environment, Economic Development, Production, and Family Agriculture [7]. This resolution defines that "PROAPE Sustainable and Organic Meat of the Pantanal aims to foster competitiveness and encourage low-environmental-impact cattle farming in the Pantanal, stimulating production based on the traditional model, with a low level of intervention in the region's natural resources, and using technological scopes for characteristic and differentiated product lines with greater added value, duly certified by independent third-party companies, accredited by Inmetro[1]". Basically, the awards for certification are:

1. For the organic system (following the organic production legislation determined by Federal Law No. 10831), an exemption of 67% of the Tax on Circulation of Goods and Provision of Services (ICMS); and

2. For the sustainable system (adoption of the ABPO protocol) exemption of 50% of the Tax on the Circulation of Goods and Services (ICMS[2]).

Cattle ranchers from the Pantanal region created ABPO. They aimed at standardizing their production to meet the requirements of the Certified Organic Livestock.

[1] Inmetro is the acronym of the Brazilian National Institute of Metrology, Standardization and Industrial Quality.

[2] ICMS is a state tax under Brazilian state administration.

This was fulfilled by embracing the statements of quality, social and ecological responsibility, and sustainable development in their activities. They established a protocol managed by the Brazilian Confederation of Agriculture and Livestock (CNA[3]), which must be adopted and applied by producers in their livestock production systems. The protocol was developed primarily for heifers raised in the Pantanal; the age range for slaughter is 2.8–3 years, the latter being the maximum age allowed [5]. The use of herbal medicines and homeopathy are the preferred breeding practices. The use of urea as fertilizer or as an element mixed in any type of supplement is prohibited. In the Pantanal, fattening is limited due to logistics issues. In addition to the strong dependence on the cyclical factors of the price of cows and steers and the abundant supply of pasture, which does not occur every year due to the more severe floods, restricting the grazing areas. The animals are taken by groups to finishing farms, where they remain for around 4 months in semi-confinement or confinement. After this period, they are slaughtered in a specific meatpacking plant prepared for deboning in accordance with the specificities of the sale. Afterwards, the meat receives its own packaging and identification of sustainable or organic origin from the Pantanal, and then the partner private company is responsible for selling and distributing it.

Considering this brief context, the aim of this chapter is to evaluate the progress of the sustainable and organic meat production system in the Pantanal, using data from farmers under the Pantanal Sustainable Livestock Program, for the period 2019–2022. We propose here an unstructured modeling approach using the K-means method to cluster these observations.

1.1 Livestock in the Pantanal

All links in the beef cattle production chain face a considerable challenge in achieving a balance between economic viability, environmental responsibility, and social acceptability to ensure the sustainability of meat production [8]. Data regarding the balance of these factors should generate information, which is fundamental and is part of the concerns of more conscious consumers [9]. This is more noticeable in biomes with peculiar characteristics, such as the Pantanal (**Figure 1**). Pantanal is located in the central-western region of Brazil. It is an immense sedimentary plain, with an area of 138,183 km^2, 65% of it in the state of Mato Grosso do Sul state and 35% in the state of Mato Grosso. It is bordered to the north by the southern formations of the Amazon rainforest (near the municipality of Cáceres); to the east, by the cerrados of the Brazilian central plateau; to the west, by the wetlands of the Bolivian-Paraguayan borders; and to the south, through the Chaco forests, on the border with Paraguay.

Pantanal of Mato Grosso, such as the sub-regions of Nhecolândia, Paiaguás, and Miranda, among others, do not have any rock formations on their surface. The slope of the region is very small, which contributes to the prolonged retention of water on the soil surface during river and rain floods. The number of rivers that run through the region is considerable, with the Paraguay River being the most important one. It flows from North to South and runs through the entire region, extending 2730 km from its sources to its confluence with the Paraná River at the border between

[3] CNA represents small, medium, and large Brazilian rural producers and is responsible for bringing together associations and political and rural leaders throughout the country.

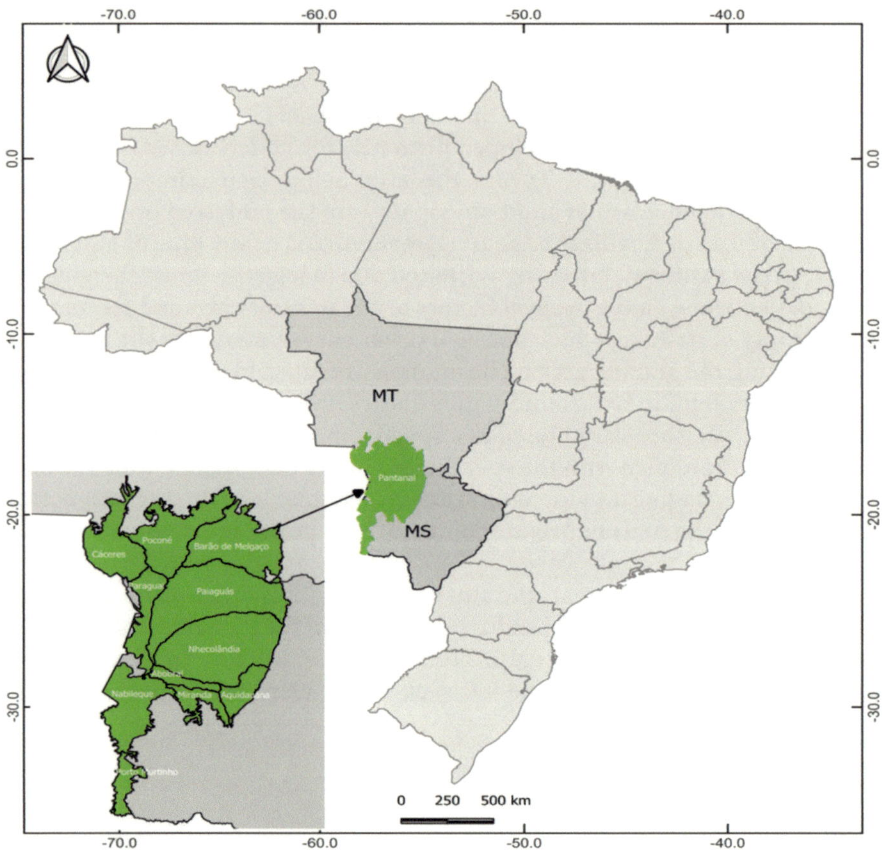

Figure 1.
Location of the Pantanal and sub-regions in Brazil. MT is the acronym for the state of Mato Grosso and MS for the state of Mato Grosso do Sul. Source: Developed by the authors.

Argentina and Paraguay, forming the Upper Paraguay River Basin. The Negro, Miranda, Aquidauana, Bento Gomes, Taquari, Nabileque, Cuiabá, APA, etc., rivers flow directly into the Paraguay River. In addition to these, the Pantanal is home to a number of other rivers, including the Piquiri, Itiquira, and São Lourenço.

The vegetation covering the Pantanal is varied. The name 'Pantanal Complex' is used to define it. The region includes cerrado, clear grassland, dirty grassland, marshes with hydrophilic vegetation, subdeciduous tropical rainforests, and others. There are several plant communities with a clear dominance of one species, with the plant community being named after the dominant species (e.g., canjiqueiral, caronal, paratudal, etc.).

The origin of cattle farming in the Pantanal dates to the seventeenth century, with two major periods identified by the type of cattle [10]. Before the beginning of the twentieth century, Iberian cattle, locally called Tucura or Bovino Pantaneiro, predominated. The second period begins with the replacement of Tucura cattle by Zebu cattle [11]. The average ranch size decreased over time. By the end of the nineteenth century, the largest ranches ranged between 100,000 and 400,000 hectares, and by the end of the twentieth century, the figures were from 1000 to 10,000 hectares for most of the ranches [10].

On the other hand, cattle have increased in the same period, from about 700,000 to over 5 million [10]. Cattle production is still the main economic activity in the region, developed in more than 50% of the total area of the Pantanal, with the predominance of calf rearing. In this context, [11] reinforced that Pantanal is a provider of calves to other areas, concentrated on cow-calf farming, and with cattle management adapted to the flood regime. Sustainable intensification is seen by [11] as the way to increase cattle production efficiency.

2. Methodology

After the PROAPE decree in November 2018, systematic data collections were carried out between January 2019 and December 2022 on the number of slaughters in the two certification protocols. We considered the count of the number of animals slaughtered as the dependent variable. We used Loess regression methodology [12] to fit the trend curves. Non-parametric regression analysis, as the Loess local regression method, estimates a function without reference to a pre-defined functional form. In local regression, a change in perspective occurs. Since the functional form is not previously established, the estimated curve takes on the central role in the analysis. Therefore, a central aspect of local regression is visualization. In other words, the "data speak for themselves." This characteristic is interesting because, often, theoretical analysis does not establish the structural form among the variables [12, 13]. The Gretl econometric analysis program was used to perform modeling using the Loess regression method [14].

In the period 2019–2022, in parallel, 78,533 head of cattle were slaughtered, distributed in 864 lots. We used data from nine slaughterhouses accredited by the Pantanal Sustainable Livestock Program. During the period, the total tax exemptions (premium in BRL) for each lot of animals slaughtered were also recorded. The data were organized and subjected to descriptive analyses to verify frequencies, trends, dispersions, and positions. The unsupervised K-means clustering method [15] was then used. The method aims to partition the observations into K groups, in which each observation belongs to the cluster with the closest mean (centroid). This results in groups in the data space, and the observed groupings minimize the variances within the cluster [15].

3. Results

Figure 2 shows the behavior of the number of slaughters and the tax exemption received by producers who chose to develop the certified Organic Protocol. The observed trend was growing until mid-2020. After this period, there was a progressive decrease, as observed in the Loess regression. The average premium per slaughtered animal was BRL 157.76.

Although the tax exemption for organic certification is 67% of the ICMS (greater than the exemption for the Sustainable Protocol, which is 50%), the number of animals in this certification has clearly been decreasing. The estimated Compound Growth Rate (CAGR) was negative, at −12.5%, meaning that there was a decrease in the number of slaughters. The conversion of traditional systems to the organic system for beef cattle in Spain, using the Global Conversion Index (GCI), and the legal requirement for European organic farming were analyzed by [2]. This author concluded that farms needed to introduce significant changes before starting the conversion process. He also showed that cross-cutting support measures are relevant,

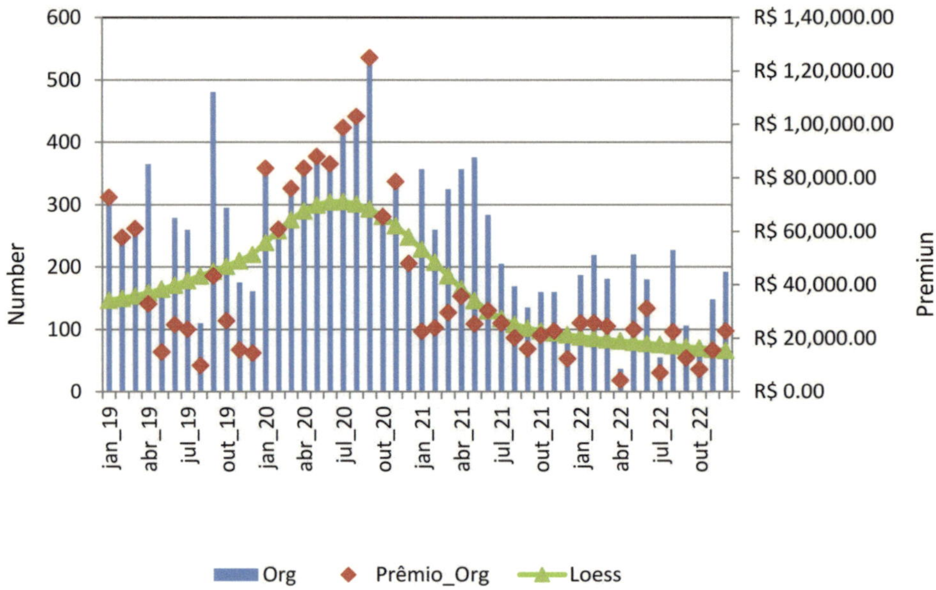

Figure 2.
Number of animals slaughtered (Org), tax exemption obtained (Prêmio_Org), and the trend over time (verified by the non-parametric Loess regression methodology) of the organic protocol in the Pantanal. Source: Developed by the authors.

such as raising consumer awareness of organic food and willingness to pay premium prices. On the other hand, in Mexico, [16] found that the complicating factors for adherence to the organic system were sustainable pasture management and ecological management, since producers do not receive consultancy and/or training for the implementation of organic process management on the property.

In the Pantanal, the increase in the scale of production will probably also be adhered to with better communication of the product to consumers and with the arrival of technological information to a greater number of producers. This is despite this protocol being stricter, especially with the prohibition of the use of urea in nutritional inputs and the obligation not to use transgenic grains. The slaughter of animals with the Pantanal sustainable meat protocol has seen a marked upward development, probably because of its flexibility in relation to veterinary and zootechnical practices, as can be seen in **Figure 3**.

After the period of organization of the production chain by producers and slaughterhouses and the pandemic period (2020 and 2021), it was possible to observe the growth of the scale of production and slaughter. The total accumulated exemption reached the value of BRL 14,597,206.75. There was an increase in slaughters in the periods of 2020, 2021, and 2022, with a total of 9265, 37,796 and 76,593, respectively. The positive CAGR was calculated at 102.2%. The average premium received by producers per animal slaughter was BRL 115.29.

The total value of the exemption passed on to the producer was BRL 8,495,587.53, with two groups observed: C1 with 716 lots of 42,871 head of cattle, and C2 with 148 lots with 35,662 head of cattle. Considering the variables number of animals and the total exemption in each lot, the two groups can be graphically observed in **Figure 4**.

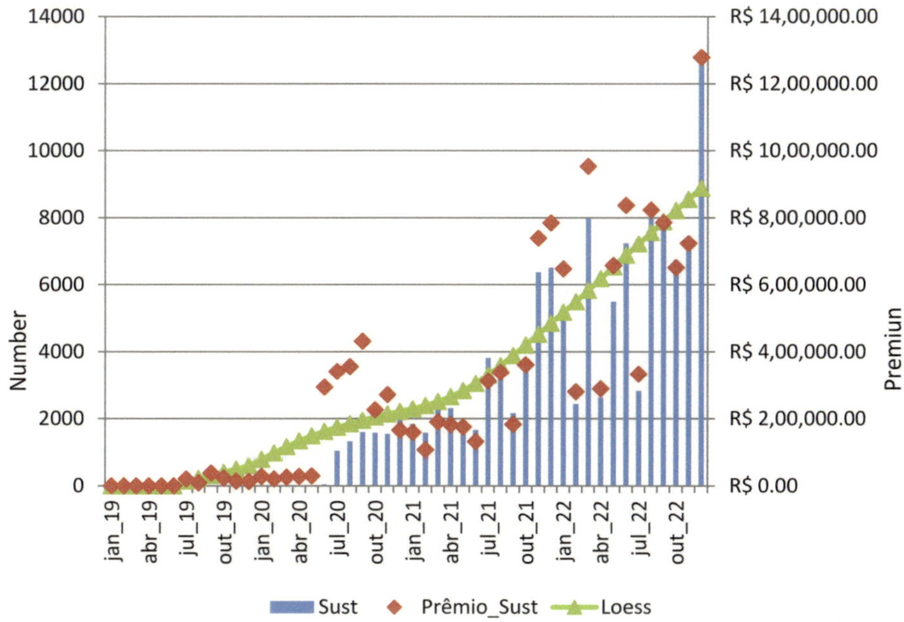

Figure 3.
Number of animals slaughtered (Org), tax exemption obtained (Prêmio_Org), and the trend over time (verified by the non-parametric Loess regression methodology) of the sustainable protocol in the Pantanal. Source: Developed by the authors.

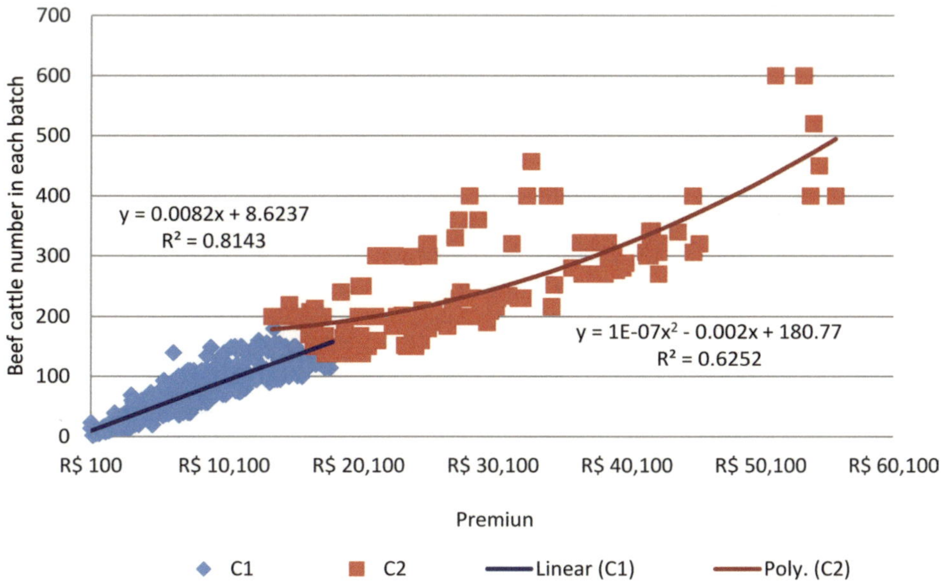

Figure 4.
Graphical representation of the two groups (C1 and C2) of lots of animals slaughtered in 2022, according to the number of heads of cattle in the lot and the premium received for each lot. Source: Developed by the authors.

As expected, the lots with the highest number of heads of cattle at slaughter were the ones that received the most tax exemption. In other words, 82.9% were slaughters with the lowest number of animals in the lot (C1), in contrast to lots C2, which represented 17.1% of the total lots in which there were heads of cattle slaughtered. On the other hand, despite the difference in the number of lots between the two groups, the total value of the tax exemption in the two groups was 52.6 and 47.4% for groups C1 and C2, respectively. This means that, despite the much larger number of lots in group C1 compared to C2, the exemptions are relatively similar, probably due to the greater number of animals slaughtered in each lot, especially males (**Table 1**).

In lots with male slaughter, in which there is a greater burden of slaughter, the financial return from the public policy of exemption was much higher (64.2%) when compared to the slaughter of females (35.8%). This fact can be a direct consequence of PROAPE. As reported [17], organic certified slaughters were mainly of females from 2013 to 2016, before the decree of November 2018. Therefore, there was a complete change in the profile of certified producers, with a significant slaughter of males; 54.1% of the slaughters occurred in the period evaluated, totaling 42,498 males (**Table 1**), in relation to the total of females (36,035 animals, 45.9%).

The marketing strategy of slaughter lots also has its own characteristics. The lots of females are smaller but more numerous. Groups C1 and C2 presented 363 and 52 lots, with 62.2 and 258.79 animals per lot and 22,578 and 13,457 head of cattle, respectively. For males, the situation is the opposite, as groups C1 and C2 presented 353 and 96 lots, with 57.49 and 231.3 animals per lot and 20,293 and 22,205 head of cattle, respectively. In group C1, there was a great number of lots with few animals, especially females. In C2, we may see the opposite: a great number of animals per lot, but a smaller number of lots, especially males. This fact may be a function of the producers' marketing strategy, which, considering the greater weight of males, opted to sell in larger lots and, thus, achieved higher premium values.

The November 2018 decree was instrumental in establishing the system for producing beef cattle with sustainable and organic certification. In 2019, producers began organizing the production chain. Despite all the changes caused by the COVID-19 pandemic, which affected all links in the agricultural production chain, the Pantanal's sustainable and organic meat chain performed well in 2020 and 2021. Even though the uncertainties experienced by economic agents led to tensions with imbalances in the market, farmers adjusted, adapting demand throughout the process. In 2022, it was possible to verify the organization and maturity of the production chain. In this sense, it is already possible to identify points that need changes to promote the

Cluster		Premium	N	Batch	N/Batch
C1	F	BRL 1,919,533.71 (22%)	22,578	363	62.20
	M	BRL 2,547,870.05 (30%)	20,293	353	57.49
C2	F	BRL 1,123,900.67 (13%)	13,457	52	258.79
	M	BRL 2,904,283.10 (34%)	22,205	96	231.30
Total		BRL 8,495,587.53	78,533	864	

Source: Developed by the authors.

Table 1.
Sex of slaughtered animals, total tax exemption (Premium), number of animals (N), lots and number of animals per lot (N/lot) by groups (Cluster).

continuous improvement in efficiency. One possibility is deepening the analyses in the future, for example, using the input-output matrix method to observe the growth and decline rates of the economic aggregates of beef cattle farming with certification (inputs, agriculture, agroindustry, and agroservices) and thus design scenarios and better management practices [18].

4. Conclusions

The sustainable and organic meat production system in the Pantanal showed significant growth after the 2018 state decree. And despite the difficulties during the pandemic (2020–2021), in 2022 producers organized the links of the chain, and the public policy implemented strengthened and provided a new option for the development of Pantanal livestock farming. However, the two certifications that have tax exemptions through PROAPE have quite different adoption by ABPO producers. The Sustainable Protocol showed significant growth in the period analyzed. This was diametrically opposite to that seen with the Organic Protocol. This is probably because the Organic Protocol presents more restrictions regarding the use of veterinary and zootechnical inputs.

Protocols are dynamic processes and must incorporate sustainability strategies and policies for livestock farming that promote system certification, protect natural resources, do not harm the environment in the long term, and can offer effective and scalable solutions. The future of sustainability and food security faces many challenges. To succeed, they must not be focused on one but on several strategies, and livestock farming itself must be part of the solution.

Conflicts of interest

The authors do not declare any conflicts of interest.

Author details

Urbano Gomes Pinto de Abreu[1]*, Eliane Gonçalves Gomes[2] and Silvio Henrique Ribeiro Balduino[3]

1 Embrapa Pantanal, Brazil

2 Embrapa Solos, Brazil

3 Associação Pantaneira de Pecuária Orgânica e Sustentável (ABPO), Brazil

*Address all correspondence to: urbano.abreu@embrapa.br

References

[1] Piorr HP. Environmental policy, Agri-environmental indicators and landscape indicators. Agriculture, Ecosystems and Environment. 2003;**98**:17-33

[2] Escribano AJ. Beef cattle farms' conversion to the organic system. Recommendations for success in the face of future changes in a global context. Sustainability. 2016;**8**:572-595

[3] Dalcin D, Souza ARL de, Freitas JB de, Padula AD, Dewes H. Organic products in Brazil: From an ideological orientation to a market choice. British Food Journal. 2014;**116**:1998-2015

[4] Gianezini M, Barcellos JOJ, Ruviaro CF, et al. Sustainability and market orientation in the Brazilian beef chain. Journal of Agricultural Science and Technology B. 2014;**4**:249-260

[5] WWF Brasil. Pecuária sustentável no Pantanal 10 anos (2004-2014): memória do projeto que reúne a cadeia produtiva de carne bovina e o WWF Brasil no desenvolvimento sustentável do bioma. Brasília, DF: WWF Brasil; 2015. Available from: https://wwfbrnew.awsassets.panda.org/downloads/10anos pecuaria_07nov15_port.pdf [Accessed: September 06, 2024]

[6] Ferreira GC, Padula AD. Gerenciamento de cadeias de suprimento: novas formas de organização na cadeia de carne bovina do Rio Grande do Sul. Revista de Administração Contemporânea. 2002;**6**:167-184

[7] Mato Grosso do Sul. Secretaria de Estado de Fazenda/Secretaria de Estado de Meio Ambiente, Desenvolvimento econômico, Produção e Agricultura Familiar. Resolução n° 074, de 22 de novembro de 2018. Dispõe sobre o Subprograma de Apoio à Produção de Carne Sustentável do Pantanal, no âmbito do Programa de Avanços na Pecuária de Mato Grosso do Sul (PROAPE), instituído pelo Decreto n° 11.176, de 11 de abril de 2003, bem como sobre a extensão do incentivo fiscal previsto na Resolução Conjunta SEFAZ/SEPAF n° 69, de 30 de agosto de 2016, aos respectivos produtores rurais. 2018. Available from: https://www.legisweb.com.br/legislacao/?id=369608#:~:text=Disp%C3%B5e%20sobre%20o%20Subprograma%20de,SEPAF%20n%C2%BA%2069%2C%20de%2030 [Accessed: September 06, 2024]

[8] Guedes AC, Torres DAP, Campos SK. Sustentabilidade e sustentação da produção de alimentos e o papel da Brasil no contexto global. In: M Buainain, E Alves, JM da Silveira, Z Navarro, editors. O mundo rural no Brasil do século 21: a formação de um novo padrão agrário e agrícola. Brasília, DF: Embrapa; 2014. pp. 117-146

[9] Capper JL. Looking forward to a sustainable future—How do livestock productivity, health, efficiency and consumer perceptions interact? Cattle Practice. 2017;**25**:179-193

[10] Abreu UGP, McManus C, Santos SA. Cattle ranching, conservation and transhumance in Brazilian Pantanal. Pastoralism Research, Policy and Practice. 2010;**1**:99-114

[11] MDA P, Bungenstab DJ, VPB E, Malafaia GC, PHN B, GRO M, et al. From traditionally extensive to sustainably intensive: A review on the path to a sustainable and inclusive beef farming in Brazil. Animals. 2024;**14**:2340. DOI: 10.3390/ani14162340

[12] Cleveland WS, Grosse EH. Computational methods for local regression. Statistics and Computing. 1991;**1**:47-62

[13] Marquetti A, Vialli L. Princípios e aplicações de regressão local. Análise Econômica. 2004;**22**:253-277

[14] Baiocchi G, Distaso W. GRETL: Econometric software for the gnu generation. Journal of Applied Econometrics. 2003;**18**:105-110

[15] Izenman AJ. Modern Multivariate Statistical Techniques. New York, USA: Springer; 2008. 731 p

[16] Cuervo-Osorio VD, Martinez BC, Quetz-Aguirre EM, et al. Assessment of the capacity of chance of Campeche livestock systems towards an organic production model. Tropical and Subtropical Agroecosystems. 2020;**23**:1-12

[17] Abreu UGP de, Oliveira LOF de, Gomes R da C, Dittmar E. Desenvolvimento da cadeia produtiva de carne sustentável e orgânica do Pantanal. Corumbá: Embrapa Pantanal; 2019. 8 p. (Embrapa Pantanal. Comunicado Técnico, 111)

[18] Malafaia GC, Rainer DM, Casagranda YG, Zevedo DB d. A mensuração do produto interno bruto do complexo da bovinocultura de corte no Brasil. Cadernos de Ciência and Tecnologia. 2021;**38**:e26777

Chapter 9

New Management Techniques for Increasing the Efficiency and Production Rates of the Natural Grazing Land and the Grazing Herds and Fed Herds, and Their Effect on Reducing Global Warming

Arieh Brosh and Aviv Asher

Abstract

The full potential for pasture and grazing animal production worldwide is not realized. Efficient herd management must address the mutual interaction between the pasture and the herd's needs. Cattle grazing' activities, forage availability, and cattle's heat production (HP) measurements can be used to calculate the grazing herd's energy-balance metrics and the actual consumed forage quality and can identify health and reproduction events. The forage availability index corrects the effect of a shortage of forage biomass. Direct individual HP and energy-performance measurements of grazing and confined cattle enable ranking them according to their efficiency. The methods for such measurements are available for managing grazing herds and their lands. A sample of animals can be used to characterize herds' energy status and grazing land. Selecting grazing and confined cattle for improved efficiency and optimization of grazing land management will increase annual forage production and soil organic matter content (soil quality). As a result, the number of heads of cattle and their production per unit area will significantly increase, and greenhouse gas emissions relative to cattle production rate will decrease. Although the technologies for measuring cattle's HP and activities are mainly commercially available, coordination between manufacturers is required.

Keywords: decision support system, heat production, energy balance, individual production efficiency, remote monitoring, grazed land management, pasture quality, pasture biomass, grazing cows, herd management and performance, global warming

1. Introduction

1.1 Animals' energy balance: Terminology and quantification methods

The energy cycle of a living organism is defined by its ability to increase its energy level (entropy decreases, order increases). Consequently, animals' maintenance requirements and production rate (including reproduction) largely depend on their energy supply. Plant consumption, which uses solar radiation as the energy source for their production, constitutes the energy source for eukaryotic animals. The ruminant's energy supply is based directly on grazed forage intake. Thus, we use energy terminology to characterize the grazing animal's needs and output and its grazing habitat's management needs and production.

Animals' energy management is the main bottleneck for their survival, reproduction, and production ability. This energy path is the primary evolutionary selection factor. Consequently, scientists have developed methods to measure animals' energy balance, starting with the measurement of oxygen (O_2) consumption. O_2 is absorbed by the cells' mitochondria to oxidize the organic substrate (diet source), thereby supplying the energy for mitochondrial ATP production, which in turn supplies most of the energy used by aerobic eukaryotic animal cells. The animal's daily O_2 consumption is a parallel term for the animal's daily energy expenditure (EE) and heat production (HP). Its energy unit is the Joule (J), the same unit used to describe an animal's diet energy input.

Initially, humans' resting metabolic rate (HP output) was measured by monitoring gas exchange (O_2, CO_2) using a respiration chamber [1]. However, it took time to understand the importance of measuring human and animal HP, and all of their energy-balance components in their natural environment, to be appreciated and significantly applied.

New methods were then developed to measure the energy-balance components of domesticated and wild animals. The doubly labeled water technique, invented in 1950, was an isotope-based method that measures the EE of subjects by monitoring the differential enrichment of two isotopic labels in the bodily fluids [1]. This technique was only applied to small animals until the early 1980s, primarily because of the expense [1]. It opened an essential window to information on the EE of domesticated and wild animals in their natural state. However, the accuracy of this technique is relatively low (8–9%), making it accurate enough to provide information on a group of animals but not to determine the EE variation among individual animals; therefore, it does not consider the individual variation in EE [1]. Moreover, the doubly labeled water technique requires significant labor and time to obtain the data; it is still quite expensive, limiting its use to research.

Following these insights, methods were developed to measure animals' energy input and output, that is, all of the animal's energy-balance components; with regard to grazing animals' energy balance, it was essential to include the complex interactions (both positive and negative) between grazing animals and their grazed lands. Novel methods were developed to measure and analyze the animal's adaptation to its grazing land and to understand ruminants' digestive mechanisms. They included measurement of the digestive tract fluid and fiber kinetics, that is, its volume, flow rate, and fluid and fiber mean retention time, and its relation to diet quality and environmental conditions [2–12]. The mutual interaction between grazing animal management and grazing land management was investigated. Further research methods were developed for real-time tracing of the grazing animals' and grazing lands' energy status. Moreover, the potential of a new scientific method to rank grazing cattle and confined cattle according to individual production efficiency without measuring their individual intake was validated [13, 14].

The main goal of the present book chapter is to present scientific research and its progression for application to commercial methods that are ready to use now and that can be applied soon. The targets of the applying techniques are: (1) to improve the combined management of grazing cattle and their grazed lands. (2) To increase the production efficiency of grazing cattle and confined cattle. Applying these methods will significantly improve grazing land production and grazing herd's production and decrease cattle's effect on global warming.

1.2 Summary description of the quantification methods

The quantification techniques to determine the animals, their habitats 'nutritional' production status, and their selection for greater production efficiency that are presented in this book chapter were:

1.2.1 Quantifying of grazing herds consumption and grazed lands production

We used and developed measurement methods to quantify the nutrition status of free ruminants and their natural environment conditions. We first targeted the animals of feedlots and cowsheds, their diets, and their cultivated grazed lands. We continued to target the natural' free-grazed animals' world habitats and their grazed ruminants. It is based on techniques [9] where we dose markers to measure digesta fluid and fibers (Rumen and Entire digestive gut) flow rate and retention time. By testing cutting grazed field forage *in vitro* digestibility and fecal 'DM flow rate, we calculate cows' dry matter intake (DMI) and forage ME quality and its multiplication of the MEI. However, this calculated ME does not represent the actual selected consumed forage ME. We improved the techniques by measuring the grazed actual selected consumed forage plant's quality using the near-infrared spectroscopy (NIRS) method [4]. We used and partly developed [5] the N-Alkenes to determine the diet botanical composition. Testing each consumed plant and fecal sample nutritional concentration values and its Indigestible Natural Detergent Fiber (INDF) internal marker concentration and its fecal sample concentrations used to calculate the actual consumed diet *in vivo* digestibility and its ME quality (see Section 2) and the actual consumed animal's MEI [5–11]. As mentioned above, the analyses of the cited techniques take time. They enable the investigation of the grazed lands and animals' nutritional status. When the herd's size and grazing areas are considered, the above data can quantify total grazing land production and recommend the optimal cattle and grazed land-stocking rate. However, the above technologies do not supply full information on the animals' energy balance and cannot used for real-time management of the animals and their natural environment.

1.2.2 Grazing cattle HP and energy balance measurement

We developed the heart rate (HR)–O_2 pulse method for HP measurements of confined and free-grazed ruminants [3] (see Section 4.4.1). We used simultaneously, along significant time, the HP measurements of grazed and confined ruminants, with the above-mentioned (Section 1.2.1) presented technologies [3, 10, 12, 15, 16]. Consequently, we got the capability to measure and calculate the free-range ruminants (feedlot, cowshed, and grazing), the entire energy balance variables and their components, that is, the selected consumed diet (fed and grazed) ME, DMI, MEI, HP, and Recovered Energy (RE).

1.2.3 Grazing cattle energy balance metrics determination by HP measurement

Following the validation of the HR–O_2 pulse method for HP measurements and the *in vivo* determination of the actual selected consumed diet quality (ME), the correlation between animal HP and MEI of confined and free-grazed ruminants was investigated. As expected, the correlation between HP and MEI was highly significant. Consequently, the regression coefficients slopes and intercept, and the other energy-dependent metrics that characterize the animal's energy status were determined (**Figure 1**); see in details Section 4.4.1. In addition, as expected, it was found that for a defined animal's conditions (productive, reproductive, breed age, and environment conditions), the animals' energy balance positively correlates to their diet ME. Thus, it can be concluded that under the broad range of the yearly season's nutritional range, the grazed herd average HP measurement value can used to calculate the herd's entire energy balance variables metrics, including the selected consumed forage ME. However, we can relate it to the herd's average state and not to the individual one, as individuals have significant variations in maintenance costs and production efficiencies. This subject will be discussed following.

1.2.4 Using the correlation between grazing cows daily grazing activities and herd's energy balance for optimize grazed land and grazed herds performances

We used published manuscripts [17–20] where grazing beef cows, HP, and daily behavior were measured continuously. Daily behavior was measured according to Ref. [21] based on GPS data. All the GPS data of cattle locations and the activity, HR, and HP data were synchronized to produce a simultaneous record at 5-minute intervals.

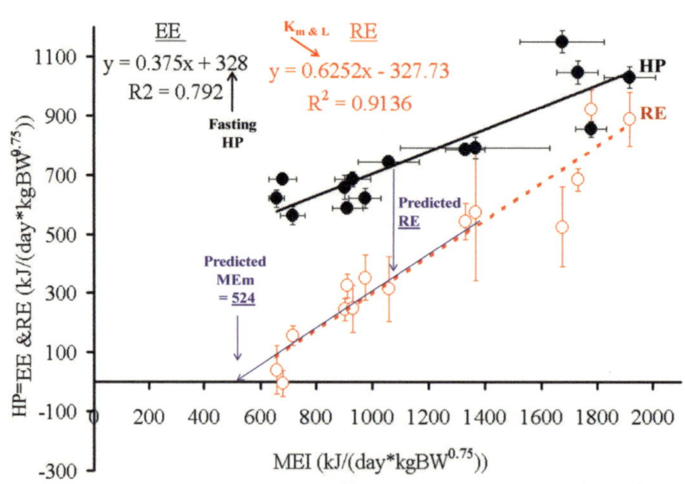

Figure 1.
Grazing cows' Energy Expenditure (EE = HP, ● ± SE) and recovered energy (RE, o ± SE) of eight cows, compared with their metabolizable energy intake (MEI) Mean ± SE. RE was defined as MEI-EE. The HP to MEI regression line was used to calculate predicted RE, MEI at maintenance conditions (MEm), fasting HP, and efficiency (K) of using MEI to maintenance (Km) and lactation (KL). Data source: Ref. ([15], Figure 7).

Partition of time spent by the cows lying down, standing, grazing, and walking without grazing, the horizontal and vertical distances walked per day, and the number of steps taken per day were summed from these records. We tested the correlation between the cow's daily grazing activity (h/day) and the cow's HP in the above studies, and we found a significantly high correlation between the cow's daily grazing time and the cow's HP and all the above-presented energy balance metrics, including the actual selected consumed forage ME.

Consequently, we conclude that cows' daily grazing time can be used to present grazing herds' entire energy balance metrics, including herd total yearly DMI and MEI consumption and, consequently, the grazed land total yearly DM and ME production. Moreover, as cows' daily activities can be remotely monitored in real-time, and the cow's daily grazing activities can be used to optimize their grazing pressure (herd stocking rate) to maintain an optimal daily forage vegetative state (ME value) and its biomass, which enables its maximal production. Indeed, it is also subjected to the grazed land soil quality and climate conditions. Increasing grazed lands forage production will improve soil organic matter deposition, that is, soil quality, its potential forage yearly production, and its carbon sequestration, which relatively decreases global warming gas production (see Section 3.4).

A software algorithm summarized, analyzed, and delineated the above data on the relationship between grazing land conditions and grazing cows' activities. This algorithm was adjusted to use an index to correct the effect of shortage in forage biomass, which affects (increases) cows' daily grazing. The above basic algorithm was applied and approved [2] for optimal herd and grazed-lands management. The tested algorithm was based on activity collars that determine cows' activities every 4 seconds using three-axis accelerometers. The collars have GPS to determine the cow's location. The accelerometer output is also used to detect health and reproductive events.

1.2.5 Using the residual heat production (RHP) index of grazing and confined cattle to select (ranking) according to their individual efficiency performance (see section 6.1)

The modern, well-accepted method used to select individual animals according to their production efficiency, using the Residual Feed Intake (RFI) index, is based on individual measurement of animals' DMI, body weight (BW), and performance (ADG, Body Condition Score (BCS) changes (see Sections 4.4.2 and 6), milk (energy) production). Applying this method is expensive, mainly because of the cost of individual DMI measurements. Its uses are limited to confinement conditions. The animal energy balance is defined by only three variables: MEI-HP = RE (Recovered Energy), the RE is accounted as (milk energy + body-retained energy), that is, the animal performance in energy unit. Thus, when animals BW and RE are measured, as used for RFI calculation, for ranking the animals according to their individual efficiency trait, the HP can replace the MEI and DMI. Following the approval of the HR–O_2 pulse to measure the HP of confined and free-grazed cattle (see Sections 4, 5, and 6), we suggest that confined and free-grazed cattle can be ranked according to their individual efficiency by using the RHP index [13, 14, 22].

2. Animal energy status

The energy-balance components are defined as dietary intake, measured by weighing the daily dry matter intake (DMI, kg DM/day). This is followed by determining diet energy digestibility using *in vitro* digestibility [13], using digestive tract microbiota, and mainly mimicking the ruminants' *in vivo* digestibility. Other researchers and we also measured the actual selected consumed natural grazed forage *in vivo* digestibility using plant sources' of Indigestible Natural Detergent Fiber (INDF) used as natural internal diet markers [13]. Another *in vivo* method uses N-alkanes internal markers [6]. The diet's energy content after its digestion, and after subtracting the animals' energy loss from eating digestion and absorption cost, the heat increment (HI) and its loss in the urine are used to calculate the diet's available energy content for animal metabolism, that is, the metabolizable energy (ME) content (MJ/kg DM) for the animals' metabolic needs.

The animal's diet ME content multiplied by the DMI is defined as the animal's metabolizable energy intake (MEI, MJ/day), that is, the available energy input from the diet. The same units present the animal's heat-emission output, the HP. Thus, the animal's available energy (input-output) = the animal's production energy content, the recovered energy (RE) which represents the daily energy retained in the animal's body (and in the produced milk when referring to lactating animals); in short, animals' energy balance is defined as MEI-HP = RE.

For grazing herds' management, the quality of the grazing pasture can rapidly be estimated from its ingredient concentration. Under most conditions, higher pasture protein content has higher digestibility and lower DM and neutral detergent fiber (NDF), which reflect an increase in grazing quality. Consequently, the forage ME value and its consumption rate, that is, MEI, increase [2].

3. The interaction between the grazing animals' management and the grazing land management

3.1 Forage quality and intake rate

In ruminants, as the dietary quality (ME concentration) increases, so does rumen microbial digestion is faster. Consequently, the animals' DM and ME intake rates (DMI and MEI, respectively) are significantly positively correlated to forage quality (ME concentration).

3.2 Optimal energy management of the grazing herd

Grazing herd management relates to the animals' energy status, which depends on forage ingredient chemical concentration and availability (the biomass of the selected consumed plants).

One of the major concerns of herd owners is the herd's annual consumption of pasture in the grazing areas, which is quantified by the total yearly DMI and MEI consumed by the grazing livestock herds. Moreover, maximization of grazing herd and grazing land production should be synchronized to interact with the grazing herd management according to the grazed plots' forage quality (forage ME) and availability.

Adjusting the grazing pressure (annual grazing area unit/animal) to optimize the grazed forage production rate can be defined as a state in which, throughout the year, the selected consumed forage (the available to-eat forage) and its vitality do not limit the grazed lands' forage production rate under the defined climate and soil conditions.

In addition, optimal grazing management should be adjusted to the grazing land's soil quality, water availability, climate (solar radiation, temperature, air humidity, wind, etc.), and topographic conditions (including slope to solar radiation) throughout the year's seasons.

This grazing land optimization will maximize the annual grazing land forage plants' ME production and the potential of yearly consumed DMI and MEI by the grazing herd.

3.3 Interaction between soil quality and optimal energy management of the grazing herd

Soil quality is positively correlated with soil organic matter (OM) content. For the grassland (Inner Mongolia steppe), the DM of herb roots (belowground biomass, B) increases with the (aboveground, A) herb biomass (shoots) increases, and the B/A ratio is variated. The slope of the allometric scaling regression between log-transformed belowground biomass and log-transformed aboveground biomass significantly differed from 1.0, and B/A ratios changed along the precipitation gradient [23]. Consequently, the optimal yearly grazing herd management that increases forage production rate (Section 3.2) will also increase soil OM content, which will significantly improve soil water capacity: Increases from 1–6% in the soil OM by weight are equivalent to approximate increases from 5–25% in the soil water capacity [24]. The increase in soil water capacity enables the grazed green forage to maintain greater vitality during the long breaks in precipitation, increasing the total annual forage production, which again increases soil OM and quality.

3.4 Influence of grazing herd management on greenhouse gas emissions and soil quality

The grazing plants' roots absorb O_2 (for their energy supply) from their aerobic soil environment. Thus, the grazing plants' roots and aboveground-withered plant leaves and stems, which are not consumed by the grazing animals, as well as the animals' feces, are digested under aerobic conditions by fungi and cellulolytic aerobic bacteria [25, 26]. This is a methane-free biochemical process. Aerobic plant fiber decomposition is significantly slower than the anaerobic digestion in the herbivore's gut. Consequently, an increase in grazing land forage production will increase the soil OM content, mainly from the roots, delay CO_2 emissions, and decrease methane emissions relative to the grazing land forage production and the grazing animal's output. This will have a significant positive effect on reducing global greenhouse gas emissions.

4. Methods used to quantify grazing ruminant production and grazing habitat production and their interactions

4.1 The challenge in quantifying the energy status of grazing animals and their grazing lands

Managing the grazing land and grazing herds throughout the seasons of the year based on quantifying the grazing land condition and the grazing animals' status is a great challenge. The grazing land's habitat conditions (soil, climate, landscape topography, etc.) mainly depend on the geographical location, current climate, and many other factors that the herd manager cannot control. The grazing herd's condition is characterized by its stage of sexual maturity, age, body weight (BW), average daily gain (ADG), expected production rate (including lactation rate), and reproductive state.

It is not easy to quantify the physical and biological conditions of the grazing land and its grazing herd managed as one unit. However, domesticated grazing herds can quantified by its yearly: (a) number of grazing animals, (b) grazing land-stocking rate, and (c) herd performance – animals' BW, expected and actual ADG, calving rate, weaning rate, etc. The quantification of animals' performance metrics in units of energy opens the way to developing scientific methods for quantifying the herd's performance in fiscal–mathematical metrics and deriving from it the grazing land's forage production performance.

To maximize a grazing animal's production rate, its management should optimized for its nutritional needs, including its energy needs, the animal's production level and reproduction state should be monitored and adjusted to plant varieties (species, genus, maturity, etc.) and its availability in the specific grazing land throughout all seasons of the year.

Quantification should be adapted to the grazing animals' breed, gender, stage of maturity, BW, productive and reproductive states, climate conditions, etc.

The well-known nutrient requirement guides of the National Research Council, Agricultural and Food Research Council, France's National Research Institute for Agriculture, Food and Environment, and the Commonwealth Scientific and Industrial Research Organization have well-documented the nutritional needs of animals. However, the references relate mainly to confined conditions. Making adjustments for grazing in a natural grazing habitat is much more complex because it is difficult to estimate the essential information on the animals' selected dietary ME in natural habitats with large variations in geographical location and distribution of plant varieties during the year.

The main barrier to optimizing the management of domesticated grazing livestock is the high labor and financial costs of accurately measuring their energy metrics. It is characterized by (a) the grazing land's forage conditions, that is, the actual consumed grazed forage's biomass and quality (forage ME), (b) the grazing land's expected forage growth rate (kg DM or MJ ME/day) per hectare (ha) when the forage is viable (green), (c) the grazing herds daily DMI (kg DM/day), (d) the grazing forage' diet' ME content, and (e) its daily intake (MEI, MJ/day). This information for the grazing land and herd also significantly depends on climatic conditions. Moreover, herd-management decisions must be made before grazing conditions adversely affect the herds and grazing land.

4.2 Methods that require long periods for measuring herd animals: Chemical laboratory tests and data analysis

Grazing animals' intake can be monitored by measuring their fecal output (kg DM/day) and correcting it to DMI (kg/day) by dividing it by the diet DM digestibility ratio {(DM fecal output, kg/day)/(DM digestibility ratio)}. The fecal output is measured by dosing indigestible markers into the tested animal's mouth at a known rate (g/day). The dilution rate of the markers in the daily fecal sample is used to calculate fecal output. The animals' selected consumed forage plants are tested by *in vitro*

DM-digestion procedure to determine the digestibility and ME value of their selected diet. The animal's daily fecal output and its digestibility ratio are used to calculate its DMI and MEI, respectively.

There are several techniques for dosing indigestible markers to measure fecal output and to calculate diet digestibility when the diet's composition and intake (DMI) are known [7]. Alternatively, calculating diet intake when the diet digestibility (tested *in vitro*) is known. The techniques are a) daily dosing of markers, which is difficult to apply manually to free-grazing animals because the intensive dosing activities interfere with their natural daily behavior. However, for dosing it is possible to use an automatic feeder with a marker dose. (b) Using a rumen bolus consisting of a controlled-release device that releases the indigestible marker chromium oxide [8]. (c) Using a single dose of two markers, an indigestible fiber marker (chromium mordant) and indigestible fluid markers (CoEDTA) [9, 10], this method is also used to analyze the digestive tract dietary fibers and fluid's kinetics and mean retention time in the rumen and in the entire digestive tract. (d) Captec (Nufarm, Auckland, New Zealand [11]) provides the slow release (at a constant release rate) of indigestible N-alkane (C32 and C36) markers. The use of this intra-ruminal controlled-release capsule combined with sampling and analyses of the forage plant species' N-alkane profile and the animals' fecal N-alkane profile enables calculating the botanical composition of the individual animal's diet, its dietary ME concentration, the daily DMI, and MEI consumptions [12].

These methods, based on indigestible marker-dilution technology, are suitable for scientific studies. They are very important for understanding and improving the management of the grazing animal and its consumed plant–ecology interactions. However, they cannot serve for commercial use – for optimizing grazing herd and grazing land management. This is because analysis of the data takes a long time, that is, they do not provide the real-time information required by herd managers. In addition, these methods are very expensive to use, and scientific knowledge is required for the analyses.

4.3 Methods that require relatively short periods for data analysis of forage nutritional composition and ME

The nutritional composition and ME content of the actual consumed forage can be detected by near-infrared spectroscopy (NIRS) analysis of the grazing land's plants compared to their spectrum in the animal's feces [4]. As presented in Section 4.2, we can use the analyzed N-alkane profile of the grazed land plants and the grazing animal's feces to determine the botanical composition of the animal's diet [5, 6]. When the nutritional value of the grazed plants is determined, the animal's selected plant diet and its nutritional value (ME) can be calculated. Following the significant correlation between the quality of the plants selected for consumption and the plant parts selected by the cows, the cows' herd' actually consumed forage ME can be measured, the herd's MEI and HP can be accurately calculated, and all of the energy-balance variables of the herd are determined [2]: MEI/ME = DMI; MEI-HP = RE.

In addition, it is possible to use the well-known method of quadrats sampling of the plot's cut forage samples for biomass and *in vitro* analysis or NIRS analysis of its quality to evaluate the grazing plot biomass and nutritional value. However, this method presents the plot's forage nutritional status and does not represent the grazing animals' actual selected consumed diet.

4.4 Real-time monitoring of the energy balance status of the grazing animals to evaluate the actual consumed (selected) forage quality (ME) and its availability, the biomass index of the forage that is selected for consumption

4.4.1 Using animal HP measurements and the regression equation variables of the correlation between cattle' energy balance values to quantify the energy status of the grazing animals and their grazing lands

The animal's energy balance variables status, MEI, HP, and RE, which mainly depend on the actual consumed forage quality [2], are significantly correlating under a wide range of grazing conditions [3]. However, these correlations [3] and regression variables were delineated under conditions of sufficient forage availability (biomass), where the changes in the forage biomass (kg DM/ha) do not affect the cattle's grazing behavior. The HP of free grazing and fed ruminants (cattle, sheep, goats, and yaks) can be measured by the HR–O_2 pulse method. The method is based on continued daily HR measurement throughout several days, multiplied by the measures, once for each measurement condition, of the net (input-output) oxygen volume consumed by the animal body for each heartbeat (the oxygen pulse, O_2 pulse) [3]. For most of the defined grazing land conditions, the measuring of the grazing animal's HP, using the heart rate (HR) O_2 pulse (HR–O_2 pulse) method [3] allowed calculating the herd' animal's MEI, RE, and DMI [3], and the actual consumed forage's ME [2]. A comment: The HR–O_2 pulse measures the individual animal's HP. However, the calculated' MEI, HP, RE, and the selected to-consume grazed forage ME content are based on the regression coefficients of the herd data analyses. Consequently, they present the herds average' energy balance metrics and not individuals (**Figure 1**).

4.4.2 Using the daily activities of grazing cattle to quantify the cattle herd's energy-balance metrics and the actual consumed forage quality (ME) and availability, that is, an index that represents the biomass of the actual consumed selected plants' forage. An index that will be used for correcting the energy-balance metrics calculations when the forage availability shortage affects grazing behavior

As an alternative to direct measurement of the cattle's HP [3], the grazed cattle herd's daily average HP can be calculated from the daily grazing duration of about 10 representative (in terms of breed, age, and reproductive state) cows [2]. Commercial collars for cows that carry three-axis accelerometer sensors and internal processors are available to determine cattle grazing behavior [2].

As mentioned in Section 4.4.1, when the herd HP is known, the herd energy-balance variables MEI, HP, RE, and the actual consumed forage ME and the cattle DMI can be calculated.

Cow collars were used to determine daily activities [2] and characterize the duration (h/day) of each type of activity – resting idle, resting with mastication, that is, for grazing cattle, it's mainly rumination (not considered in the study [2]), grazing, and walking. In addition, the collar counted the daily number (N) of grazing bouts (grazing without stopping) and calculated the above daily average Grazing Bout Duration (GBD, s), which is used as an index to identify the actual consumed forage's availability [2]. Forage biomass, daily activities and the GBD index (**Figure 2**) are used to calculate the herd average energy balance variables (**Figure 3**).

These calculated energy balance metrics and the GBD forage availability index enable the farmer to manage the grazing herds and their grazing lands and maintain

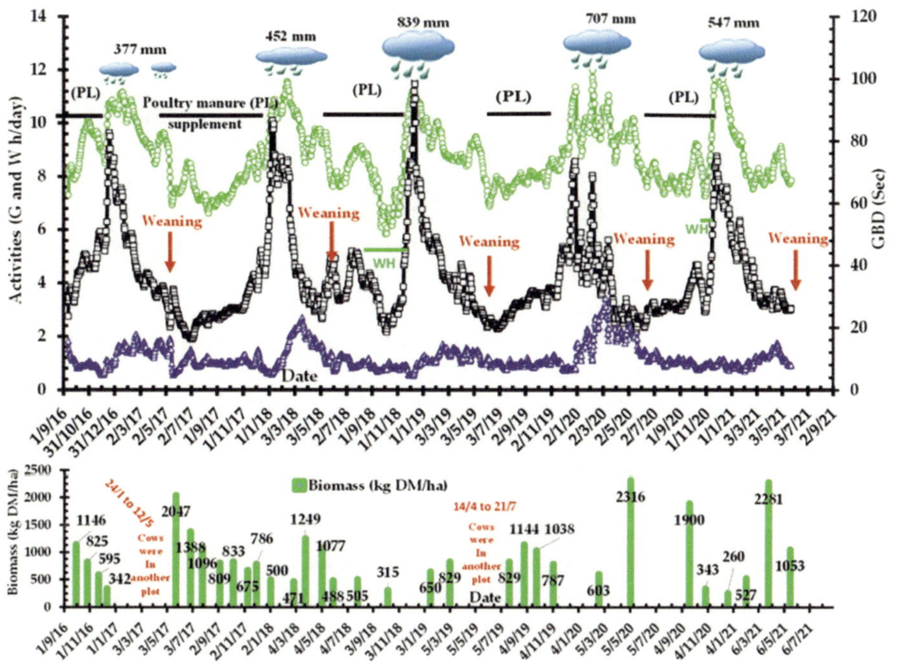

Figure 2.
The study plot, biomass (kg DM/ha, direct sampling), yearly seasonal winterrain, duration of poulty manure (PL-) and of wheat hay (WH-) supplementation, and cows' weaning events. Cows' daily walking (W, h/day, △) and grazing (G, h/day, ○), and the average daily grazing bout duration (GBD, □). Dates are given as day/month/year. Weaning dates: 9/5/17, 24/5/18, 2/6/19, 9/6/20, and 25/5/21. Data source: Ref. ([2], Figure 5).

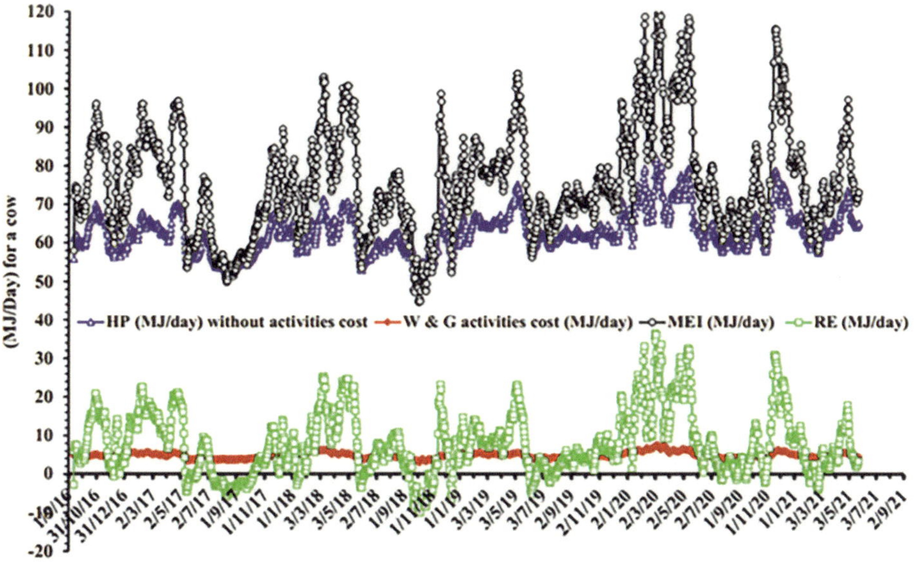

Figure 3.
Studied cows' energy balance variables (MJ/day) as calculated by the HMS. Metabolizable energy intake (MEI, o), heat production (HP, △), energy cost of walking (W) and grazing (G) activities (activity cost, ♦) and recovered energy (RE, □). Dates are given as day/month/year. Data source: Ref. ([2], Figure 6).

an optimal grazing biomass state. The herd and grazed land management information is supplied by the Herd Management System (HMS) using a unique algorithm developed by MOOnitor LTD. The herd's calculated algorithm information, including the GBD, is based on the cows' daily activities. Maintaining the optimal state of forage plant vitality and availability (selected consumed biomass) during the green seasons will maximize the yearly and/or tested period photosynthesis rate of the selected consumed plants under the actual local climatic conditions. Thus, it will maximize the grazing lands' yearly biomass production with greater ME quality of the consumed forage.

In addition, these data values can be used to calculate the total yearly DMI and MEI of the actual consumed forage in that year by the grazing cows on their unit area of grazing land, that is, yearly kg DM and MJ ME per ha.

The calculated annual forage consumption and the estimated forage left over on the grazing land, estimated from the GBD values, can be used to calculate the grazing land plot's total annual production of forage DMI and MEI per unit area [2].

Integrating the herd consumption and grazing land production and testing yearly changes in the animals' body condition scores (BCS) to calculate changes in the herd's energy balance can serve to calculate the recommended stocking rate for each specific grazing land location [2]. However, as the climate can change throughout the year, this general stocking rate recommendation should be based on several years of measurements and adjusted to each year's specific climate conditions. As the metrics are continuous variables, about 10 cows can used to represent the grazing herd's energy status and the grazing land's energy availability (quality (ME) and biomass).

4.4.3 Using daily activity measurements to identify grazing cattle's health and reproductive events

As presented in Ref. [2], the cows' daily activities can be used to identify reproductive and health events. These events are not continuous variables like the energy and biomass metrics. Consequently, to identify them, we have to use about 30 cows with collars per herd. Analysis of the daily activities of these 30 cows can give early alerts for the following events. (a) The herd has a low proportion of cows in estrus before the introduction of breeding bulls. (b) The proportion of conceiving cows after a defined number of days from the introduction of the breeding bulls to the cows' herd is too low. (c) The proportion of abortion events in the herd is too high. (d) A high proportion of cows have mostly simultaneous health alerts, indicating an epidemic.

4.4.4 Additional potential information that can be received from the cattle collars: The yearly stocking rate calculation. Cows and breeding bulls' individual reproduction control

When the cattle collars are equipped with GPS [2], the cows' daily grazing locations throughout the year can be recorded. The cow's daily locations (plots and subplots), daily intake, and other performance info are accumulating and supplying the herds' user. That information calculates plots and subplots of cows' average daily and yearly DMI and MEI forage consumption. When the grazed herd number of cows and their suckling calves is recorded, the grazing plot and an optimal recommended yearly stocking rate can calculated [2]. Using individual collars with GPS or another location ID enables the treatment of any health events, including calving problems.

Indeed, using collars with GPS on the herd's cows and their breeding bulls will give essential information on each bull and help quickly identify and treat any problem in this crucial breeding season and in the calving season.

5. Future challenges for commercial automated monitoring of cattle's HP

As presented in Section *4.4.1*, the HR–O_2 pulse method [3] enables scientists to measure the HP of grazing and confined cattle herds using several representative animals. The HP data are used to calculate the grazing herd's MEI and RE, as well as the selected diet's ME [3]. These metrics are used to calculate the herd's daily DMI and the total forage consumed by the cattle annually from its grazing land.

Calculation of daily HP is based on the animal's daily HR measurement over several days and the O_2 pulse measurement over a relatively short time. Both metrics should be measured automatically for commercial use.

A new review manuscript [27] discusses, with a significant positive conclusion, the potential of using new technology for continuous HR measurements. Another recent manuscript [28] presents a test prototype of a nose ring device for HR measurements; however, the technique has not been tested throughout a full day. For commercial use, nose ring HR sensors with minimal connection accessories have significant potential advantages. When this sensor is used, the HR sensor weight should continue touching and lying on the nose's soft live tissues. This press may damage the soft tissues. Thus, this proposed technology must be tested for a significant extended period.

We suggest assessing the potential of using a system such as C-Lock Inc.'s GreenFeed [29] to measure the cattle's O_2 consumption in real time and simultaneously using the animals' HR sensor measurement to calculate the O_2 pulse. The daily HP is then calculated by multiplying the daily HR by the O_2 pulse [3].

The GreenFeed system is typically configured to offer a small amount of pelleted bait attractant (supplemental feed) to entice the animals to visit the measurement units multiple times per day. During the animal's visit, its gas fluxes methane (CH_4), carbon dioxide (CO_2), and optionally, O_2 and hydrogen (H_2) can be measured [29].

Animals' HR and O_2 consumption change during the day according to their activities, including their grazing activities and eating under confined conditions, diet digestion and absorption, etc. [3]. We did a new analysis of the daily HR patterns throughout the day of grazing cows publication [15]. The actual HR (beats/min) was measured once a minute. The daily variation was based on 2-h averages. Analysis of the average distribution of the grazing cows' HR throughout the day showed that the daily proportion of maximum to minimum HR on a range of forage qualities across seasons [15] was 1.21 to 1.29. Certainly, the averaging over 2 hours significantly decreased the amplitude of the daily variation. Thus, we can conclude that to represent grazing cows' daily O_2 consumption through direct measurements, these measurements should be repeated quite often during the day.

The subject of the effects of cows' daily activities and the cattle' daily variation in their HP, that is, the O_2 consumption, was cited and referred to by Ref. [29]. They found that 40 spot samples are needed to represent grazing cattle's daily O_2 consumption. However, in their O_2 consumption sampling discussion, the authors did not relate to the daily pattern of the grazing cattle activities and their HP. This pattern depends on the diurnal daily conditions, forage quality, and availability [3]. The daily pattern of sunlight and ambient temperature significantly affects the daily time pattern of cattle grazing, resting, and ruminating, substantially affecting the cattle's daily HP pattern, that is, the cattle's O_2 consumption pattern. Moreover, topographic distance limitations will significantly limit the measurement system's location availability around the grazing area of measurement systems like GreenFeed for most grazing area conditions. Consequently, matching the daily direct O_2 consumption sampling to the daily HP pattern will be substantially challenging.

In addition, and very important! When the measurement of animals' HP (and consequently the entire animals' energy balance) is based on an external equipment unit that measures several animals per day, we should be safe that the measurement procedure will not bias the HP calculation compared to the natural conditions. Thus, the following subjects should be investigated: (1) How many animals can be sample per day? (2) The effect of animals' internal interactions in the equipment access. (3) The animals' time needs to be in the measurement unit. (4) The animal's time needs to arrive (walk from the daily grazing area) to the measurement unit. (5) How much do all the above four subjects bias the animal's daily behavior patterns, and consequently, it biases the energy balance measurement?

Regarding the daily O_2 consumption measurement for determining herd and grazing land energy status, where only a sample of animals (10 is recommended by Ref. [1]), we assume there is enough time to use any system like the GreenFeed system. However, matching the daily O_2 consumption measurement sampling to the daily HP pattern is still supposed to be substantially challenging in most of the world's grazing areas because of the topographic access limitation of the measurement of the O_2 consumption site to the grazing area.

When most or all the herd's animals are subjected to be measured (see following for efficiency selection), we assume that using a similar system will be challenging; that is, many measurement units should be used for each herd, which can affect the animals' daily behavior.

However, using a system like GreenFeed for O_2 consumption sampling, combined with cattle individually equipped with HR sensors and using the HR–O_2 pulse method [3] to measure cattle daily HP, can be the optimal solution for measuring the HP of all the cattle individuals of the herds.

Applying the direct daily repeated O_2 consumption sampling method for confined cows is supposed to be more straightforward. However, we must still validate that it accurately represents the actual daily pattern of O_2 consumption and that the number of measurement units is sufficient and does not affect daily behavior and HP measurements.

Repeated addition of supplemental feed to attract the animals in a system such as GreenFeed is likely to interfere with their daily grazing behavior and bias the accuracy of the daily O_2 measurement. In addition, giving too much supplemental feed may affect the cows' daily grazing time and their grazed feed intake. However, under confined conditions, assuming there are enough O_2 consumption measurement systems, the daily repeated direct measurements of O_2 consumption are supposed to represent the tested animal's daily HP. Nevertheless, application of this method to confined cattle still needs to be validated with respect to its ability to present daily variations in O_2 consumption as affected by eating, digestion, absorption, resting time, and the animal's daily behavior.

In addition, the number of animals that can use one unit of measured O_2 without significantly interfering with their daily behavior should be accounted for. However, when the HR–O_2 pulse method is used, the pressure on the O_2 consumption measurement unit and its possible adverse effects decrease significantly, as the O_2 pulse is mainly stable during the day [3].

The recommended duration of the O_2 consumption measurement for O_2 pulse calculation and HR measurements for representing daily HP, their recording frequency, and other details that can affect the calculated HP's accuracy have been well-discussed in the literature (discussion in Ref. [3]), and they should be strictly taken into account when trying to apply the measurements in a commercial system.

6. Selection of grazing and cowshed cattle according to their individual production efficiencies, applying commercial methods

The authors suggest (hypothesis) of evolutionary' and genetic' positive linkage between the greater protein deposition in cattle body tissues and the lower HP of the efficient cattle.

The Residual Feed Intake (RFI) efficiency index is the preferred index for ranking cattle according to their feed efficiency. However, it requires direct and accurate individual measurement of the cattle's DMI, which needs very expensive systems and can only applied under confined conditions. However, most of the world's cattle are grazing cattle; mostly beef, but also a significant amount of grazing dairy cattle. The animal's energy balance is defined by three metrics: HP + RE = MEI. The individual variation of the diet digestibility did not relate to cattle ranking for RFI [13]. Thus, for a defined, measured diet and RE of a group of cattle, the individual variation in the cattle HP directly represents variation in their MEI, which also directly represents the variation in the DMI. Consequently, individual HP measurements can replace the individual MEI (or DMI) for efficiency calculation and for ranking grazed' cattle and confined' cattle according to their individual efficiency [13].

The use of cattle and other animals' DMI values and RFI index for calculating and ranking them according to their efficiency trait are well characterized by the commercial aspects of their individual efficiency variation. However, this descriptive index does not relate directly to the biological essence (mechanism) that can explain individual variation among animals' efficiency. Moreover, scientifically, the animal's efficiency trait is characterized by the three energy balance variables: MEI, HP, and RE. However, when an animal's RE is represented by the ADG, when growing animals are tested, the animal's efficiency ranking calculation is also substantially depends and is biased by the fat-to-protein ratio in the body's tissues daily gain [13]. Based on animals' energy metabolism and allometric mathematics, the comparisons, the MEI, HP, and RE metrics interrelation should be weighted by their ratios to the metabolic BW ($BW^{0.75}$); see **Figure 1** ([15], Figure 7).

When the RFI index is used, we assume that the DMI variations directly represent the MEI variation. The above is supported by the finding that for a group of growing calves tested on the same diet, the individual variation of diet digestibility was minor and unrelated to the RFI [13]. The use of DMI without weighting it to $BW^{0.75}$ can be supported by using (choosing) a group of calves with minimal variation of age and BW for the efficiency-tested group. However, when growing calves are used, there is a significant bias between calves ADG and calves RE [13]. It is well explained by the considerably greater protein-to-fat tissue ratio in their ADG of growing calves [13]. The low RFI (greater efficiency trait) for growing calves does not represent a relatively greater RE. When a mature cows' herd (beef and dairy) is used, we try to minimize the changes in the tested herd's individual ensemble, which can affect the group's social hierarchy. Consequently, the right adjustment of the energy balance metrics as a ratio to $BW^{0.75}$ becomes more essential. Thus, we considerably suggest adjusting (calculating as ratio to $BW^{0.75}$) all the three energy balance variables (MEI,

HP, RE (ADG, milk energy) and their alternative, DMI instead of MEI, when RFI is used). We should also be aware of the ADG's limitations for mature cows; see Section 6.2 for the gut fill and the effects of the last trimester of pregnancy, and Sections 6.1 and 6.4 also discuss the using of BCS changes to represent body-retained energy.

Asher et al. [13], in the Appendix section, discuss and refer to publications related to the biology mechanism that can explain inter-animal variation of production efficiency. The study [13] used Holstein intact growing bull calves. As was present by other studies (see above [13], Appendix), we concluded that the highly efficient growing calves gain less fat and more protein in their body weight. In the third period of the study, when we used more mature and higher BW bull calves. The low, medium, and high RFI groups' RFI were – 1.50, 0.23, and 1.27 kg DM/day, respectively, and have a mid-tested period BW of 565, 576, and 557 kg. The calves, respectively, daily HP were 873, 888, and 914 (MJ/kg $BW^{0.75}$); the HP of the low RFI group was smaller than that of the high RFI (SEM ±7.17, $P = 0.02$). The calves, respectively, daily MEI were 1204, 1429, and 1567 (MJ/kg $BW^{0.75}$). The daily MEI of low RFI was significantly smaller than that of the high RFI (SEM ± 12.0, $P < 0.001$).

The energy content of protein and fat is 5.23 and 32.54 (kJ/kg live tissue), respectively ([13], Appendix p. 1005, Section 2). Thus, the fat tissue weight unit has 6.22 times more energy content than the protein tissue.

On the energy-based calculation (MJ to MJ), the HI of protein deposition is 17.5 times greater than that of fat deposition ([13], see Appendix, p. 1005–6; Section 3). The respectively 17.5/6.22 = 2.81, that is, the cost (HI) of the deposition of kg protein tissue is 2.81 times greater than of fat kg tissues.

Moreover ([13], see Appendix Section 4, p. 1006), the energy required for the daily maintenance of protein and fat tissues are 192.9 and 20.7 (kcal × kg $BW^{-0.75}$ × d^{-1}), respectively, which is 807.1 and 86.6 (kJ × kg $BW^{-0.75}$ × d^{-1}); that is, for the same animal BW, the maintenance cost of protein live tissue is 9.32 times of the fat live tissue.

The bull calves' digestive tract fill volume did not affect the efficiency trait of the ranking RFI animals. It was presented by the final not significantly carcass weight (end of study [13] third period, see Table 7).

The scientific publication (cited and discussed in Ref. [13], see Appendix section) showed that the protein-to-fat tissue ratio and its energy aspects are: (1) The low RFI ranking calves have significantly greater protein tissue proportion than the high FRI calves. (2) The energy cost (HI) of the deposition of protein tissues is greater than that of fat tissues. (3) The maintenance energy cost of protein tissues is greater than that of fat tissues. Thus, it was substantially expected that the HP (kJ × kg $BW^{-0.75}$ × d^{-1}) of the efficient ranked calves would be greater than that of the low efficient calve.

In agreement with the above expectation of greater HP of the more efficient calves with a greater body protein ratio, in the above study [13], the third period, the HP/MEI ratio of the low, medium, and high RFI groups were 0.74, 0.67 and 0.61, respectively. The above low RFI group ratio was significantly ($P < 0.001$) greater than the high RFI (SEM ±0.02). However, the calculated residual heat production (RHP, kJ × kg $BW^{-0.75}$ × d^{-1}) of this [13] third' period, low RFI calves group (−10.37) was significantly ($P < 0.05$) more efficient than that of the high RFI calves group (5.32), which is the opposite of the HP/MEI ratio. Thus, based on the study [13], we can conclude that the cattle's lower maintenance cost is directly related to the greater energy efficiency trait.

Here in the present review chapter, we suggest that there is a significant link (probably a genetic and a biochemical pathways link) between the greater efficiency trait (as categorized by the lower RFI and lower RHP and the presented high proportion of body protein and the efficient animal's relative lower HP). We suggest that the greater efficiency traits are related (suggested linked to each other) to an individual's greater energy efficiency of protein tissue production, a lower protein turnover (slower protein decomposition, which was supported by the finding of lower urea concentration in blood serum of low RFI calves compare to high RFI calves (study [13] unpublished data). The above are combined with a lower protein maintenance requirement. As the animal's essential and main biological' bottleneck of survival and reproduction are related to the protein biochemical pathways activities and energy supply under shortage of its sources, it is more than logical that the evolution for increasing animals' efficiency in both pathways (protein and energy) are linked to each other.

Based on the above-discussed subjects, we suggest that the cattle' HP metrics as criteria for calculating and ranking cattle efficiency are much more important than a practical index for replacing the DMI. We suggest that the efficient energy utilization from ME to the net energy of maintenance and production, especially the greater body protein ratio linked to a lower body HP of the more efficient cattle, is significantly related to the biological mechanism of improving animal efficiency. Thus, using animals' HP and its interaction with the animals' protein metabolism to characterize (select) animals according to their production efficiency has a significant potential to yield additional scientific, including genetic knowledge and insight into the potential for genetic selection of animals according to their production efficiency. This hypothesis should be deeply tested (biochemically and genetics).

6.1 Using the residual heat production (RHP) index of grazing and cowshed cattle to select cows according to their individual efficiency performance

The daily HR calibrated to O_2 consumption, using the HR–O_2 pulse method, that is, the HP measurement, was validated on large numbers of domesticated ruminants' cattle, sheep, goats [3], and yaks [16] under a wide variety of measurement conditions (grazing, confined cowshed cattle, and growing feedlot bull calves [3, 13, 16].

The individual RE of the cattle can measured by using cattle BW, BW' gain or loss (ADG), milk production (for lactating cows), and changes in BCS [2, 3]. The RHP as an efficiency index is defined as: {RHP = animal (actual measured HP) – (expected HP) [13]}.

The similarity between the RHP and the RFI efficiency index calculations was testing using individual efficiencies ranking of confined-growing bull calves [13]. The calves were classified into three efficiency groups and ranked according to their RHP and RFI efficiency calculation.

The ranking consistency of the efficiency indexes, including the RHP in three tested periods, was analyzed using the total agreement test, and the P – Significance of Kappa value was used. The consistency of the calf's RHP efficiency ranking between periods 1 and 2, 2 and 3, and 1 and 3 was not significant ([13], see Table 8).

When the RHP calculation groups average values ($kJ \bullet BW^{-0.75} \bullet d^{-1}$) of the individuals' bull calves that were ranked to groups by the RFI index in each tested period were compared to the group average of the RFI ranking ([13], see Table 5), the deviation average range between RHP and the RFI efficiencies groups in the three periods was similar. However, only in the third period (high-energy diet, high BW calves) the differences between the high RHP and the low RHP calculated groups (where RFI ranked their calves) was significant ($P < 0.05$).

6.2 Time duration needed for individual efficiency ranking using the RHP index

The length of time taken for the ranking depends on two criteria: (a) the number of days needed to measure and calculate the animal's performance in energy units, the RE and (b) the number of days needed to measure the HP of the tested animal accurately.

For criterion (a), a relatively short period of 2 weeks is needs for lactating dairy cows in weeks 16–30 of lactation [30]. The tested period should be a relatively stable reproductive period. For cows, it is not recommended to test efficiency during the breeding period (high activity), the last trimester of pregnancy, or during the calving seasons, where the fetus and pregnancy-supporting tissues change the cow's weight significantly, affecting its ADG. Those tissues have substantially more protein and lower energy content than for normal ADG, where changes in the cattle's BCS are used to estimate the energy content of the animal tissues' gain or loss [31] (see also [2], p. 7 for more details). Calving events significantly decrease cows' weight without any connection to their energy balance. Following calving, the cow's rumen volume increases [7]. Thus, the cow's ADG cannot be used to calculate the animal's retained energy.

The nutritional status, which substantially affects the animal's production rate, must be relatively stable during the testing period. For grazing cows, a seasonal decrease in forage quality will significantly increase the digestive gut's fill volume, which can increase the tested animal's BW [32, 33] without any connection to the animal's actual energy balance.

Regarding criterion (b), as the daily HR is measured continuously during the day and calculated for the entire day (24 h), it is generally an accurate value that well represents the HP on the tested day. Thus, for the HR measurement and its calibration to HP by the O_2 pulse measurement, 4 days are needed [15] to represent each period of the HP measurements. However, as daily conditions, nutrition state, climate, etc., can change between days, it is important that all the individuals' performance will be measured simultaneously, and all the HP measurements will be measured simultaneously. In addition, the cattle HP measurement period will be part of the cattle' performance' measurement period.

For grazing cattle, a shorter time between repeated HP measurements should be used when the changes in nutritional conditions (grazing forage ME) are greater, that is, during the transitional seasons, autumn for winter, and spring for summer (see **Figure 2**).

6.3 Efficiency calculation based on weight units compared to energy units

Whether the animal's efficiencies are ranked, using the RFI when individual DMI is measured or by RHP when the HP is measured, the cattle maintenance requirements and performance should be used to complete the calculations. Cattle's maintenance requirements and performance are calculated from cattle BW, ADG, BCS changes, and milk energy (when lactated cows are used). When the RFI efficiency index is used, the cattle maintenance cost and performance are compared to the cattle's daily DMI. When the RHP efficiency index is used, the cattle maintenance and performance translate to the cattle's daily RE and are compared to the cattle's daily HP.

6.4 Discrepancies in the cattle performance estimation

When the efficiencies indexes calculations are based on tissue weight unit ([13], see the Appendix chapter), we have to be aware that the protein tissue has 5.23 MJ/kg tissue, and the fat tissues have 32.55 MJ/kg tissue, that is, the kg fat tissue have 6.224 times more energy than of protein tissue. However, the ME spent for the tissue production, the heat increment (HI) for protein and fat depositions, with a content of 1 kJ, are 5.56 and 1.26 kJ, respectively, that is, 4.41 times greater per energy unit of protein than of fat.

The maintenance energy requirements of protein and fat tissues are 807.1 and 86.6 (kJ • kg $BW^{-0.75}$ • day^{-1}), respectively. This means that, for metabolic BW, the maintenance requirement of protein tissue is 9.28 times greater than that of fat tissue.

The protein tissue decomposition' rate is part of the maintenance energy cost of protein tissue turnover, that is, the cost of depositing new protein tissue to replace the protein tissue decomposition ([13]; see the Appendix chapter, of Section 6).

The protein-to-fat ratio deposition rate, which accumulates in the BW and affects the whole-body protein-to-fat ratio, is significantly affected by diet ME' quality, the cattle's age and breed (maturity age variation) [34], and its individual variation.

The change in the digestive tracts fill volume affects the proportion of the accumulated protein and fat to the body weight (percentage of BW). These are affected by diet quality, cattle age [33], pregnancy stage, and calving [7]. The above-presented information should be used to correct (adjust) the individual efficiency calculation, whether the cattle's maintenance and performance requirements are compared to DMI and the presented weight units for the RFI calculation or compared to the cattle HP by the RE for the RHP calculation.

The individual measurements of the above tissues and compartment (weight and volume) for correcting the above-presented discrepancies are not practical to use for efficiently selecting commercial confined' growing and dairy cattle and grazing herds' cattle. However, considering the above possible deviation errors, we should plan for the efficiencies tested periods to be carried out under stable conditions where the discrepancies in the cattle performance estimation will be minimal. Furthermore, we should be aware that for growing cattle, the efficiency variation, whether selected according to the RFI or RHP, is related per se to the fat-to-protein ratio in tissue gain and its accumulation in body tissues, which, by definition, also affects the calculated energy efficiency.

Testing the individual efficiency of mature (not growing animals) is mainly relevant for cows. The effect of pregnancy stage and calving on body composition and digestive tract volume should be considered (see above, this section and Section 6.2). Cows' performance should be measured from daily milk energy (MJ/day) and changes in cows' BCS [2]. Thus, it is significantly more straightforward to measure when dairy cattle are tested (the cows are milked daily), whether in a cowshed or under grazing conditions. For grazing beef cows, the individual's milk production can measured by using the Weigh-Suckle-Weigh Milk Production Estimates method [35].

Alternatively, it is possible to test grazing beef cows' energy balance and individual efficiency after calves weaning, taking about a two-week break (we did not test it) for mental (weaning stress) and physiological recovery. In well-managed herds, the cows in this time are in the second pregnancy trimester when the effect of the fetus size and pregnancy-supporting tissues change on the cow's weight and body composition changes are minor. The efficiency test should carried out in a stable energy balance' period, where forage quality and availability are stable and enable a positive' moderate cows' energy balance and a relatively stable BW. Under those conditions, most of the MEI is emitted as heat, which is the measure HP; consequently, the error effects in the cattle RE estimation are minor.

Thus, the preferred conditions for cows' efficiency measurements are to use mature cows when their protein tissue weight changes are relatively minor. It primarily regards dairy cows in cowsheds and for grazing beef and dairy cows.

Under grazing conditions for non-suckling cattle where the ADG is relatively moderate or mainly stable, the individual efficiency of growing calves and heifers can be tested (see [14] for heifers). Indeed, only the RHP efficiency index can be used for grazing cattle.

The RHP repeatability was well validated when grazing heifers were compared under two significantly different energy-balance conditions [14]. The RHP was successfully used to test the efficiency variation of two breeds of lactating dairy cows [22], and under cowshed conditions when growing bull calves are tested in confined feedlot [13].

7. The potential of using grazing and confined cattle's daily behavior metrics and cattle's daily HR to select cattle according to their production efficiency

7.1 General background

In the past, and probably still today, indirect selection was conducted to increase cattle efficiency. Cattle selection for increasing cattle production will dilute the maintenance energy-to-MEI ratio as the RE-to-MEI ratio rises.

Direct efficiency selection is mainly done using the RFI index based on direct DMI measurements. The RHP efficiency index is based on individual cattle's HP and RE measurements. By definition, HP + RE = MEI. Therefore, when we use the RHP efficiency index, we calculate and use all cattle's energy balance metrics.

Holstein bull calves growing in feedlots were ranked according to their efficiency trait (using several efficiency indexes) [13]. Calves' BW (kg) in the mid-periods of the efficiency testing of low and high RFI calves, respectively, were 309 and 304 in period 1, 392 and 386 in period 2, and 565 and 557 in period 3. When the calves were fed diets with high, low, and high ME, respectively [13], in each period, the calves ranked into three efficiency RFI subgroups: low, medium, and high. The low RFI and high RFI values (kg DM/day) during the three periods were, respectively, −0.94 and 0.80; −0.63 and 0.63; −1.5 and 1.27. The calculation (low RFI BW gain-to-feed intake ratio)/(High RFI BW gain-to-feed intake ratio) in the three tested periods were very high: 1.224, 1.416, and 1.460, respectively.

The significantly high ratio value of the low to the high RFI groups in their gain-to-feed ratio justifies testing the possibility of monitoring alternative metrics logically and significantly related to the above conventional efficiency determination metrics, that is, alternatively to daily DMI, the daily HP will use. Those suggested metrics are much more accessible, faster, and cheaper to measure.

When RFI is used, the suggested direct alternative metrics to the DMI are daily grazing time duration, daily time in the feed bunk duration, daily time head-down duration, and the number of daily eating' bites, that is, the cattle's metrics related to cattle daily feed consumption activities. When RHP is used, the suggested direct alternative metric to the HP is the daily HR.

7.2 Using cattle activities metrics to select cattle according to their production efficiency under confining and grazing conditions

In an invited review manuscript [36], the authors cited three manuscripts under confinement conditions where growing steers, heifers, and pregnant beef females were tested; they found that the eating rate of the selected low RFI groups was slower than that of the high RFI groups. The authors [36] noted that most studies on cattle efficiency were conducted under confinement conditions, even though beef worldwide is primarily based on grazing cattle. They reported 'that there was no consistent feeding behaviors-related traits observed between divergent RFI phenotypes of heifers and lactating beef cows that grazing on pasture' [36].

However, a study that characterized the feeding behavior traits of steers that consumed a high-concentrate diet under confined conditions and categorized them into low, medium, and high RFI groups found significant statistical differences in the direct measurement of 14 feeding behavior criteria [37]. The daily meal duration (min/day) in the feed bunk and the daily head-down duration (min/day) were positively correlated to the cows' DMI [37]. When we calculate (their data) the daily meal duration and the daily head-down duration of the (low RFI)/(high RFI) ratios were respectively 0.873 and 0.673, while the respective daily intake ratio of the RFI groups was 0.836, which is almost identical to the above daily meal duration in the feed bunk. Similarly, under confinement conditions, the eating behavior of pregnant cattle selected for low and high RFI groups was tested [38]. They tested 13 criteria related to cattle daily behavior, including the direct DMI. Only six criteria were found related to cattle behavior in the feed bunk. Only the daily feed bunk visit duration (min/day) was directly related to the RFI ($P < 0.01$). The DMI of the low RFI to the high RFI ratio was 0.827, and the respective RFI group's daily bunk visit duration ratio was 0.745. The manuscript [38] authors cited other studies with similar significant findings that cattle daily feed bunk visit duration is positively correlated to the RFI. The manuscript [38] authors suggested, 'that grouping feeding bouts into meals is the most biologically relevant method to examine feeding behavior'. Another confinement study [39] tested Simmental and Simmental × Holstein-Friesian beef cows offered a grass silage diet and selected the cows to high (H), medium (M), and low (L) RFI groups. The cow's respectively DMI and total daily eating duration grading were H > M > L. We calculate the DMI ratio of the H/L and M/L RFI of the above groups. We found, similarly, that the daily feeding duration ratios of the H/L RFI = 1.222 are greater than those of the M/L RFI = 1.111, which are almost identical to the above RFI groups' DMI ratios, which were 1.252 and 1.157, respectively. Therefore, from the above-cited studies, it looks that, for the same group of cattle on the same diet, under confinement conditions, the individuals' distribution between the cattle to the RFI groups and daily eating time duration well represents the individuals' distribution in their daily DMI.

7.2.1 A general comment about the feed bunk infrastructure and the measurement conditions

The above-presented measurements of the cattle eating behavior are based on measurements located in the feedthrough (feed bunk). The cattle behaviors time duration out of the feed trough was excluded from the measurements. At least 2/3 of the cattle's daily time duration under confinement is eating-free. When we tested cattle activities by using collars (MOOnitor L.T.D [2] under confinement conditions, we observed (unpublished data) significant recorded times duration where the cows identified as eating (grazing) while they were not eating. Indeed, cattle consuming from the feed bunk are identified as eating (grazing). However, when the cows were standing with their head in the feed bunk and did not eat, the time duration criteria in the feed bunk categorized it as eating, while the collar identified it as resting.

When the number of feed bunks is less than the animals' number (we used one feed bunk for 4.3 calves), we observed and recorded the calves' eating duration vs. the time duration in the feed bunk. We found (not published) that higher-hierarchy calves stay longer without eating than lower-hierarchy calves. Thus, we suggest that under confinement conditions, the daily duration in the feed bunk can be significantly greater than the actual daily eating time.

Thus, we suggest that when cattle daily eating time duration is used under confinement conditions to estimate the individual eating time, we should be aware of possible faults in its measurement accuracy. Moreover, the cattle should have free daily access to the feedthrough lot, which is essential for representing the individual cattle's variation of DMI from its eating duration time.

7.3 Using cattle direct daily HR measurements instead of using HP for efficiency selection

Cattle HR values tested under a range of feedlot diets' qualities [13] and grazing forage qualities [3, 15] correlated well with cattle HP. However, the individual' measured O_2 pulse multiplied by their daily HR should be used to calculate the individual daily HP [3]. Cattle O_2 pulse varies by individuals and affected by diet quality and other factors [3]. In addition, there are individual variations between cattle HR when they are in the same nutrition and reproductive status, whether under confinement or grazing conditions. We recorded it but did not investigate if it was only an individual variation per se or if it related to individual variation in cattle HP. This means that cattle with lower HP have lower HR at the same nutritional and reproductive status. Thus, this subject should be deeply investigated. As we wrote above, as the variation in cattle efficiency and intake values relative to its performance as measured by RFI and intake to average daily gain (ADG) is very high, we strongly recommend testing (firstly scientifically) the possibility of using Residual Heart Rate (RHR) instead of RHP as an efficiency index (see above, Section 7.1).

7.4 Improving the comparison of using cattle behavior for efficiency' selection to the use of cattle RFI index for efficiency selection

To test the potential of using cattle daily behavior metrics for efficiency selection, we suggest relating them to the cattle BW and cattle performance in the same way that cattle DMI is related to the RFI index calculation.

Firstly, the primary potential behavior's metrics that are logical to select and test have to relate proportionally and directly to cattle's daily DMI, that is, using metrics like daily meal duration (actual eating) in the feed bunk and animal's daily time when it head was down when we related to the feed bunk used by confined cattle [37]. For grazing conditions, we suggest using the following: daily grazing duration, head-down duration, daily eating bite number, and daily bite times duration. However, the head-down duration can be used only when the animals graze on grass forage and do not browse on shrubs and trees.

Each activity criterion can directly replace the DMI as the regression's dependent variable (Y) in the calculation test (as it is used for the RFI calculation). At the same time, the cattle' maintenance and performance variables (BW, ADG, BCS change, or milk, or the calculated RE) have to be used (as usual when RFI is calculating), as the independent variables (x1, x2...) in the regression equations, paralleling to its uses for the RFI calculation. For example, the tested efficiency of (actual daily grazing duration – predicted daily grazing duration, the dependent variable) = Residual Daily Grazing Duration (RDGD), which can be tested in the face of the RFI index. For the significant test, we suggest ranking the cattle individual's RDGD and comparing it, using the total agreement test, to the individual efficiency ranking of the RHP under grazing conditions, see [13, 14].

Under confinement conditions, behavioral metrics can validated by comparing the cattle's behavior-tested trait to the cattle RFI when an individual feed intake measurement system in a feedlot or cowshed is automatically available. However, the cost of individual feed intake measurement systems is high, and their availability is relatively rare. Alternatively, we suggest using the individual HP and the RHP efficiency index for the above comparison. HP can be measured using direct O_2 consumption measurements [29] or the HR–O_2 pulse method [3]. Combining the two systems by measuring daily HR by individual sensors and oxygen consumption using a system like the GreenFeed system [29] for the O_2 pulse calculation is possible. However, for growing cattle with high ADG, when the ratio of fat-to-protein in gain and total accumulated body tissue is high and probably significantly variated between individuals throughout the growing period, the RHP does not fully match the RFI; see Section 6.1 of the present book chapter.

7.5 The potential of using a system for direct O_2 consumption, like the GreenFeed, for daily HP measurement for selecting grazing and confined cattle according to its production efficiency, using the RHP index

Under grazing conditions representing the majority of the world-domesticated cattle, as much as we know, only the RHP efficiency index can be used as a practical reference [14] for testing the potential of using cattle behaviors for selecting individuals according to their efficiency traits. Cattle's HP can be measured using the HR–O_2 pulse method [3]. However, from the scientific and practical aspects, HP can be measured directly using an O_2 consumption measurement system like GreenFeed [29]. However, the variation in the daily pattern of cattle grazing and its HP is relatively high [3]. Consequently, the frequency and the timing of daily samples of cattle O_2 consumption should be relatively high and reflect the daily O_2 consumption changes.

Moreover, the effect of attracting the cattle by supplementing feed to enter the oxygen-consumption measurement system on the daily grazing behavior and the total daily grazing time should be estimated.

For cattle's individual efficiency' ranking, the measurement conditions, whether grazing or confined, are not supposed to affect its efficiency ranking order significantly. The efficiency-testing period is not supposed to represent all yearly grazing conditions. Thus, it is possible to adjust the testing period to grazing conditions (topographic, yearly season, forage biomass) where the frequency of visits and the time needed for the repeated daily direct oxygen-consumption measurement will not significantly shift the cattle efficiency ranking.

Consequently, a system such as GreenFeed [29] may be used for cattle efficiency ranking. Nevertheless, the above possible limitations should be considered. In addition, the effect of the cattle's social ranking (pressure, competition) on entering the oxygen-consumption unit should considered.

Alternatively, the HP measurement to rank the cattle according to RHP can use the oxygen-consumption measurement system [29] for the O_2 pulse measurement, while the cattle HR sensors and the HR–O_2 pulse method [3] will be used for daily HP measurement. This procedure may enable efficient use of the direct O_2 consumption measurement system for a significantly greater number of cattle. It will save significant labor and time for the grazing cattle HP measurement and its efficiency ranking.

Under confinement conditions, for a similar diet ME and daily MEI, the beef cow's daily eating time is about four times shorter than that of grazing cows. Throughout the day, the changes (the variation) in cattle HP are significantly more minor than for grazing cattle [Brosh unpublished data]. The location of the O_2 consumption measurement system is close to the measured cattle throughout the entire day. Thus, under confinement conditions, the number of cattle heads and their repeated measurements throughout the day can be significantly greater than that of grazing cattle. Consequently, direct O_2 consumption measurement systems seem better to use in commercial HP measurement for cattle efficiency selection using the RHP index. However, combining the leading diet quality relative to supplemental attraction to visit the O_2 consumption measurement system should be considered.

8. The potential of applying the above-presented technologies

8.1 The topics that have been implemented and those that are close to implementation at the research level

The technology of using the daily HR–O_2 pulse method to measure grazing and confined ruminants' daily HP was well-validated on cattle, sheep, goats, and yaks. The above technologies of using grazing herds' daily activities to manage grazing herds and their grazing land and improve their annual production were validated on grazing beef cattle herds and are fully ready for commercial application. The use of cattle's HP and RHP efficiency index for ranking cattle according to their efficiency was validated on grazing cattle, dairy confined cattle, and partly on growing' calves and mainly ready to apply. Improving commercial sensors for continuous daily HR measurements combined with the currently available O_2 measurement system will open the potential for cattle direct selection according to their individual efficiency traits. Using confined and grazing cattle activities traits that relate directly to daily feed consumption for ranking cattle efficiency, like using the RFI, looks promising. However, it should be intensively validated. The above cattle efficiency trait can be validated by comparing animal individual efficiency selection (ranking order) to those selected by the RHP efficiency index.

8.2 Commercial application in the near future

Indeed, there are no significant technology barriers to the worldwide application of the above technologies. The contribution of applying the above technologies to the cattle producers, their service companies, and the entire worldwide cattle industry will increase their efficiency and profits significantly. Moreover, increasing production efficiency will reduce the greenhouse gas emissions of the cattle industry (relative to their production rate). The real-time information on the grazing cattle (beef and dairy) and their grazed land energy status will significantly improve their production and their grazed lands' soil quality and yearly photosynthesis rate of a vast world range of accessible grazed natural habitats. It will decrease (relative to its production rate and absolutely) the yearly greenhouse gas emissions. To apply those technologies, cattle producers and their organizations need a good awareness of the potential of technology, which should be supported by the technology industry and government institutions (environment, agriculture, industry, etc.).

8.3 Future commercial application

Regarding the other domesticated ruminants industry, and in the future, the wild ruminants too. Currently, the unique algorithm [2] that uses animals' activities to manage grazing herds and their grazed land is available only to grazing cows. The calculations and the technologies use basic energy calculations for the ruminants, their diet, and their grazing lands management. We do not see any significant barrier to adapting the above [2] algorithms to the other ruminants. We assume that the algorithm adaptation for each of the other ruminants will focus on coefficient changes (regression slopes, intercepts, correction factors, etc.). The [2] basic algorithm calculates its metrics in energy daily units to the animal's metabolic body weight. Consequently, the software currently allows the adaptation of animal body weight variation.

9. Conclusions

9.1 General background

Grazing land's forage production, its quality, and the yearly potential of forage production are defined as the grazing' land's potential available energy (ME) output, which is the grazing herd's available energy (MEI) input, which is directly related to the grazing herd production. Real-time remote measurement of those metrics enables quantifying the optimal number of animals per grazing area, which will optimize (maximize) the grazed land's and the grazed herd's production rate and the herd's offspring weaning percentage.

The grazing land's forage available energy production depends on its habitat, including climate conditions, and the grazing herd's management.

Optimal herd management will increase forage quality (ME), production (yearly biomass), and grazing soil OM content, that is, soil quality. Older research methods can quantify the metrics related to grazing lands and herd production interactions, and supply the necessary knowledge to improve management. However, only real-time knowing of the above metrics can provide optimal management of the grazing herds and their grazing lands to maximize their yearly production.

9.2 Grazing cattle's energy balance, the quality, and availability of their actual selected' consumed forage. The uses of it for herds and grazing lands management: Potential measurement methods (see above, sections 4.4 to 4.4.3 of this book chapter)

9.2.1 The use of the cattle' heat production (HP) measurement, the HR: O_2 pulse method

The grazing cattle's individual' real-time continuous HR measurement and short time of oxygen-consumption measurement transmitted to the user using the HR–O_2 pulse method can be used to measure the individual cattle's daily HP and its daily changes. The individuals' daily HP can be used to calculate the cattle herd's energy balance metrics (MEI, HP, and RE) and the consumed selected forage quality (ME), as well as the herd's yearly total DMI and MEI, that is, the annual forage production of the grazing lands that used by the cattle herd.

The calculations of the above-presented herd's energy-balance metrics are biased up when the forage availability is too low, and the cattle invest more time and energy on daily grazing.

9.2.2 Using grazing cattle activities measurements

Real-time remote monitoring of grazing cattle's daily activities can detect the herd's (not individual) current energy-balance metrics and the grazed forage quality and availability. The above-presented bias (Section 9.2.1) effect of low forage availability on calculating the energy balance metrics can be corrected using the GBD values, that is, the forage availability index, determined from the cattle's daily activities. Changes in cattle's daily activity duration can identify cow health and reproduction events. Consequently, the cows' daily activities can be used to optimize the management of the grazing herds and their grazing lands and to increase their production.

9.3 Cattle (grazing and confined) selection according to their individual efficiency traits by using their individual' HP and the RHP index (see above, sections 5 and 6 of this book chapter)

As the animal's energy balance is defined by HP + RE = MEI, and the *in vivo* diet digestibility was not found to affect the RFI, by definition, the reliability of the RHP as an efficiency index is at least equal to that of the RFI. Whether confined or grazed, cattle's energy status conditions do not affect their ranking into efficiency subgroups as long as all individuals are tested under the same conditions. Those conditions also relate to the cattle's energy and reproductive status. The tested group's individual ensemble should be constant, as changes in the animals' social ranking interactions can affect the individual's behavior, performance, and efficiency ranking. To calculate the expected HP, the individual animal's RE should be measured under the same conditions as the HP is measured. The select period for the cattle RHP measurement and calculation does not have to represent the cattle's energy status throughout the year. It is preferable to test the cattle's efficiency under yearly and reproductive period conditions, in which the expected errors in the measurement of the variables that affect cattle energy balance and that are used to calculate the cattle's expected HP, that is, the individual's RE, are minimal. It is preferable that the length of the testing period will be minimal (but all tested variables measurement should be accurate). For Israel's Mediterranean climate conditions, for example, the recommended period can be the beginning of the summer when forage availability is relatively high (and will not change significantly during the tested period), and its quality is close to maintenance conditions. The cows are about 2 weeks month after weaning when the cows are in the second pregnancy trimester. Another possible condition is to test the RHP of grazing heifers, where cattle performance is significantly related to the heifer's growing period. It includes the situation when the heifer's RE is near zero [14]. The RHP method for selecting cattle according to their individual efficiency traits can also be used for confined-growing cattle, lactating beef, and dairy cattle, as long as their performance measurements (energy of gain, energy of milk, i.e., the RE) are reliable.

For efficiency index calculations, when cattle performance is directly presented by energy units, that is, as the recovered energy (RE), it is strongly recommended that all the used cattle energy efficiency metrics (HP, RE, MEI (or DMI when they represent the MEI variation)) be weighted to a $BW^{0.75}$ ratio metrics.

9.4 The potential genetic and biochemical evolution linkage between the greater protein deposition in cattle body tissues and the lower HP of the efficient cattle (see above, section 6 of this book chapter)

We suggest that the cattle's HP metric as a criterion for calculating and ranking cattle efficiency is much more important than a practical method for replacing the DMI metric. We suggest that the efficient energy utilization of ME to the net energy for maintenance and production and adapting to a greater protein ratio in body tissue gain is a tremendous evolutionary advanced linkage. This should be significantly related to the biological mechanism of improving animal efficiency. Thus, using animals' HP and its interaction with the animals' protein metabolism to characterize (select) animals according to their production efficiency has a significant potential to yield additional scientific, including genetic knowledge and insight into the potential for genetic selection of animals according to their production efficiency. This hypothesis should be deeply tested (biochemically and genetics).

9.5 The readiness for the application of the technologies (see above, sections 4.4 to 4.4.4 and 5 of this book chapter)

9.5.1 Immediate ready to use

There is no significant technology barrier to applying grazing cow's activities for yearly management of grazing cattle herds and their natural free or cultivated grazed lands. Only a sample of cows with collars is needed to represent the herds' energy, reproductive, and health statuses. Applying the use of the RHP efficiency indexes to rank individuals' efficiency in grazing and confined beef and dairy cows is mainly ready to use. Improving and coordinating for better economic' use of the individual HR sensors and O_2 consumption systems for O_2 pulse and HP calculation will help. As the measurement period for individual efficiency ranking of a group of cattle is relatively short (2 weeks for confined dairy lactated cows), the HR sensors and the O_2

consumption system can pass between cow groups throughout the year and can cover a significant number of cows' groups for efficiency ranking.

9.5.2 Promising technologies that should be validated

The potential of using cattle daily grazing duration (grazing cattle), daily eating duration (confined cattle), and other mentioned behaviors related to daily feed consumption for efficiency ranking looks promising. It was validated on the group's basis average, that is, the daily grazing duration, the daily feed bunk duration, etc., of the low RFI selected cattle group average, is significantly shorter than those of the high RFI selected cattle group average. However, the potential of using cattle behaviors for efficiency selection will be better validated by relating (calculating) the cattle behavior criteria to their maintenance cost and performance as done for the RFI test. Those individual cattle's rankings (behavior basis) have to be compared to the cattle's RFI ranking. Moreover, the selected behavior criteria for representing individual cattle intake variation have to represent daily intake variation as much as possible. For example, the cattle's daily time duration in the feed bunk with head down (can tested by feed bunk sensors) is a significantly more reliable criterion than the cattle's daily time duration in the feed bunk.

9.5.3 Future application on the other ruminant species

The presented uses of sensing cattle activities and cattle HR are open to using worldwide for any ruminant species and its grazed land natural habitats and confinement conditions. Indeed, the technology adaptation will start for domesticated ruminants, but there is no barrier to using it to control and save the world's wild ruminants and their natural habitats. Adaptations should be made to the animal grazing activity metrics for each animal species to determine the animals' and their grazed land energy status.

Conflict of interest

The authors declare that they have no conflicts of interest regarding the submission and publication of this book chapter.

DOI: http://dx.doi.org/10.5772/intechopen.1007816

Author details

Arieh Brosh[1*] and Aviv Asher[2]

1 Beef Cattle Section, Newe Yaar Research Center, Agricultural Research Organization, Ramat Yishay, Israel

2 Department of Animal Science, Northern R&D, MIGAL – Galilee Research Institute, Kiryat Shmona, Israel

*Address all correspondence to: ariehbrosh5@gmail.com

References

[1] Speakman JR. The history and theory of the doubly labeled water technique. American Society for Clinical Nutrition. 1998;**68**(suppl):932S-9388S

[2] Asher A, Brosh A. Decision support system (DSS) for managing a beef herd and its grazing Habitat's sustainability: Biological/agricultural basis of the technology and its validation. Agronomy. 2022;**12**:288. DOI: 10.3390/agronomy 12020288

[3] Brosh A. Heart rate measurements as an index of energy expenditure and energy balance in ruminants: A review. Journal of Animal Science. 2007;**85**:1213-1227. DOI: 10.2527/jas.2006-298

[4] Landau S, Glasser T, Dvash L. Monitoring nutrition in small ruminants with the aid of near infrared reflectance spectroscopy. (NIRS) technology: A review. Small Ruminant Research. 2016;**61**:1-11

[5] Brosh A, Henkin Z, Rothman SJ, Aharoni Y, Orlov A, Arieli A. Effects of faecal n-alkanes recovery in estimates of diet composition. The Journal of Agricultural Science (Camb.). 2003;**140**:93-100

[6] Dove H, Mayes RW. Plant wax components: A new approach to estimating intake and diet composition in herbivores. Journal of Nutrition. 1996;**126**:13-26

[7] Brosh A, Holzer Z, Aharoni Y, Levy D. Intake, rumen volume, retention time and digestibility of diets based on poultry litter and wheat straw in beef cows before and after calving. Journal of Agriculture Science (Cambridge). 1993;**121**:103-109. DOI: 10.1017/S0021859600076851

[8] Costigan P, Ellis KJ. Analysis of faecal chromium derived from controlled release marker devices. New Zealand Journal of Technology. 1987;**3**(2):89-92

[9] Aharoni Y, Brosh A, Holzer Z. Double-marker digestive tract model. Journal of Theoretical Biology. 1994;**166**:407-417

[10] Aharoni Y, Brosh A, Orlov A, Shargal E, Gutman M. Energy balance of grazing beef cows in Mediterranean pasture, the effects of stocking rate and season: 1. Digesta kinetics, faecal output and digestible dry mater intake. Livestock Production Science. 2004;**90**(2004):89-100

[11] Dove H, Mayes RW. The use of plant wax n-alkanes as marker substances in studies of the nutrition of herbivores: A review. Australian Journal of Agricultural Research. 1991;**42**:913-952

[12] Brosh A, Henkin Z, Orlov A, Aharoni Y. Diet composition and energy balance of cows grazing on Mediterranean woodland. Livestock Science. 2006;**102**:11-22

[13] Asher A, Shabtay M, Cohen-Zinder Y, Aharoni J, Miron R, Agmon I, et al. Consistency of feed efficiency ranking and mechanisms associated with inter-animal variation among growing calves. Journal of Animal Science. 2018;**2018**(96):990-1009. DOI: 10.1093/jas/skx045

[14] Marín MF, Naya H, Espasandin AC, Navajas E, Devincenzi T, Carriquiry M. Energy efficiency, reproductive performance, and metabolic parameters of grazing Hereford heifers. Livestock Science. 2024;**279**:1-10. 105389. DOI: 10.1016/j.livsci.2023.105389

[15] Brosh A, Aharoni Y, Shargal E, Choshniak I, Sharir B, Gutman M. Energy balance of grazing beef cows in Mediterranean pasture, the effects of stocking rate and season: 2. Energy expenditure estimated from heart rate and oxygen consumption, and energy balance. Livestock Production Science. 2004;**90**(2004):101-115

[16] Luming M, Ding YP, Wang A, Brosh JQ, Chen MJ, Gibb ZH, et al. Seasonal heat production and energy balance of grazing yaks on the Qinghai-Tibetan plateau. Animal Feed Science and Technology. 2014;**198**(2014):83-93. DOI: 10.1016/j.anifeedsci.2014.09.022

[17] Brosh A, Henkin Z, Ungar ED, Dolev A, Shabtay A, Orlov A, et al. Energy cost of activities and locomotion of grazing cows: A repeated study in larger plots. Journal of Animal Science. 2010;**2010**(88):315-323. DOI: 10.2527/jas.2009-2108

[18] Brosh A, Henkin Z, Ungar ED, Dolev A, Orlov A, Yehuda Y, et al. Energy cost of cows' grazing activity: The use of heart rate GPS methods for direct field estimation. Journal of Animal Science. 2006;**84**:1951-1967

[19] Aharoni Y, Henkin Z, Ezra A, Dolev A, Shabtay A, Orlov A, et al. Grazing behavior and energy costs of activity: A comparison between 2 types of cattle. Journal of Animal Science. 2009;**87**:2719-1731

[20] Aharoni Y, Dolev A, Henkin Z, Yehuda Y, Ezra A, Ungar ED, et al. Foraging behavior of two cattle breeds – A whole-year study: I. Heat production, activity and energy costs. Journal of Animal Science. 2013;**91**:1381-1390. DOI: 10.2527/jas.2012-5400

[21] Ungar ED, Henkin Z, Gutman M, Dolev A, Genizi A, Ganskopp D. Inference of animal activity from GPS collar data on free-ranging cattle. Rangeland Ecology & Management. 2005;**58**:256-266

[22] Aharoni Y, Brosh A, Kafchuk E. The efficiency of utilization of metabolizable energy for milk production: A comparison of Holstein with F1 Montbeliarde × Holstein cows. Animal Science. 2006;**82**:101-109

[23] Liu Y, Xu M, Li G, Wang M, Li Z, De Boeck HJ. Changes of aboveground and belowground biomass allocation in four dominant grassland species across a precipitation gradient. Frontiers in Plant Science. 2021;**12**:1-13. DOI: 10.3389/fpls.2021.650802

[24] Hudson BD. Soil organic matter and available water capacity. Journal of Soil and Water Conservation. 1994;**49**(2):189-194

[25] Bautista-Cruz A, Aquino-Bolaños T, Hernández-Canseco J, Quiñones-Aguilar EE. Review cellulolytic aerobic bacteria isolated from agricultural and Forest soils: An overview. Biology. 2024;**2024**(13):102

[26] Wilson DB. Three microbial strategies for plant cell wall degradation. Annals. New York Academy of Sciences. 2024;**1125**:289-297. DOI: 10.1196/annals.1419.026

[27] Nie L, Berckmans D, Wang C, Li B. Is continuous heart rate monitoring of livestock a dream or is it realistic? A review. Sensors. 2020;**20**:2-27. DOI: 10.3390/s2008229

[28] Salzer Y, Lidor G, Rosenfeld L, Reshef L, Shaked B, Grinshpun J, et al. Technical note: A nose ring sensor system to monitor dairy cow cardiovascular and respiratory metrics. Journal of Animal Science. 2022;**2022**(100):1-8. DOI: 10.1093/jas/skac240

[29] Dressler EA, Bormann JM, Weaber RL, Rolf MM. Characterization of the number of spot samples required for quantification of gas fluxes and metabolic heat production from grazing beef cows using a GreenFeed get access arrow. Journal of Animal Science. 2023, 2023;**101**:skad176. DOI: 10.1093/jas/skad176

[30] Asher A, Shabtay A, Haim A, Aharoni Y, Miron J, Adin G, et al. Time required to determine performance variables and production efficiency of lactating dairy cows. Journal of Dairy Science. 2014;**97**:4340-4353. DOI: 10.3168/jds.2013-7265

[31] NRC. Nutrient Requirements of Dairy Cattle. 7th ed. Washington, DC, USA: National Research Council, National Academy Press; 2001. p. 24. Available from: http://www.nap.edu/catalog/9825 [Accessed: December 20, 2021]

[32] Aharoni Y, Brosh A, Holzer Z. Effects of fill volume of diets on digestive tract kinetics and fattening pattern of growing Holstein-Friesian bull calves. Journal of Animal Science. 1995;**73**:2418-2427

[33] Brosh A, Aharoni Y, Levy D, Holzer Z. Effect of diet energy concentration and of age of Holstein-Friesian bull calves on growth rate, urea space and fat deposition, and ruminal volume. Journal of Animal Science. 1995;**73**:1666-1673. Available from: http://jas.fass.org/content/73/6/1666

[34] National Research Council (NRC). Nutrient Requirements of Beef Cattle. Washington, DC: National Research Council (NRC), National Science Education Standards. The National Academy Press; 1996

[35] Williams JH, Anderson DC, Kress DD. Milk production in Hereford cattle. I. Effects of separation interval on weigh-suckle-weigh milk production estimates. Journal of Animal Science. 1979;**49**(6):1438-1442. DOI: 10.2527/jas1979.4961438x

[36] Kenny DA, Fitzsimons C, Waters SM, McGee M. Invited review: Improving feed efficiency of beef cattle the current state of the art and future challenges. Animal. 2018;**12**(9):1815-1826. DOI: 10.1017/S1751731118000976

[37] Parsons IL, Johnson JR, Kayser WC, Tedeschi LO, Carstens GE. Characterization of feeding behavior traits in steers with divergent residual feed intake consuming a high-concentrate diet. Journal of Animal Science. 2020;**98**(7):1-9. DOI: 10.1093/jas/skaa189

[38] Hafla AN, Carstens GE, Forbes TDA, Tedeschi LO, Bailey JC, Walter JT, et al. Relationships between post weaning residual feed intake in heifers and forage utilization, body composition, feeding behavior, physical activity, and heart rate of pregnant beef females. Journal of Animal Science. 2013;**2013**(91):5353-5365. DOI: 10.2527/jas2013-6423

[39] Fitzsimons C, Kenny DA, Fahey AG, McGee M. Feeding behavior, ruminal fermentation, and performance of pregnant beef cows differing in phenotypic residual feed intake offered grass silage. Journal of Animal Science. 2014;**2014**(92):2170-2181. doi. DOI: 10.2527/jas.2013-7438

Section 7

Agroforestry and Extensive Animal Farming

Chapter 10

Agroforestry and Animal Husbandry

Jacopo Goracci and Francesca Camilli

Abstract

Agroforestry (AF) entails deliberate agricultural practices where farmers integrate a combination of trees, grasses, shrubs, crops, and livestock within their farming systems to create more resilient and sustainable agricultural landscapes. This strategic combination enhances soil fertility, conserves water, and mitigates climate change by sequestering carbon. Additionally, it fosters diverse habitats for wildlife and promotes biodiversity. Animal husbandry within AF systems optimises land use efficiency, providing animals with access to forage, shade, and shelter. This holistic approach enhances resilience against environmental challenges, ensuring diversified and consistent agricultural production over time, ultimately boosting farm productivity and profitability. This chapter underscores the multifaceted benefits of integrating AF and animal husbandry, offering valuable insights into viable agricultural practices for achieving food sovereignty and providing ecosystem services.

Keywords: silvopasture, biodiversity, agroecosystem, agroforestry systems, livestock

1. Introduction

1.1 Agricultural intensification and abandonment of traditional practices

Since the mid-twentieth century, agricultural intensification, spurred by population growth and the corresponding need for more food, has led to the abandonment of agricultural practices that had been applied for centuries. These traditional methods included the integration of crops with trees and the management of livestock [1]. Up until that time, much of the Italian rural landscape was significantly influenced by agrosilvopastoral systems, resulting in diverse, mixed-use landscapes in areas typically organised into small holdings [1].

To streamline agricultural operations, expand arable land, and facilitate mechanisation, trees that bordered or crossed fields—forming alleys that separated different crops or provided shelter for animals—were removed. This led to the segregation of agriculture, forestry, and animal husbandry into distinct, isolated sectors, thereby also simplifying the agroecosystems.

1.2 Landscape simplification and environmental implications

Additionally, the simplification of rural landscapes has occurred due to a decrease in the diversity of cultivation arrangements and the gradual disappearance of key elements such as hedgerows, tree alleys, buffer strips, ponds, ditches, various types of trees (whether lined, grouped, or isolated), field borders, windbreaks, terraces, stone or earth walls, flowering borders, riparian buffers, and patches of natural habitats that serve as shelter, feeding, or nesting areas for wildlife. In mountainous areas, a different form of landscape homogenisation has emerged as people moved to urban centres and abandoned agricultural activities. This socio-economic shift has brought about significant changes also in environmental conditions of mountain areas, notably, the unchecked expansion of woods and forests into areas previously managed as lands and grasslands.

The simplification of the rural landscape has multifaceted implications. Ecosystem services rely on the maintenance of biological diversity for their persistence [2]. The loss of these natural services due to biological simplification can lead to significant economic and environmental costs [3]. Historically, Italian landscapes have been characterised by a high diversity of cultivations and land uses, often on small patches, due to traditional management practices and the fragmented nature of land ownership. This has resulted in landscape mosaics with complex and diverse structures, crucial components of bio-cultural diversity that support a variety of animal and plant species linked to traditional agricultural practices [1].

1.3 Environmental and agricultural challenges and solutions

The intensification of agricultural production, the promotion of monocultures, and the simplification of agroecosystems have exacerbated environmental issues. The reduction in farming system diversity has consequences at the soil, landscape, and farm scales [4]. According to research, separating livestock from crops and trees disrupts the carbon cycle and its interactions with water, nitrogen, phosphorus, and sulphur cycles [5]. Intensive livestock management practices contribute to increased greenhouse gas emissions such as methane (CH_4) but integrating livestock with crops and trees can mitigate this impact by reducing the global warming potential of CH_4 and nitrous oxide (N_2O) [6].

In summary, the move towards simplified agricultural systems has broad implications for both biodiversity and environmental sustainability, highlighting the importance of integrating diverse agricultural practices to maintain ecosystem services and mitigate environmental impacts. In 2016, global greenhouse gas emissions were 49.4 billion tons CO_{2eq}, with 18.4% of these emissions caused by agriculture, forestry, and land use [7]. In 2021, greenhouse gas emissions in Italy were 40 million tons CO_{2eq} in the agricultural sector, representing 10% of the country's total emissions, while emissions in the forestry and land use sector were -12.90% (**Figure 1**) [7].

Global future scenarios still depict a growing population and the need to achieve food and nutritional security. The United Nations estimates that the world's population is expected to increase by nearly 2 billion people in the next 30 years, from the current 8 billion to 9.7 billion by 2050 and could peak at nearly 10.4 billion in the mid-2080s. This seems to conflict with the need to reduce the negative environmental footprint of agriculture [5].

Agroforestry works on renewing the integration of agriculture, forestry, and husbandry and could meet the need to support both productivity and the environment.

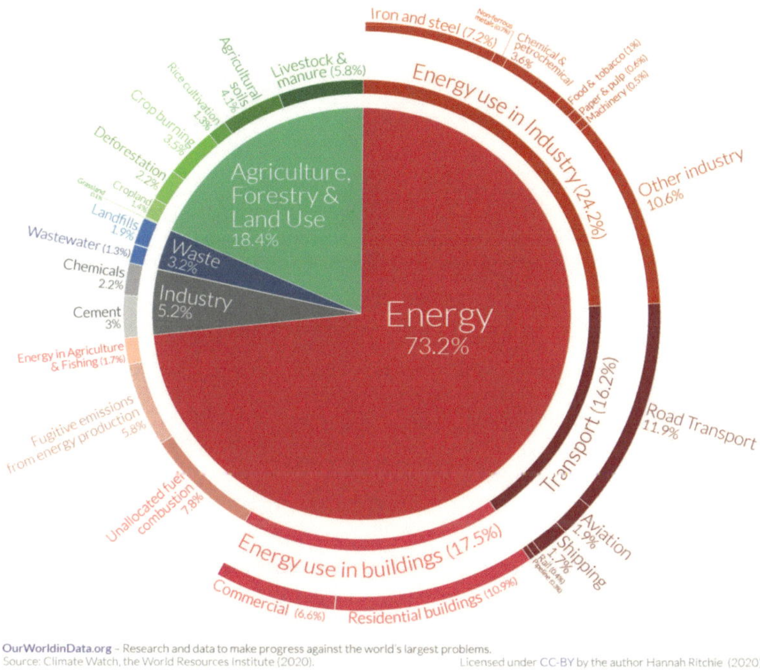

Figure 1.
Global greenhouse gas emissions by sector [7].

Agroforestry is defined as "the practice of deliberately integrating woody vegetation (trees or shrubs) with crop and/or animal systems to benefit from the resulting ecological and economic interactions" [8]. It can be applied at a plot and the farm scale and at the landscape level [9].

Agroforestry comprises silvopastoral systems where plants, i.e. grasses and leguminous herbs, are deliberately intercropped with shrubs and trees as fodder for animal nutrition and other uses [10].

Despite being a traditional land use system, agroforestry has shown to respond in a feasible and sustainable way to current challenges [11], namely by mitigating climate change and the impacts on natural environments, societies, and food production systems [11].

1 billion hectares of agricultural land with tree cover (> 10%) occupy about 7% of the global land area (**Table 1**) [17].

In 2018, silvopastoral systems were the prevalent agroforestry practice with Spain showing the highest percentage (34%), followed by Greece (16%), France (11%), Italy (7%), and Portugal (7%). These countries together formed about 61% of the whole European agroforestry area and 75% of the total silvopastoral area. The area of grazed grasslands (103,238 km^2) was slightly less than that occupied by silvopastoral systems (113,717 km^2), which include grazed woodlands and shrublands as well as grasslands with sparse tree cover [18]. Within silvopastoral systems, agrosilvopastoral practices were most prevalent in Portugal (51%), Spain (45%), and Italy (4%) [10].

Also the evaluation of "permanent grassland" should be considered. In the EU-27, there are 54.051 million hectares estimated by combining the land-cover evaluation from satellite images (Corine) and field sampling on a 2×2 km grid (LUCAS), while

Country/ Region	Total agroforestry area (km^2)	Agroforestry/ total country area (%)	Silvopastoral area (km^2)	Main tree species	Livestock	References
Spain	33230	6.60	18	Cork oak, Holm oak	Cattle, Sheep, Goats	[12, 13]
Greece	16089	12.28	10	Olive trees, Various oaks	Sheep, Goats	[12, 14, 15]
France	12103	2.20	7	Poplar, Walnut	Cattle, Sheep	[12, 13]
Italy	9170	3.00	4	Turkey oak, Chestnut	Cattle, Sheep	[12, 16]
Portugal	5029	5.40	3	Cork oak, Holm oak	Cattle, Goats	[12, 16]
Total (EU28)	114621	2.60	56	Various	Various	[12, 13, 16]

Table 1.
Estimated values of agroforestry areas in some of the EU Mediterranean countries.

47.964 million hectares of grassland areas is the estimate provided by farmers through the "Farm Structure Survey" [19].

According to [20], the extent of permanent grassland remained stable from 2005 to 2020. However, livestock numbers decreased by 10%, with sheep numbers falling by 20%. These trends are observed without specific attribution to intensive or extensive farming practices, including silvopastoral systems. To understand how silvopastoral systems can contribute to preserve the environment and the landscape, while guaranteeing economic sustainability, this chapter provides an overview of the possible integration of trees and livestock in different types of silvopastoral systems, particularly, focusing on animal grazing management in the wooded areas.

Special attention is given to the Maremmana autochthonous cattle breed to highlight the importance of the role and functions of autochthonous breeds in preserving biodiversity and actively contributing to the resilience of the ecosystem.

2. Methodology

Bibliographical research was performed using the Web of Science and Google Scholar platforms combining key words such as: "silvopastoral" or "silvopasture"; "wood" or "forest"; "agroforestry"; "Mediterranean"; "Maremmana cattle breed"; "extensive systems", "land use"; "sustainability".

As the literature referred to the niche fields like the topics of this publication is not abundant, no restrictions in terms of time periods were considered. Therefore, to highlight the research findings related to silvopastoral systems, older papers were also included, as they serve as a foundational baseline, offering valuable insights into the broader interactions between animals, grasslands, forests, and human activities.

Data from grey literature was also used, as it is important in supporting dissemination and knowledge transfer, and it also helps to cross-fertilise different disciplines [21].

Data was organised through the chapter designed to introduce the topic, show the methodology, and discuss the different types of silvopastoral management as the result of the bibliographical research. Notably, Section 3 was dedicated to the relationships between animal grazing and woods/forests. One part of this section specifically illustrates the connections between silvopastoral management activities and the different components of the silvopastoral system (soil health, forest vegetation, animal welfare and behaviour, meat quality, and carbon footprint). Here, a focus was put on the management of the Maremmana cattle breed, and a particular mention was provided to the activities conducted in Tuscany within two recent rural development projects: NEWTON (Available from: https://gonewton.it/) and MARRUCA. Observations on the relationships between silvopastoral management and fire prevention and economic challenges close the result section. Final reflections close the chapter.

3. Results and discussion

3.1 Managing woods for grazing

The interaction between animals, plants, and non-living environmental components such as soil and climate forms an ecosystem. When this ecosystem is primarily managed by humans and includes domesticated animals and plants, it is known as an agroecosystem. On small-scale and backyard farms, human and community systems can integrate with the agroecosystem, creating a larger, interconnected system.

Within agroecosystems, resources are both competed for and synergistically utilised. Often, a stable and sustainable biotic community—balancing animals and plants—is established. The equilibrium includes the cycling of carbon, nitrogen, and minerals, as well as the flow of energy through soil, plants, and animals. Excess plant material serves as animal feed, while animal waste or by-products provide nutrients for plants. Sustainable farming practices leverage the natural interactions to enhance the ecological, biological, and socio-economic sustainability of farms and agriculture as a whole.

Silvopastoral systems can be implemented in many different settings (**Table 2**).

In native woodlands, a coppice or tall tree system should be implemented according to the following guidelines [29]:

1. In uneven-aged stands, grazing should be consistently banned due to continuous regeneration.

2. In even-aged stands, animals should be excluded during the juvenile stages, from seedlings to young saplings.

3. In coppices, grazing should be avoided between the time of cutting and when the suckers have reached diameters that grazing livestock can no longer damage (5–6 cm, corresponding to 14–15 years).

4. In compound coppice, seedlings grow slower than suckers and can be more easily damaged by livestock.

5. In dense forests with reduced or absent vegetation cover, grazing should be minimised.

6. In sparse forests, where herbaceous and shrub cover is abundant, good grazing conditions can be maintained, including clearings with a good sward.

Type of design	Description	References
Single-row plantings	Trees are spaced approximately 2.5–3.5 m apart within each row, and 5 to 10 m apart between rows.	[22, 23]
Double-row plantings	Stagger tree rows with 2.5–3 m between trees and rows.	[23]
Multiple-row spacing	This system has two to three rows of trees spaced fairly close together (2.5 × 3 m or 3 × 3 m), with an alleyway of 6 to 12 m between sets of tree rows for forage production.	[24, 25]
Triple and quadruple rows	Trees are planted in three or four rows with specific spacing to optimise tree growth and forage production. These rows can be staggered or uniformly spaced.	[23, 26]
Block plantings	A typical silvopasture block planting spaces trees 3.5 × 3.5 or 4.5 × 4.5 m apart.	[23, 24]
Directly in native woodland	Coppice or tall tree system.	[27, 28]

Table 2.
Overview of different design types, detailing planting distances for each method, from single-row plantings to block plantings, and describing the spacing between trees and rows.

In marginal areas, conifers fit silvopasture, as they can grow in different sites, fast react to intensive management, and facilitate the light to get the forest ground [30]. Moreover, compared with hardwood trees, they are less likely to be damaged by livestock.

Non-rational animal husbandry in forests can lead to a regression of the forest, with the most valuable wood species often being the least hardy and rustic, risking their disappearance in favour of more robust but less valuable species.

The design of firebreaks is covered in Paragraph 6.

3.2 Managing animals in forests

Silvopasture can be considered a multilayered system intentionally integrating trees, domesticated animals, and forages: each component benefits from its relationship with the others, enabling multiple yields from the same land. Compared to crop-based systems, livestock-based systems can have a net positive influence on the depth of the water table and water quality, as runoff in pasture has been shown to be one-fifth of that in cropland [31].

Soil and climate are of paramount importance in the management of forest grazing. Clay soils, which are easily compacted especially in wet seasons, are the least suitable (e.g. Turkey oak native forests, *Quercus cerris* L.), whereas the permeable soil of cork forests, for example, appears less damaged by animal presence and trampling [29]. Some authors assert that forest environments can support a wild mammal load per hectare ranging from 10 kg [29] to 100–110 kg live weight [32]. Therefore, regardless of how extensive and low the stocking rate of animal breeding in the forest may be, this will have a greater impact compared to the natural regeneration potential of woods. However, such values cannot be universally applied in all environments and seasons; they must be related to the richness of vegetation, the regenerative capacity of ecosystems, soil fertility, climatic trends, rotational plans, and possible food competition between domestic and wildlife in the area [33].

Dividing the forest into sectors allows livestock to consume the available grass without constantly moving in search of food. Unnecessary transfers cause trampling and burial of grass, leading to a strong reduction in palatability. For adult cattle, the trampled area amounts to 60 m^2 per km of walking [34]. Therefore, herd permanence on the same plot must be well-defined. In grasslands, a resting period of 20–30 days seems optimal, allowing sufficient recovery of grazed vegetation and limiting soil compaction due to trampling, especially in circumscribed areas where cattle preferentially rest (i.e. watering, foraging, and resting points) or frequently travel (trails). Additionally, plot size and herd size are important factors as, in pastures, shorter grazing periods ensure high utilisation rates and optimal ingestion levels (**Figure 2**) [33].

Concerning wooded areas, the regeneration times need to be longer than those studied in pastures. For instance, after completing thinning in a Turkey oak forest, the resting time required to ensure proper regeneration from seeds (acorns) is 20–30 years. During this period, the presence of farm animals in the forest is strictly prohibited to allow for the sprouting and growth of future tall trees. An oak tree can produce up to 3 million acorns during its lifetime, providing tons of protein, fat, carbohydrates, tannins, and important secondary compounds for feed. Besides, a mature oak tree can drop up to 700,000 leaves each year, forming a soft mantle on the ground that quickly becomes an excellent habitat for numerous beneficial animals and insects. This system also facilitates other important ecological functions: up to 3 months calves of all Italian breeds (Chianina, Podolica, Piemontese, Marchigiana, Romagnola, and Maremmana) are reddish, for example, are born reddish (a colour

Figure 2.
Examples of silvopastoral management: (a) Maremmana cattle breed in Turkey Oak forest (Quercus cerris l.) in Maremma region (Tuscany, Italy); (b) Pezzata Rossa cattle breed in summer transhumance on a pine forests in Friuli Venezia Giulia (Italy); (c) Cinta senese pig breed grazing under an oak tree in South-Tuscany (Italy); and (d) Sarda pig breed browsing in cork oak forest (Quercus suber l.) in Sardinia (Italy). Source: owned by Jacopo Goracci.

called "fromentino" from *fromentum*, wheat in Latin), to get competitive advantage of camouflage in the woods, especially in the late winter/early spring period. Stopping grazing also contributes to the spread of nemoral species which are essential for improving the morphological and functional characteristics of the humus type and the mineralisation of litter by biological agents [35].

At the beginning of the grazing period, animals tend to consume a daily ration of forage that far exceeds their actual needs and assimilation capacities, leading to a quantitative and qualitative degradation of the food available at the end of the season [36]. Heavy grazing affects soil porosity, reducing its permeability and water capacity, with negative repercussions on the hydrogeological and regulatory functions of the forest [35]. Compared to species such as goats and sheep, cattle cause less damage by trampling, allowing for better exploitation of dense coppices. Including forest in a proper animal rotation also helps contain the environmental microbiome and the possible spread of diseases typical of high-stocking rate housing systems [37].

Forage growing under the shady, low-wind environment near trees tends to mature more slowly, be lower in fibre and more digestible than that growing out in the open. Livestock appreciate a varied diet and frequently eat a bit of tree foliage daily. However, this minimal browsing can be harmful if animals stay in the silvopasture for a long time, so proper management is essential. The age and experience of the animals are more likely to influence their tendency to browse or strip bark from trees, than their breed. Browsing damage can sometimes be mitigated by removing or managing these animals. Browsing by livestock is unlikely to kill young trees unless it is both severe and repeated multiple times [36]. Wood pasture is particularly appetising to cattle in the late summer period: leaves and shoots, while having less nutritional value compared to spring, contain more astringent tannin and cellulose substances that seem to balance an overly watery and laxative diet based on late meadow shoots. Early autumn rains also moisten and soften the lignin-rich grasses and shrubs of the undergrowth making them much more palatable [29].

An interesting behaviour of grazing animals, particularly related to breeds like Maremmana cattle, is the cooperation between conspecifics. This happens, for example, when some animals to bend the plants to lower their canopies to the ground, to allow others to feed on them [35]. Younger livestock are more likely to damage trees compared to older and more experienced animals [30]. Finally, drinking water management is a key aspect of silvopasture and often proves to be a factor influencing the animals approach to different grazing areas, notably, in summer or drought periods. In fact, the water source must be at a certain distance from grazing animals that need to reach water regularly. The livestock drink more when water is available at a distance under 250 m. When water is located close to the grazing area (within 250 m), cattle will drink individually rather than as a herd. Pasture utilisation tends to become less uniform if cattle need to travel more than 250 m to water, although this value must be related to the species and breed being reared. If adequate watering facilities are limited, lanes and alleyways can be used to access a single water source serving several paddocks; however, this can result in mud and trampling around the water source (**Table 3**) [36].

3.3 Management of silvopastoral systems components

This paragraph depicts the connections between silvopasture and the different components of an agroecosystem. Much of the information collected is the result of two projects run in Maremma, Tuscany (Italy).

Characteristic/Benefit	Description	References
Biodiversity enhancement	Increases plant and animal biodiversity by providing diverse habitats and resources.	[38–40]
Soil health improvement	Enhances soil structure, fertility, and microbial activity through the integration of trees, which contribute organic matter and prevent erosion.	[14, 38, 41]
Carbon sequestration	Sequesters carbon in biomass and soil, mitigating climate change.	[39–41]
Water regulation and quality	Improves water infiltration, reduces runoff and erosion, and enhances water quality by filtering pollutants.	[38, 41, 42]
Microclimate regulation	Modifies local microclimates by providing shade and windbreaks, which can reduce temperature extremes and protect crops and livestock.	[13, 38, 40]
Diversified production	Produces multiple outputs (e.g., timber, fodder, fruit, and livestock), reducing economic risks and increasing resilience.	[14, 39, 41]
Pest and disease control	Reduces pest and disease incidence through enhanced biodiversity and natural predator habitats.	[13, 14, 38]
Economic benefits	Provides additional income streams from diverse products and can reduce input costs through ecosystem service provision.	[13, 40, 41]
Sustainable land use	Promotes sustainable land management practices that integrate agriculture and forestry, preserving land productivity and ecosystem functions.	[14, 39, 41]
Greenhouse gas emission mitigation	Reduces emissions of methane and nitrous oxide from livestock through integrated management practices.	[14, 40, 41]

Table 3.
Summary of the key features, components, and benefits of silvopastoral systems, providing an overview of how these systems function and their ecological, economic, and environmental advantages.

The relationships between soil health and silvopastoral practices were the focus of monitoring activities conducted within the EIP-AGRI Operational Group (RDP of the Region of Tuscany 2014–2022) “NETWork for agroforestry in TOscaNa - NEWTON.” The aim of the Operational Group was to promote agroforestry and the environmental sustainability of agroforestry systems.

NEWTON activities were based at the Tenuta di Paganico (Civitella Paganico, Grosseto), an agrosilvopastoral farm of about 1500 hectares where a long-term agroforestry system is conducted, through traditional pasture and transhumance land. Here, silvopastoral systems have historically been a typical land-use type [43]. At the farm, oak forests include an over-80-year-old wood of Turkey oaks. This species is important and widespread in the area, commonly grazed by the autochthonous Maremmana cattle due to the palatability of the acorns [43]. The Maremmana cattle are a very hardy breed well adapted to the Mediterranean forests [37].

As reported by [44], NEWTON aimed at evaluating the impact of diverse grazing regimes on the two main agroecosystem components: as regards to soil, chemical (soil organic carbon; total nitrogen), physical (bulk density and soil hydraulic conductivity), and biological (QBS—biological quality of the soil) parameters were analysed. As regards to vegetation, the monitoring of the tree structure, shrub component, natural regeneration, and growth of the dominant storey was performed.

MARRUCA project (LEADER measure 16.2, GAL FAR Maremma, Tuscany region) includes the study, development, and evaluation of new indicators for the

Maremmana beef supply chain. The integrated approach encompasses the assessment and enhancement of environmental sustainability, agricultural and forestry land use, and animal welfare. The innovation spans the entire supply chain, focusing on efficiently organising the meat production from Maremmana breed animals raised in a silvopastoral system. The goal was to achieve environmental and economic improvements at both product and land levels. The life cycle assessment (LCA) was utilised as an innovative tool to evaluate the environmental sustainability of the companies involved in the Maremmana beef production chain. This assessment defined a framework to possibly develop a future Environmental Product Declaration (EPD) or environmental product certification.

3.3.1 Silvopasture and forest vegetation

Studies conducted [43, 44] highlight the need to consider the role of grazing management in the silvicultural treatment for the stand regeneration, as grazing can affect the complexity of the forest and its potential regeneration, besides impacting on preserving biodiversity in a Turkey oak stand in a sub-Mediterranean area.

In fact, the results of NEWTON activities carried out to monitor the tree, shrub, and natural regeneration components of the agroecosystem at the Tenuta di Paganico suggest that the grazing activity of Maremmana cattle is responsible for a simplified stand structure, the reduction of vertical layering, while maintaining a regular horizontal spatial distribution, and constraining the development of the lower tree layers and the shrub vegetation. The theses differed only in the dominated and shrub layers and in the natural regeneration as regards to density, height, and coverage values and were inversely related to grazing intensity. Significant differences between grazed and control areas were found in the overstorey, only for medium-sized trees [43, 44].

3.3.2 Silvopasture and soil health

The contribution of grazing to carbon sequestration is still debated, as the knowledge emerging from the literature is lacking [45]. The effects of grazing intensity in terms of regimes, season, and species are still understudied and need a broader transdisciplinary approach [46]. As silvopastoral systems share resources in time and space [47], when analysing a Mediterranean grazed wood pasture, it is necessary to consider different factors such as annual and seasonal variations in climate, the impact of grazing animals, disturbances due to natural events, and the consequences of human use of these landscapes over time [48].

The type of grazing management can affect physical and biological characteristics of soil. Regarding physical features, soil hydraulic conductivity (Ksat) shows a strong trend as a function of animal treading as Ksat values are lower in areas more prone to animal treading and dependent on the barn distance [44]. Moreover, grazing management based on the regulation of animal load in the forest, in a long-standing agroforestry system, can also influence soil biodiversity and its ecological equilibrium. As shown by [49], lower soil respiration levels can be found in areas where animals graze intensively. The soil biological parameters were measured through the Soil Biological Quality index (QBS-ar) highlighting higher levels of biodiversity by mesofauna detection and fungal metabarcoding analyses in medium and low grazing intensity areas [49].

3.3.3 Silvopasture, animal welfare, and behaviour

The goal of intensive livestock farming has traditionally been to create a production system that completely controls the animal's living conditions, including feeding, microclimate, rest, reproduction, and so on. In this system, the animal has no freedom of choice.

Studying the behaviour of cattle in loblolly pine (*Pinus taeda*) silvopasture versus open pasture landscapes revealed that animals utilised the land more evenly in the silvopasture. This difference was mainly attributed to the reduced solar radiation recorded in the silvopasture. Grazing was the dominant behaviour in the silvopasture while loafing was more common in the open pasture. The shade in the silvopasture appeared to reduce heat stress for cattle grazing during warm weather [50]. In the warmest months of the year, cattle spent most of the daylight hours lying down or loafing (engaging in activities other than grazing or lying) in wooded areas. Grazing primarily took place in shaded areas of grasslands near woods or wooded riparian habitats. During the cooler months of March and October, cattle preferred grasslands and woods, showing the least preference for riparian habitats [51]. In addition to reducing solar radiation, trees can also influence other microclimatic conditions, as well as forage productivity and quality [51, 52].

Animal behaviour exhibits significant variation depending on species, breed, and the management system applied. Scientific literature related to cosmopolitan breeds, often selected for specific traits fitting the widespread intensive management practices, is dominating and, typically, investigates the animal behaviour under controlled and high-input conditions, reflecting the demands of modern agriculture. In contrast, the research on heritage breeds in extensive systems is scarce.

Local breeds adapted to extensive systems are often overlooked in scientific research despite their importance in traditional and sustainable farming practices. These breeds have evolved in specific geographical and environmental conditions, exhibiting behaviours that are integral to their survival and productivity in less intensive farming systems.

Understanding the behaviour of local breeds in extensive systems is crucial for several reasons. Firstly, it provides insights into how these animals interact with their natural environment, including grazing patterns, social dynamics, and responses to environmental stimuli. Secondly, it fosters sustainable management practices that capitalise on the inherent behavioural traits of these breeds, promoting their conservation, utilisation in diverse agricultural landscapes, and increasing system performances.

Dominance hierarchies in social animals are closely tied to behaviours essential for managing social interactions, such as affiliative behaviours and social play, which are crucial even from early stages of life. A study on the behavioural interconnections in a group of Maremmana beef cattle within a silvopastoral system [53] revealed a clearly linear dominance hierarchy within the group, where older animals held higher hierarchical ranks. Affiliative behaviours were predominantly directed upwards in the hierarchy: dominant individuals received more affiliative behaviours, while subordinates initiated these interactions, suggesting that affiliative behaviours may serve subordinates in bonding with dominants to potentially gain benefits within the group. Social play, primarily involving play fighting, occurred most frequently among dominant individuals. In this context, affiliative behaviours appear to serve a communicative role, possibly mitigating competition during play fighting. Overall, under natural conditions, young beef cattle exhibited a full range of adult behavioural repertoire, and their agonistic, affiliative, and playful behaviours are intricately linked in shaping social dynamics. Therefore, it can be highlighted that extensive biodiverse farming environments provide an ideal setting to observe and understand the behavioural strategies domestic animals employ to establish cohesive social groups. Such insights are basic to improve the management and welfare of domestic ungulates in free-range running.

Thus, animal internal states can manifest through social behaviours shaping group dynamics, i.e. allogrooming, social play, and agonistic encounters. In extensively bred domestic species, interactions with humans during mandatory health checks can disrupt group dynamics, especially if animals are not accustomed to human handling. In a study involving young Maremmana heifers and steers under silvopastoral system in the inland rural areas of Tuscany [54], changes in social behaviours relative to the presence and timing of health checks were examined. Aggression levels did not fluctuate significantly in response to health checks. However, a marked decrease was observed in time spent playing and affiliating with each other on the day immediately following the routine health check compared to control days when no health checks occurred. Subsequently, social play and affiliative behaviours significantly increased on the following day, potentially reflecting a rebound effect aimed at compensating for reduced social interactions and restoring social equilibrium. These findings suggest that social interactions in Maremmana cattle reared in the silvopastoral system are indicative of their internal states. Specifically, social play and allogrooming appear to function as coping strategies for these animals in response to disruptive events like health checks. Understanding these behavioural responses can provide insights into the well-being and social dynamics of cattle managed under semi-extensive conditions, opening the door to future inspiring research.

3.3.4 Silvopasture and meat quality

There is no extensive bibliography on the relationship between card quality and silvopastoralism. Estimating the nutritional potential of a pasture and the intake quota is particularly challenging because such factors are closely tied to the plant's phenological stage, the season, and the climatic pattern [55]. Species like arbutus, holm oak, lilac, and broom are grazed year-round, whereas others such as oak and elm are only grazed from spring to fall [35]. To date, limited research from Italy has explored ruminant meat quality within silvopastoral systems, with only one PhD thesis [56] addressing this topic. The study compared two genotypes, Aubrac and Maremmana, under different rearing systems (feedlot and pasture with forest access) during the calf finishing stage. The research found that the rearing system did not significantly impact most evaluated parameters, except for the n6/n3 ratio ($p < 0.001$), which was lower in beef samples from grazing steers. This reduction was attributed to higher levels of α-linolenic acid (+151.93%) and its derivatives, namely C20:5n3 (+112.83%) and C22:5n3 (+72.11%), in the pasture-based diet. Pasture grazing is known to enrich α-linolenic acid content, leading to higher n3 fatty acid concentrations in beef from grazing cattle.

Furthermore, beef samples from the grazing system exhibited significantly higher levels of C18:3n-3 ($P < 0.01$). This increase correlated with elevated intermediate products from rumen biohydrogenation of C18:3n-3, including C18:1 trans11 (+73.34%), C18:1 trans16 (+37.09%), and CLA9:2 cis trans (+37.07%).

Pastures, forages, and crop residues often constitute a larger portion of ruminants' diets compared to feedlot-based systems. Utilising more forage is advantageous to meet increasing food demands, given its perceived health benefits in quality meat for human consumption. Therefore, optimising the use of lignocellulosic feedstuffs holds significant long-term importance.

Integrated livestock-forest systems offer various advantages including providing shaded areas that enhance animal comfort. High temperatures from excessive heat stress can adversely affect animal performance, thermal comfort, and behaviour, ultimately impacting carcass and meat quality. This influence is shown in two main ways: (a) elevated temperatures alter muscle metabolism, affecting meat quality; and (b) animals develop heat dissipation strategies in response to high temperatures, which can persist post-slaughter and affect carcass composition and meat quality [57].

Research in Brazil on Angus-Nellore heifers [58] found that a silvopastoral system with lower tree density did not significantly impact animal performance and carcass characteristics when supplemented. However, higher tree density reduced weight gain per hectare, carcass weight gains per hectare, and stocking rate. Systems with trees improved microclimate variables and thermal comfort indices compared to treeless systems. Additionally, tree introduction did not compromise carcass and meat quality standards required for human consumption.

The influence of a silvopastoral system on animal welfare and productive performance was examined using Canchim (3/8 Nellore, 5/8 Charolais) beef heifers in meat production systems [59]. In the first experiment, the silvopastoral system lowered body temperature and enhanced heifer welfare and performance, improving thermal comfort, dry matter intake, body weight gain, and in vitro embryo production, if compared to rotational grazing system where shade was not available. In the second trial, the silvopastoral system raised the body weight without altering the metabolism or the general reproductive performance, while also improving the recovery rate of oocytes in primiparous cows.

Comparing silvopastoral systems to exclusively pastoral systems pose challenges in determining effects on meat quality due to varying definitions and research methodologies regarding "quality". To identify significant and potentially distinguishing differences, focusing on qualitative "systemic" aspects of quality rather than quantitative measures such as fatty acids, fat, and protein may be more insightful. This approach could involve studying sensory characteristics, as sensory properties significantly influence meat acceptability among consumers [60]. Sensory analysis by trained panels is crucial for elucidating differences perceived by humans, with eating quality—encompassing flavour, tenderness, and juiciness—often considered the most critical factor influencing repeat consumer purchases [61].

3.3.5 Silvopasture and carbon footprint

Silvopasture systems significantly enhance ecosystem services, contributing to environmental benefits and increased overall stability in the face of climate change. One of the primary services is carbon sequestration which is facilitated by the growth of aboveground biomass and the presence of deep-rooted perennial grasses and trees. Silvopasture systems are estimated to sequester between 0.55 and 1.9 Mg ha^{-1} C per year. Converting degraded and abandoned lands to silvopasture globally has the potential to absorb 26.6 gigatons of CO_2 equivalent by 2050 [62]. However, managing silvopasture systems requires careful consideration of tree selection, grazing intensity, the optimal mix of trees, and forage and nutrient management to maximise carbon sequestration and nutrient cycling.

Research suggests that silvopasture exhibits higher levels of soil organic carbon, increased water content, and overall better soil quality compared to conventional pastures. Additionally, silvopasture systems show lower temperatures in both soil and among cattle. To achieve equivalent net productivity (including trees, nuts, forage, and animal weight), conventional pasture systems would require approximately four times more land area per hectare than silvopasture. Temperate silvopasture systems enhance ecosystem services by improving soil quality and promoting animal welfare without compromising productivity, making them a viable option for sustainable production in the face of climate change [62].

Building on these experiences, the MARRUCA project conducted a LCA of the beef hamburger production chain of the Maremmana cattle, from breeding to processing. The analysis involved three livestock farms in the province of Grosseto: Tenuta di Paganico (TdP), Aia della Colonna (AdC), and Tenuta di Alberese (TdA), which manage the breeding phase in an agroforestry system [63]. Differences emerged in the fattening phase: TdP practises pasture fattening, TdA uses barn fattening with self-produced forages and feed, while AdC primarily relies on farm forages and commercial feed for barn fattening.

The LCA results indicated that pasture fattening at TdP results in higher CO_2 emissions per kg live weight than the other systems, with TdP at 27.8 kg CO_{2eq}, AdC at 16.8 kg CO_{2eq}, and TdA at 23.6 kg CO_{2eq}. For a single hamburger, the carbon footprint is 8.2 kg CO_{2eq} for TdP, 5.6 kg CO_{2eq} for AdC, and 7.0 for TdA. Enteric fermentation is the main source of impacts, accounting for over 70% of the total. Although pasture fattening generates more emissions per kilogramme of meat produced, farms with forest areas can offset these emissions through carbon uptake by trees. Annual carbon uptake was calculated using national forest inventory tables, which show annual carbon growth and uptake rates by stand type.

The results show that AdC can offset about 30% of its emissions while TdP and TdA achieve full offset, resulting in net impacts of 0 kg CO_2 equivalent per hamburger produced. This highlights the potential to market a low-impact product linked to local resources, distinguishing it from conventional products from intensive breeding of specialised breeds typically found in the refrigerated sections of large-scale retailers. The analysis of the carbon footprint and offsetting potential underscores the possibility of achieving zero-emission meat production through extensive and agroforestry management. This demonstrates that pasture fattening, despite its higher emissions, can be made sustainable through offsetting with forested areas.

3.4 Silvopasture and fire prevention

Forests and garrigue are vegetation types common in Mediterranean pastoral systems, known for their high flammability [64, 65]. Historically, fires have been managed using firebreaks, but a regional survey in Tuscany, Italy, identified several issues with these firebreaks, such as incorrect placement, low efficiency due to the colonisation of flammable bushes, and excessive slopes leading to erosion [66]. To address these problems, tree lines of low-flammability species have replaced some firebreaks, and others are actively managed to prevent the colonisation of flammable weeds. Forage species have been proposed to cover the soil and control vegetation regrowth through animal grazing. Subterranean clover has proven effective in trials due to its prostrate growth habit, good palatability, and autumn-spring growth cycle, which reduces fuel biomass in the summer after grazing [67, 68]. Additionally, grazing by sheep, cattle, and game animals has been shown to manage firebreaks effectively by reducing weed encroachment and shrub density, particularly at forest edges, which are prone to transmitting fires to trees [67]. Research conducted in the Mediterranean forests of northeastern Spain [69] observed that semi-feral *Alberes* cattle predominantly consume flammable woody vegetation, particularly during the winter months. This behaviour significantly contributes to reducing the risk of forest fires. Although husbandry is declining in Tuscany, it still supports the grazing management of some firebreaks and mixed pasture-forest vegetation. Mowing is used where animals are no longer available.

Firebreak efficiency is maintained by keeping sward height short through frequent mowing or grazing. Short species, such as some prostrate legumes that compete with tall grasses, can be sown [70]. A study conducted in 2001 compared four swards (*Hedysarum coronarium* HC, *Subterranean clover* SC, *Trifolium michelianum* TM, and native species NS) and two management treatments (sheep grazing and monthly mowing), focusing on evaluating fire protection and forage production over three years in the Tuscan Maremma [66]. In this Mediterranean climate, the study assessed the effects of sheep and cattle passage through coppice wood and firebreak lines oversown with SC on vegetation evolution in open spaces and under forest cover, considering the availability of herbaceous biomass and the removal of shrubs to protect the wood from fire (**Figure 3**).

Sheep utilised the herbaceous vegetation of firebreak lines more effectively than cattle, limiting weed development and leaving a higher seed yield of subterranean clover. Over three years, sheep ingested more seeds (41%) compared to cattle (18.3%), but seed digestion was 93% in sheep and only 22% in cattle. Returned seeds showed significant improvement in germinability. The stock of hard seeds in the soil increased the self-reseeding rate from 0.37 in the first reseeding to 0.68 in the second,

Figure 3.
Example of silvopastoral management in the Tenuta di Paganico Farm (Italy): firebreak line dividing a tall tree system of Turkey Oaks (Quercus cerris l.) with central swale (wooded ditch on contour line with function of both water and soil harvesting) and feed supplementation sites in the forest (hay). Source: owned by Jacopo Goracci.

with mixed grazing or sheep grazing proving more efficient. Although herbaceous production under the woodland canopy was not significant, an increase was observed when grazing was allowed in the firebreak lines. Shrubs were utilised more by cattle (81%) than by sheep (51%).

SC yield was generally low but increased fourfold each year when cattle were allowed to move through both the wood and firebreak lines, and it doubled under the same conditions with sheep grazing. Seed utilisation was very low, with digestion and germinability of returned seeds following the same trend as in the firebreak lines. The rate of the first self-reseeding was 0.32, increasing to 0.43 the following year, with higher values observed with cattle grazing.

Regarding animal behaviour, sheep mainly utilised seeds in summer and improved the vegetation in open spaces, while cattle provided better dispersion of SC plants within the wood and better control of shrub regrowth. These complementary actions suggest that mixed grazing or having cattle follow sheep grazing could provide more balanced management of silvopastoral systems.

Sward height is crucial for fire and erosion prevention, with clovers being more suitable due to their creeping form and low height, particularly SC. For these species, no management could be hypothesised, especially where steep slopes or absence of husbandry make tillage and grazing impossible. HC height was excessive when only grazed.

Fire prevention was partially incompatible with good dry matter production. HC requires good management, which is difficult to ensure in firebreaks, and native sward palatability is too poor to benefit from animal intake. Clover swards had biomass concentrated in the lower canopy layers and provided good protection against wildfire and erosion. Native and HC swards posed a higher risk of flame spread due to the high presence of stems in the upper layer.

3.5 Economic challenges for a sustainable and profitable silvopastoral system

Implementing sustainable management practices in silvopastoral systems within Mediterranean regions can yield numerous ecosystem services across biological, social, and economic dimensions [43]. These systems benefit local communities by conserving landscapes and reducing fire risks. Socio-economically, they support animal welfare, produce high-quality meat, and offer a plentiful supply of wood and non-wood products.

By combining trees, forage, and livestock, silvopastoral systems create a sustainable land management approach that produces marketable goods while maintaining long-term productivity. This strategy lowers economic risk through diversified product outputs, many of which have established markets. It also minimises costs and enhances marketing flexibility by distributing management expenses between timber and livestock components. These systems optimise land use efficiency, providing a steady income from livestock sales and selective tree and timber product harvesting. Effective forage management enhances livestock nutrition, promoting growth and productivity. The tree component can yield a wide range of products: from sawtimber to pulpwood, firewood, pine straw; from game, nuts, and fruit to ornamental flowers, greenery, mushrooms, and various secondary products [30].

Moreover, silvopastures improve wildlife diversity and contribute to better water quality. The forage cover helps to prevent soil erosion by water and wind while adding organic matter to the soil. These systems create aesthetically pleasing, park-like landscapes that serve important production purposes. Compared to concentrated livestock operations, silvopastoral systems mitigate environmental concerns related to water quality, odours, dust, noise, disease transmission, and animal welfare issues.

4. Conclusions

The sustainability of individual farms hinges on resource-based, economic, environmental, and social issues faced by farm families. The broader agriculture and food system, encompassing everything from farm producers to end consumers, involves economic, environmental, ecological, societal, and cultural aspects at both local community levels and on a wider societal scale. Animals and their products are crucial in the diet, providing essential nutrients and serving as major income sources for farms. Sustainable management of agroecosystems requires balancing competition and synergy in using fundamental resources like land, water, energy, air, labour, technology, capital, and organisation. Ideally, a stable biotic community can be established, balancing plants, animals, and humans. Livestock-based farming systems can address environmental issues such as declining soil fertility, soil erosion, water contamination, and land abandonment. Economically, animals can add value to livestock-crop-forest systems and help buffer trade and market disruptions in agriculture. Diversifying farm enterprises is vital for risk management and includes on-farm strategies, like growing various crop species, and off-farm measures, such as reallocating production resources to non-agricultural activities. However, animal productive efficiency is crucial for the viability of individual farms and agriculture as a whole, including silviculture.

Success in livestock-based systems can be significantly influenced by energy use, particularly the type of energy employed. Organising an effective forage system using "foraging chains," which include forests, natural pastures, meadows, and grasslands of varying growth periods in rotation, is a viable strategy for sustainable farm management. This approach fosters a symbiotic relationship between livestock, human, and forestry activities. Within this system, livestock, agriculture, forestry, and other human activities interact symbiotically, ensuring farm productivity and economic stability while enhancing the landscape's aesthetic and environmental value.

Silvopastoral systems, with their numerous interconnected components, can be improved through an effective One Welfare approach [71, 72], which emphasises that health encompasses more than the absence of disease and includes the overall psycho-physical well-being of the environment, animals, farmers, and communities. The One Welfare approach focuses on the ethics of the relationships between humans, animals, and the environment, seeking solutions that do not disadvantage any component and focusing solely on health risks creating a hierarchy among components (not only between humans, animals, and the environment but also between health and welfare). Thus, agroforestry management aims to promote this approach by designing systems that extend beyond short-term agricultural cycles to broader timescales, such as the growth and development of forestry systems.

Although agroforestry is recognised as a sustainable land use practice, it is not widely acknowledged by many farmers or policymakers [10]. Various organisations and movements, representing the voices of farmers and researchers at international (ICRAF, EURAF) and national levels (AIAF), promote actions to support agroforestry and silvopastoral systems [73–75]. However, these efforts need more consideration at both political and technical management levels.

Silvopastoral systems are particularly relevant in Mediterranean regions. Preserving and innovating these systems through interdisciplinary research and new technologies is crucial. The new approach to rural landscapes should be based on the "dynamic conservation" of the environment, balancing change with preservation, and integrating human activity with nature, transforming the countryside into a "cultural landscape".

Author details

Jacopo Goracci[1]* and Francesca Camilli[2]

1 Italian Agroforestry Association (AIAF) and Tenuta di Paganico Farm, Paganico (GR), Italy

2 Italian Agroforestry Association (AIAF) and Institute of BioEconomy, National Research Council (CNR IBE), Firenze, Italy

*Address all correspondence to: jacopogoracci@hotmail.com

References

[1] Agnoletti M, Santoro A. The Italian National Register of historical rural landscapes. In: Hernik J, Walczycka M, Sankowski E, Harris BJ, editors. Cultural Heritage - Possibilities for Land-Centered Societal Development. Environmental History, Vol 13. Cham: Springer; 2022. pp. 15-34. DOI: 10.1007/978-3-030-58092-6

[2] Altieri MA. Biodiversity and Pest Management in Agroecosystems. Binghamton, NY: Food Products Press (an imprint of The Haworth Press); 1994. p. 185. DOI: 10.1016/0378-4290(94)90020-5

[3] Altieri MA. The ecological role of biodiversity in agroecosystems. Agriculture, Ecosystems and Environment. 1999;**74**:19-31. DOI: 10.1016/S0167-8809(99)00028-6

[4] Lemaire G, Franzluebbers A, de Faccio Carvalho PC, Dedieu B. Integrated crop–livestock systems: Strategies to achieve synergy between agricultural production and environmental quality. Agriculture, Ecosystems and Environment. 2014;**190**:4-8. DOI: 10.1016/j.agee.2013.08.009

[5] Lal R. Managing soils and ecosystems for mitigating anthropogenic carbon emissions and advancing global food security. Bioscience. 2010;**60**(9):708-721. DOI: 10.1525/bio.2010.60.9.8

[6] Lal R. Integrating animal husbandry with crops and trees. Frontiers in Sustainable Food Systems. 2020;**4**:113. DOI: 10.3389/fsufs.2020.00113

[7] Ritchie H, Rosado P, Roser M. CO_2 and greenhouse gas emissions. Published online at Our World InData.org. 2023. Available from: https://ourworldindata.org/co2-and-greenhouse-gas-emissions [Online Resource]

[8] Burgess PJ, Rosati A. Advances in European agroforestry: Results from the AGFORWARD project. Agroforestry Systems. 2018;**92**:801-810. DOI: 10.1007/s10457-018-0261-3

[9] Mosquera-Losada MR, Santos MGS, Gonçalves B, Ferreiro-Domínguez N, Castro M, Rigueiro-Rodríguez A, et al. Policy challenges for agroforestry implementation in Europe. Frontiers in Forests and Global Change. 2023;**6**:1127601. DOI: 10.3389/ffgc.2023.1127601

[10] Murgueitio E, Calle Z, Uribe F, Calle A, Solorio B. Native trees and shrubs for the productive rehabilitation of tropical cattle ranching lands. Forest Ecology and Management. 2011;**261**(10):1654-1663. DOI: 10.1016/j.foreco.2010.09.027

[11] Pantera A, Mosquera-Losada MR, Herzog F, et al. Agroforestry and the environment. Agroforestry Systems. 2021;**95**:767-774. DOI: 10.1007/s10457-021-00640-8

[12] den Herder M, Burgess PJ, Mosquera-Losada MR, et al. Current extent and trends of agroforestry in the EU28. Agroforestry Systems. 2017;**91**(3):537-552. DOI: 10.1007/s10457-016-0059-6

[13] Mosquera-Losada MR, Santiago-Freijanes JJ, Rois-Díaz M, et al. Agroforestry in Europe: A regional analysis. Land Use Policy. 2018;**76**:735-745. DOI: 10.1016/j.landusepol.2018.03.017

[14] Eichhorn MP, Paris P, Herzog F, et al. Silvoarable systems in Europe – Past,

present and future prospects. Agroforestry Systems. 2006;**67**(1):29-50. DOI: 10.1007/s10457-005-1111-7

[15] Papanastasis VP, Mantzanas K, Dini-Papanastasi O, Ispikoudis I. Traditional agroforestry systems and their evolution in Greece. Agroforestry Systems. 2009;**76**(1):1-11. DOI: 10.1007/s10457-008-9181-3

[16] Rigueiro-Rodríguez A, McAdam J, Mosquera-Losada MR. Agroforestry in Europe: Current status and future prospects. Springer Science & Business Media. 2009. pp. 89-109. DOI: 10.1007/978-1-4020-8272-6

[17] Rubio-Delgado J, Schnabel S, Burgess PJ, Burbi S. Reduced grazing and changes in the area of agroforestry in Europe. Frontiers in Environmental Science. 2023;**11**:1258697. DOI: 10.3389/fenvs.2023.1258697

[18] Rubio-Delgado J, Schnabel S, Lavado-Contador JF, Schmutz U. Small woody features in agricultural areas: Agroforestry systems of overlooked significance in Europe. Agricultural Systems. 2024;**218**:103973. DOI: 10.1016/j.agsy.2024.103973

[19] EUROSTAT Permanent agricultural grassland in Europe. 2018. Available from: https://ec.europa.eu/eurostat/statistics-explained/index.php?title=Permanent_agricultural_grassland_in_Europe [Online Resource]

[20] Bertomeu M and Lawson G. Permanent grassland definitions in the EU. EURAF Policy Briefing #29 (v1-31.12.23). 2023. DOI: 10.5281/zenodo.10449117

[21] Falavigna G. Il ruolo della Grey Literature e della letteratura scientifica nella diffusione della conoscenza. Quaderni IRCrES-CNR. 2019;**4**(1):45-58. DOI: 10.23760/2499-6661.2019.003

[22] Ares A, Reid W, Brauer D. Tree spacing in silvopastures: Effects on forage and tree production. Agroforestry Systems. 2003;**57**(3):279-291. DOI: 10.1023/A:1024856208332

[23] Garrett HE, Kerley MS, Ladyman KP, Walter WD, Godsey LD, Van Sambeek JW, et al. Hardwood silvopasture management in North America. Agroforestry Systems. 2004;**61**(1-3):21-33. DOI: 10.1023/B:AGFO.0000028987.09206.6b

[24] Clason TR. Silvopastoral practices sustain timber and forage production in commercial loblolly pine plantations of Northwest Louisiana, USA. Agroforestry Systems. 1999;**44**(2-3):293-303. DOI: 10.1023/A:1006267114962

[25] Jose S. Agroforestry for ecosystem services and environmental benefits: An overview. Agroforest Systems. 2009;**76**:1-10. DOI: 10.1007/s10457-009-9229-7

[26] Anderson LS, Sinclair FL. Ecological interactions in agroforestry systems. Agroforestry Abstracts. 1993;**6**(2):57-91

[27] Nair PKR, Kumar MB, Nair VD. An Introduction to Agroforestry. Springer Science and Business Media. Cham: Springer; 1993. DOI: 10.1007/978-3-030-75358-0

[28] Sharrow SH, Clason TR, Sharrow TR. Silvopastoral Practices. North American Agroforestry: An Integrated Science and Practice. Madison, WI: ASA; 1998. pp. 105-131

[29] VV.AA. Aspetti agro-forestali e zootecnici del pascolamento nel bosco, Atti Conv. A.P.D.S.A.F. 1979. 18 Novembre, Monticano, Firenze. p. 80

[30] Klopfenstein NB, Allen KK, Avila FJ, Heuchel SA, Martinez J, Carman RC, et al. Proteinase inhibitor II gene

in transgenic poplar: Chemical and biological assays. Biomass and Bioenergy. 1997;**12**(4):299-311. DOI: 10.1016/S0961-9534(97)00004-4

[31] Baker FH, Raun NS. The role and contributions of animals in alternative agricultural systems. Journal of Alternative Agriculture. 1989;**4**:121-127. DOI: 10.1017/S0889189300002940

[32] Susmel L, Viola F. Pascolo in foresta. Economia Montana. 1981;**5**:17-20

[33] Cocca C, Campanile G. Pascolo in bosco solo se controllato. Ağrı. 2005;**7**:21-22

[34] Gusmeroli F. Il piano di pascolamento: strumento fondamentale per una corretta gestione del pascolo. Quaderni SoZooAlp. 2004;**1**:27-41

[35] VV.AA. Pascolo e bosco, Atti Tavola Rotonda A.I.S.F. 1984; 12 Marzo, Firenze. p. 50

[36] Hamilton J. Silvopasture: Establishment and Management Principles for Pine Forests in the Southeastern United States. USDA National Agroforestry Center: Lincoln; 2008. p. 72

[37] Lucifero M, Jannella GG, Secchiari P. Origine, evoluzione, miglioramento e prospettive della razza bovina Maremmana, Ed. Edagricole. 1977. p. 44

[38] Poudel S, Pent G, Fike J. Silvopastures: Benefits, Past Efforts, Challenges, and Future Prospects in the United States. Agronomy. 2024;**14**:1369. DOI: 10.3390/agronomy14071369

[39] Nair PKR. Carbon sequestration studies in agroforestry systems: A reality-check. Agroforestry Systems. 2011;**86**(2):243-253. DOI: 10.1007/s10457-011-9434-4

[40] Garrett HE, RL MG. Alley Cropping Practices. North American Agroforestry: An Integrated Science and Practice. Lincoln, Nebraska, USA: ASA Pub. USDA National Agroforestry Center; 2000. DOI: 10.2134/agronmonogr54.c5

[41] Shibu J. Agroforestry for ecosystem services and environmental benefits: An overview. Agroforestry Systems. 2009;**76**(1):1-10. DOI: 10.1007/s10457-009-9229-9

[42] Montagnini F, Nair PKR. Carbon sequestration: An underexploited environmental benefit of agroforestry systems. Agroforestry Systems. 2004;**61-62**(1-3):281-295. DOI: 10.1023/B.0000029005.92691.79

[43] Pacini A, Pelleri F, Marini F, Maltoni A, Mariotti B, Mazza G, et al. Impact of cattle density on the structure and natural regeneration of a Turkey oak stand on an agrosilvopastoral farm in Central Italy. Journal of Forestry Research. 2024;**35**(1):22. DOI: 10.1007/s11676-023-01680-9

[44] Manetti MC, Marini F, Pelleri F, Pacini A, Chiarabaglio PM, Bergante S, et al. Impact of cattle grazing on soil and stand structure in a Mediterranean silvopastoral system in Tuscany. In: 6th European Agroforestry Conference, EURAF 2022, Nuoro, 16th–20th May. EURAF ed. 2022

[45] Pulina A, Hernández-Esteban A, Rolo V, Seddaiu G, Roggero PP, Moreno G. A remote sensing approach to assess the stability of forage production from legume-rich mixtures oversown in Mediterranean wooded grasslands. In: 6th European Agroforestry Conference, EURAF 2022, Nuoro, 16th–20th May. EURAF ed. 2022

[46] Lecegui A, Olaizola AM, Varela E. Disentangling the role of management

practices on ecosystem services delivery in Mediterranean silvopastoral systems: Synergies and trade-offs through expert-based assessment. Forest Ecology and Management. 2022;**517**:120273. DOI: 10.1016/j.foreco.2022.120273

[47] Sales-Baptista E, Ferraz-de-Oliveira MI. Grazing in silvopastoral systems: Multiple solutions for diversified benefits. Agroforestry Systems. 2021;**95**:1-6. DOI: 10.1007/s10457-020-00581-8

[48] Franca A, Re GA, Sanna F. Effects of grazing exclusion and environmental conditions on the soil seed bank of a Mediterranean grazed oak wood pasture. Agroforestry Systems. 2018;**92**:909-919. DOI: 10.1007/s10457-018-0203-0

[49] Maienza A, Lumini E, Biancofiore G, Baronti S, Ugolini F, Fabrizio U, et al. Ecological relations of soil fauna under long term agroforestry system in temperate climate: Results from Central Italy. In: 6th European Agroforestry Conference, EURAF 2022, Nuoro, 16th–20th May. EURAF ed. 2022

[50] Karki U, Goodman MS. Cattle distribution and behavior in southern-pine silvopasture versus open-pasture. Agroforestry Systems. 2010;**78**:159-168. DOI: 10.1007/s10457-009-9250-x

[51] Kort J. Benefits of windbreaks to field and forage crops. Agriculture, Ecosystems and Environment. 1988;**22-23**:165-190. DOI: 10.1016/0167-8809(88)90017-5

[52] Bird PR. Tree windbreaks and shelter benefits to pasture in temperate grazing systems. Agroforestry Systems. 1998;**41**:35-54. DOI: 10.1023/A:1006092104201

[53] Bagnato S, Pedruzzi L, Goracci J, Palagi E. The interconnection of hierarchy, affiliative behaviours, and social play shapes social dynamics in Maremmana beef cattle. Applied Animal Behaviour Science. 2023;**260**:105868. DOI: 10.1016/j.applanim.2023.105868

[54] Francesconi M, Pedruzzi L, Bagnato S, Goracci J, Ripamonti A, Mele M, et al. Social play and affiliation as possible coping strategies in a group of Maremmana beef cattle. Journal of Ethology. 2023;**42**:41-52. DOI: 10.1007/s10164-023-00801-5

[55] Sottini E, Geri G. Composizione chimica, valore nutritivo ed utilizzazione pascoliva dei germogli e delle foglie di alcune essenze tipiche della macchia mediterranea. Informatore Agrario. 1971;**3**:27-38

[56] Ciucci F. Productive, Qualitative and Metabolic Response of Maremmana and Aubrac Steers Maintained in Feedlot or Grazing Systems. PhD Thesis. Department of Agricultural, Food and Agro-Environmental Sciences, University of Pisa; 2020

[57] Gregory NG. How climatic changes could affect meat quality. Food Research International. 2010;**43**:1866-1873. DOI: 10.1016/j.foodres.2009.05.018

[58] de Oliveira NF, Andrighetto C, Aparecida da Luz P, et al. Performance, carcass and meat quality of Angus-Nellore heifers finished in integrated livestock forest systems. Journal of Animal and Feed Sciences. 2021;**30**(1):33-41. DOI: 10.22358/jafs/133446/2021

[59] Lemes AP, Garcia AR, Pezzopane JRM, Brandão FZ, Watanabe YF, Cooke RF, et al. Silvopastoral system is an alternative to improve animal welfare and productive performance in meat production systems. Scientific Reports. 2021;**11**(1):14092. DOI: 10.1038/s41598-021-93609-7

[60] Horsfield S, Taylor LJ. Exploring the relationship between sensory data and acceptability of meat†. Journal of the Science of Food and Agriculture. 1976;**27**:1044-1056. DOI: 10.1002/jsfa.2740271111

[61] Maltin CA, Warkup CC, Matthews KR, Grant CM, Porter AD, Delday MI. Pig muscle fiber characteristics as a source of variation in eating quality. Meat Science. 1997;**47**(3-4):237-248. DOI: 10.1016/S0309-1740(97)00055-7

[62] Amorim HCS, Ashworth AJ, O'Brien PL, Thomas AL, Runkle BRK, Philipp D. Temperate silvopastures provide greater ecosystem services than conventional pasture systems. Scientific Reports. 2023;**13**(1):18658. DOI: 10.1038/s41598-023-45960-0

[63] Finocchi M, Cella FG, Villani R, Ripamonti A, Goracci J, Cappucci A, et al. Carbon balance of two beef cattle farms: Integration of LCA methodology and offset tools in agrosilvopastoral system. In: Book of Abstract EURAF Conf. EURAF ed. 2024. DOI: 10.11118/978-80-7509-982-2

[64] Pardini A, Talamucci P, Margaritelli L. Tentativi di incremento dell'offerta di pascolo in un bosco ceduo attraverso concimazione, semina e trasemina. Monti e Boschi. 1987;**3**:47-52

[65] Talamucci P, Pardini A, Argenti G, Staglianò N. Functioning of a sylvo-pastoral system based on different resources, including firebrake lines utilized by sheep. Cahiers options Méditerranéennes. FAO-CIHEAM. 1995;**12**:179-182

[66] Lombardi P, Argenti G, Pardini A. Revegetated firebreaks management in Tuscan Maremma. The future of the green Mediterranean. In: Euro-Mediterranean Conference. 2001. Alghero 1-2 JUNE. Sardinia Region ed. 2001. pp. 100-107

[67] Etienne M, Dersko M, Rigolot E. lmpact du paturage sur les arbustes dans les amenagements sylvopastoraux a objectif de prevention des incendies. Cahiers Options Méditerranée. FAO-CIHEAM. 1995;**12**:217-220

[68] Pardini A, Piemontese S, Argenti G, Staglianò N, Talamucci P. Mantenimento di bande parafuoco con trifoglio sotterraneo pascolato ai fini della riduzione della fitomassa combustibile e delle perdite di suolo in un'area forestale toscana. Rivista di Agronomia. 1995;**XXIX**(Suppl. 3):427-433

[69] Bartolomé J, Plaixats J, Piedrafita J, Fina M, Adrobau E, Aixàs A, et al. Foraging behavior of Alberes cattle in a Mediterranean Forest ecosystem. Rangeland Ecology and Management. 2011;**64**(3):319-324. DOI: 10.2111/REM-D-09-00160.1

[70] Nassiri M, Elgersma A, Lantinga EA. Vertical distribution of leaf, dry matter and radiation in grass-clover mixtures. In: Grassland and Land Use Systems. Grado (Italy): 16° EGF Meeting; 1996. pp. 269-274

[71] Pinillos RG, Appleby MC, Manteca X, Scott-Park F, Smith C, Velarde A. One welfare - a platform for improving human and animal welfare. The Veterinary Record. 2016;**179**(16):412-413. DOI: 10.1136/vr.i5470

[72] Santiago-Freijanes JJ, Pisanelli A, Rois-Díaza M, Aldrey-Vázquez JA, Rigueiro-Rodrígueza A, Pantera A, et al. Agroforestry development in Europe: Policy issues. Land Use Policy. 2018;**76**:144-156. DOI: 10.1016/j.landusepol.2018.03.014

[73] Center for International Forestry Research and World Agroforestry. CIFOR-ICRAF. 2023 [updated 2024 Aug 10; accessed 2024 Sep 11]. Available from: https://www.cifor-icraf.org/

[74] European Agroforestry Federation. EURAF. 2023 [updated 2024 Aug 10; accessed 2024 Sep 11]. Available from: https://euraf.net/

[75] Italian Agroforestry Association. AIAF. 2023 [updated 2024 Aug 10; accessed 2024 Sep 11]. Available from: https://www.agroforestry.it/

Chapter 11

Efficiency and Sustainability in Ruminants' Livestock Systems: A Case Study of Dry-Land Pastures in the Montado Ecosystem

Emanuel Carreira, Flávio Silva, João Serrano, Ana Geraldo, Paulo Infante, Rui Charneca, Cristina Conceição and Alfredo Pereira

Abstract

The Montado (Portugal)/Dehesa (Espanha), a unique agro-sylvo-pastoral ecosystem in the Mediterranean region, offers a valuable model for sustainable ruminant livestock production. This chapter explores the efficiency and sustainability of this system, with a particular focus on dryland pastures and grazing practices. Extensive livestock systems in the Montado, defined by their low-input and low-output approaches, are highly dependent on rainfed pastures. The Mediterranean climate, with its irregular rainfall patterns and increasing aridity, poses significant challenges to pasture productivity and animal performance. This chapter explores various grazing systems, including continuous and deferred grazing, and their implications for pasture health, animal welfare, and ecosystem services. By analyzing the relationship between grazing management, pasture quality, and environmental factors, we seek to identify strategies for enhancing the efficiency and sustainability of ruminant livestock systems in the Montado. Ultimately, this research contributes to a deeper understanding of the potential of agro-sylvo-pastoral systems to balance agricultural production with environmental conservation, providing valuable insights for sustainable livestock practices in Mediterranean regions.

Keywords: agro-sylvo-pastoral systems, bovine, grazing, Mediterranean climate, Montado, sheep

1. Introduction

Agriculture serves various purposes, primarily focusing on producing plants and animals to ensure global food security and meet the demand for diverse products. It encompasses a wide range of activities essential for food production, including cultivation, domestication, horticulture, and arboriculture. Livestock management and grazing practices, such as agro-sylvo-pastoral systems and transhumance, are integral components of agriculture [1]. In addition, agricultural activities must ensure the sustainability of ecosystems, while respecting the environment, and preserving soils and biodiversity. Agro-sylvo-pastoral systems are an indispensable example of how to sustain the territory, playing a crucial role in combating desertification and preventing rural fires [2]. The agro-sylvo-pastoral systems, characterized by their reliance on extensive rainfed pastures, form the backbone of traditional ruminant production in the Mediterranean. These extensive systems have always been characterized by diversification as the best way to adapt to natural constraints and to make small-scale agriculture efficient. In this context, crop rotation has emerged as a structuring factor in Mediterranean agriculture. As the most demanding and environmentally aggressive crops are usually associated with essential human foodstuffs, they must be followed in space and time by others who can restore the physical environment. These are usually of indirect interest for human consumption (pastures for grazing animals), so that the final balance of the process is positive, especially from an environmental point of view, that is, renewable resources such as soil and water [3].

The Montado, a good example of an agro-sylvo-pastoral ecosystem, is inextricably linked to the landscape of the south of Portugal (Alentejo), which is the result of the temporal interaction between natural and cultural factors. It relies on the extensive grazing of livestock, including sheep, cattle, and Iberian pigs, on slender soils supporting perennials/annual pastures interspersed and a mix of tree cover,

Figure 1.
Diagram representing the interrelationships between the components of the Montado and the services provided by the ecosystem. Diagram created with Canva [6].

predominantly *Quercus suber* (cork oak), with some *Quercus rotundifolia* (holm oak). Currently, this ecosystem occupies around 1 million (1/3 of its territory) hectares in Alentejo [4]. The Montado is associated with diverse practices, both related to multiple production activities, characteristic of complex forest-pastoral systems, and with many other traditional activities: hunting, beekeeping, mushroom picking, or activities that have recently become more relevant, such as hiking, other outdoor sports, or bird watching [5]. The multifunctionality of this agro-sylvo-pastoral system and its importance related to the area in the south of the country mean that a huge part of the region's intangible heritage, such as gastronomy, traditions, singing, imagery, and legends, refer to and are rooted in areas of Montado and particularities of this system. And so, along with the landscape, the identity of the southern region of the country also passes through the Montado. **Figure 1** represents the interrelationships between the components of the Montado (soil, pasture, trees, and animals) and the services that it provides to the environment, animals, and society.

2. Methodology

A bibliographic research was conducted utilizing the Web of Science and Google Scholar search engines. In addition, national agricultural and land use databases, along with data collected from our research group over several years on the experimental fields of the University of Évora, were employed to augment the research information.

This chapter investigates the efficiency and sustainability of ruminant livestock systems within the Montado ecosystem, a distinctive agro-sylvo-pastoral landscape situated within the Mediterranean region. By examining the relationship between grazing practices, pasture management, and environmental factors, the chapter aims to provide insights into the potential of these systems to contribute to sustainable agriculture.

3. Extensive livestock systems

From 1989 to 2023, the number of farms in Portugal decreased significantly, although the average size of farms increased (**Figure 2**) [7]. According to the National Institute of Statistics (INE) [7], each farm has currently an average of 14.8 hectares of useful agricultural surface (**Figure 2**).

In Portugal, there are 3.861 million hectares of useful agricultural land, with permanent pastures occupying 54.4% of this area, followed by permanent crops with 23.3% and arable land with 22% (**Figure 3**) [7]. However, many permanent pastures (75%) are considered poor (**Figure 4**), lacking improvements such as species introduction, fertilization, or the application of correctives, even though extensive livestock production—central to Mediterranean agriculture—primarily depends on these rainfed pastures.

Extensive livestock production, a basis of Mediterranean agriculture, primarily relies on rainfed pastures. Due to the limited water resources, these systems are particularly vulnerable to fluctuations in temperature and rainfall, which can significantly impact biomass yields and animal productivity. Extensive livestock systems mean that producers have lower relative feed costs, while also appealing to consumers who prefer food obtained from animals produced in "green pastures" [8].

Livestock production systems based on undercover pastures in Montado are characteristic and distinctive, and their products can be certified with Protected

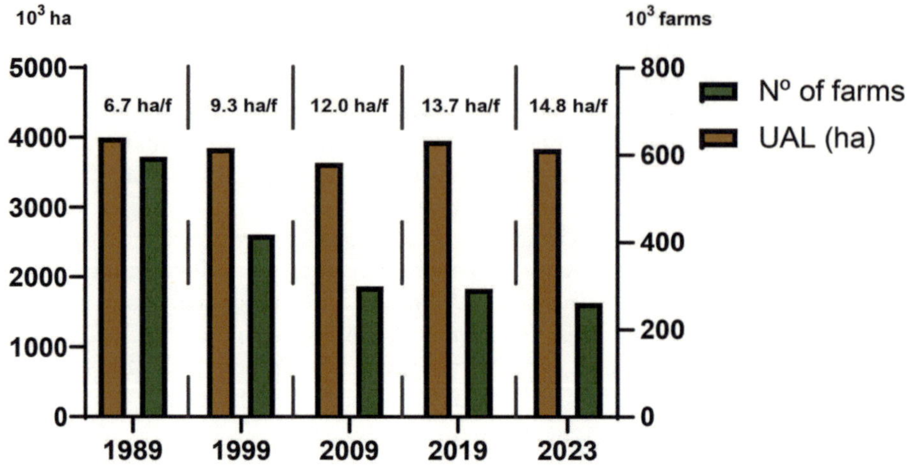

Figure 2.
Evolution of the useful agricultural area and number of farms from 1898 to 2023 in Portugal. UAL-useful agricultural area. Adapted from Ref. [7].

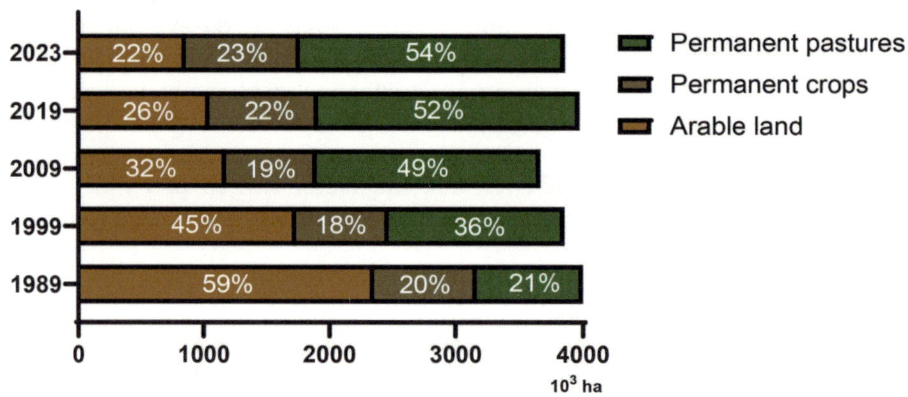

Figure 3.
Evolution of the composition of the useful agricultural area in Portugal between 1989 and 2023. Adapted from Ref. [7].

Designation of Origin (PDO). The differentiation comes from the dietary scheme that characterizes the extensive livestock production practiced on Montado farms and which gives them distinctive organoleptic qualities [5]. According to these authors, in 2010, extensive livestock farming supported more than half of the country's production of cattle (65%) and sheep (56%), with a low level of investment and application of technological developments, based on extensive grazing of permanent pasture, with low stocking rates.

A forage unit (FU) obtained from grazed grass costs only 15 to 20% of the same FU obtained from concentrate feed. In addition, the meat or milk obtained from grass is of a much higher quality than that obtained from systems with high consumption of concentrates, as the meat obtained from grass is rich in vitamin E, CLA (conjugated linoleic acids), and Omega 3, unlike that obtained from high doses of concentrates, which are cholesterol-inducing and possibly carcinogenic [9].

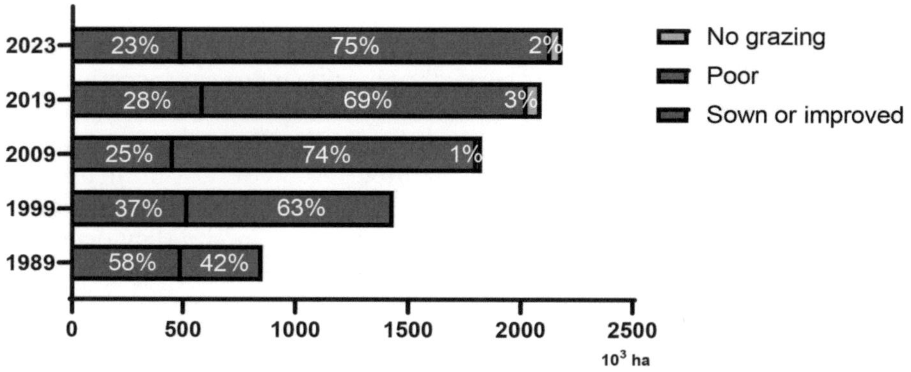

Figure 4.
Evolution of permanent pastures by type and form of installation from 1898 to 2023 in Portugal. Adapted from Ref. [7].

Since permanent pastures are arguably the less expensive feed for animals, their improvement is highly valuable [2]. However, the relationship between the number of animals on farms and the quantity and quality of pasture is more balanced in spring. During the rest of the year, several times there is a deficit and, since it is not common practice to reduce the number of animals, farms resort to producing and conserving fodder or purchasing other external food resources [5]. However, producers are increasingly buying feed produced abroad, since farms dedicate almost the entire area to the production of permanent pasture. Additionally, it is often cheaper to buy feed than to produce it, due to the large investment in machinery and labor.

Over the last 20 years, extensive livestock farming in the Montado has undergone important changes. The increase in fenced areas and in the European financial support for cattle, among other factors, has led to a consistent increase in the cattle population, which is currently the dominant zootechnical species, based on indigenous breeds (i.e., Mertolenga and Alentejana) and on exotic breeds, especially the Limousin, Charolais, and Aberdeen Angus where cross-breeding with indigenous breeds is a frequent strategy [5]. In 2024, the cattle population in Portugal was 1,534,000 animals, thus animals an increase of 5.7%, compared to 2023, when the number of animals was 1,445,897 [7]. In 2024, Portugal's cattle population was 1 534 000 animals, an increase of 5.7% compared with 2023, when the number of animals was 1 445 897. The decrease in the number of goats and sheep, due to the lack of skilled labor, has also led to an increase in cattle, as they can be easier and cheaper to manage. Also, lower rainfall and increased production factors costs between 2019 and 2023 also contributed to a decrease in the number of goats (−18.5%) and sheep (−6.7%) in these years, as well as a reduction in the number of farms (−12.7 and −8.7%, respectively) [7]. However, the average flock size per farm remained like in 2019, with 52 sheep and 15 goats per farm. According to data from INE [7], the national sheep population fell by 30.5% between 1999 and 2023 (2,929,765 animals in 1999 vs. 2,036,008 animals in 2024), nearly 1 million fewer animals than in 1999 (2,929,765 sheep).

3.1 Grazing systems

Pastures are generally managed by establishing stocking rates, with relatively low grazing pressures, allowing animals to choose their diet [10]. Producers aim to improve the profitability of their production systems while maintaining sustainability and biodiversity. However, they often fail to do so, as examples of this include situations of overgrazing and a decrease in pasture quality. This requires integrating knowledge of the species biology and correctly adjusting management actions [11]. Grazing tends to enable adequate animal welfare and the production of safe and healthy products [12].

Aiming for the most efficient use of the multiple resources available in the Montado, different livestock species can and should coexist through appropriate grazing management. This can be mixed and continuous or intermittent grazing, over an appropriate periodic crop (plots or leaves), rotating the animals through various plots. Grazing by cattle in the cork oak forest plays a very important role in maintaining it, as it counteracts the proliferation and development of shrub species such as cistus (*Cistus ladanifer*) and sargassum (*Cistus salviifolius*), among others [13].

Grazing animals are fundamental to the balance of agro-sylvo-pastoral systems, since they transform herbaceous production (dryland pasture) into products of unrivaled quality and contribute to maintaining the balance between herbaceous and shrub layers. On the other hand, the animals also act as firebreaks, thus helping to prevent fires, which damage ecosystems. Added to this are the beneficial effects of maintaining soil fertility and reducing production costs [14].

Continuous grazing (CG) involves grazing the same plot during the grazing season, year after year [15]. According to Teague and Dowhower [16], CG has limitations because it allows for selectivity, with consequences for the heterogeneity of the pasture, with overgrazed and undergrazed areas occurring simultaneously. Overgrazing can lead to soil degradation and loss of biodiversity, while undergrazing can lead to a greater preponderance of less palatable species with lower food value and loss of habitat, with the overlapping of a shrub layer [17].

In CG systems, excessive stomping is detrimental and damages not only the pasture but also the soil [18]. On the other hand, according to Barriga [19], in CG systems, there is a return of nutrients to the soil through feces and urine. CG offers several benefits over rotational grazing (RG), particularly regarding lower infrastructure costs associated with fencing and watering systems. This simplified management approach also allows animals to select their diet, potentially leading to improved performance [8, 15]. Intermittent grazing (deferred grazing on several fences) is also sometimes used, presenting quite attractive productive, ecological, and economic results [11].

RG is a grazing management system where animals are moved between fenced pastures, allowing each area to rest and regenerate before being grazed again [8]. However, most of the time, there is no pre-defined frequency for the animals to remain in each pasture plot. This system, while affording animals a degree of pasture choice, effectively balances animal nutrition with pasture health, allows animals to select their forage while ensuring that their nutritional needs are met without compromising the quality and abundance of the grass [18].

Hussain et al. [20] found that pastures benefit significantly from RG. Their study concluded that biomass accumulation is higher in areas grazed rotationally by sheep compared to ungrazed sites, especially in the spring. On the other hand, Zang et al. [21] noted that grazing under high biotic loads can lead to pasture degradation due to excessive defoliation and trampling, even though it can be effective in controlling weeds. To maintain a balanced pasture-grazing relationship, it is essential to adopt an integrated and well-managed approach. RG enhances animal productivity per hectare through sustainable intensification, achieved not by overgrazing but by employing effective management practices. A possible setback is that it increases labor and

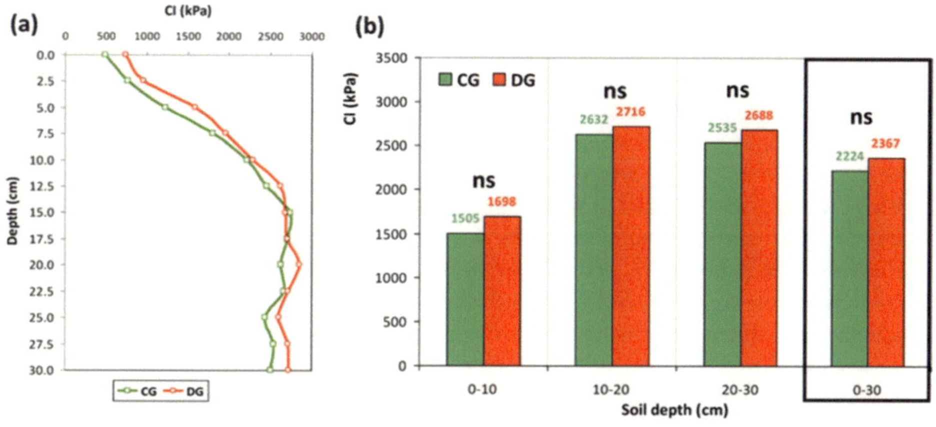

Figure 5.
Average cone index (CI) compared between continuous and deferred grazing at (a) 0–30 cm soil depth and (b) 0–10 cm, 10–20 cm, 20–30 cm, and 0–30 cm soil depths. ns - no significance. Kindly provided from Ref. [22].

additional land could be needed [8]. In semi-arid regions such as the Alentejo in Portugal, RG is recommended, with short grazing periods and long rest periods [18]. In theory, this could work, but in practice, it is not adoptable due to climatic asymmetries in both temperatures and water, both of which limit grass growth. If the grass grows at different speeds, RG at fixed times is not the most appropriate and is therefore not used. If the grazing is rotational, but with variable periods of stay depending on the availability of pasture, it is called deferred grazing (DG). The frequency and duration of grazing in each paddock depend on factors such as forage type, animal stocking rate, and environmental conditions. This management optimizes forage growth, improves soil health, and prevents overgrazing promoting biodiversity, reducing erosion, and enhancing the sustainability of grazing lands.

The DG consists of placing high stocking rates on the pasture, with the duration of the grazing period being a function of the available biomass in the pasture. The availability of pasture biomass can be estimated by measuring the height of herbaceous plants. Holechek [15] states that DG allows areas preferred by the animals not to be damaged as much as in CG, in terms of the vigor and production of the plants in these areas. DG can minimize the harmful effects of selective overgrazing, due to the overuse of areas preferred by animals [11]. A disadvantage often pointed out to DG is excessive trampling, due to the high animal loads per hectare. However, according to a study by Serrano et al. [22], as can be seen, no significant differences were observed in soil compaction when comparing two grazing systems with sheep (CG - 1 AUE/ha vs. DG - 2 AUE/ha) in the Montado ecosystem (**Figure 5a,b**).

4. Characterization of Montado/Dehesa ecosystem

The Montado, dominated by cork oaks and holm oaks, is the most extensive agroforestry system in Portugal and has the greatest incidence in the country's large southern area [23]. The Montado/Dehesa (Portugal/Spain) is a multifunctional agro-sylvo-pastoral ecosystem [24], where agricultural, livestock, and forestry activities are balanced and combined [5], consisting of a herbaceous stratum where permanent pastures predominate, an arboreal stratum with a special incidence of cork oaks

(*Quercus suber*, L.) and holm oaks (*Quercus ilex ssp.* Rotundifolia, Lam), grazed by animals (sheep, cattle, goats, and pigs) on an extensive basis [25]. This ecosystem is highly complex due to the interrelationships between its four fundamental components – soil, pasture, trees, and animals – and the climate [26].

The Montado, as a non-climax community, naturally progresses into a Mediterranean forest in the absence of human intervention. Thus, anthropogenic activities play a crucial role in maintaining the balance of this ecosystem. These activities help preserve the characteristic Montado landscape, which typically includes an herbaceous layer, a tree layer, and occasionally a shrub layer, often grazed by livestock. However, it should be noted that humans are also contributed to the degradation of certain components of this ecosystem, due to practices associated with agriculture and livestock production during certain periods of recent history.

The Montado occupies significant areas with acidic, thin and rocky soils, degraded because of erosion and nutrient loss [27]. Most of the soils where the Montado is located suffer from structural and fertility limitations, being thin, stony, acidic, poor in phosphorus and nitrogen, with imbalances in terms of micronutrients (particularly the magnesium/manganese ratio) [25].

4.1 Mediterranean climate: Irregularities vs. climate change

The Mediterranean climate is characterized by hot and dry summers and rainy winters with mild temperatures, being the only climate on the Earth that has the particularity of having a dry summer, typical of summer with high average temperatures for several months, which is associated with a high degree of irregularity in the occurrence of precipitation, both within and between years [28]. It is the only climate in which the maximum annual temperatures overlap with the minimum levels of precipitation. The designation of Mediterranean climate is since the Mediterranean Sea basin is its largest area of influence, although it is also present in California, Chile, South Africa, and Australia [3]. In addition to the irregular annual distribution of rainfall, the Mediterranean climate is also characterized by marked inter-annual irregularity, with the occurrence of years with abundant and scarce rainfall, in a bimodal frequency distribution, which means that the average annual rainfall values are not the most frequent [29]. In the Mediterranean region, annual accumulated rainfall ranges from 300 to 800 mm [30]. As an example, Évora weather station reported an average annual rainfall of 624 mm, a minimum of 203 mm and a maximum of 1186 mm, between 1871 and 2007. The values of average monthly precipitation and average minimum temperature and average maximum temperature for Évora, in the period between 1871 and 2007, are represented in **Figure 6**. As can be seen, during the summer months (June, July, August, and September), the precipitation values are residual, with the temperature values being the highest.

Lopes et al. [32] highlight that various observations and projections suggest the Mediterranean region and southern Europe, including Portugal, are more vulnerable to climate change compared to northern Europe. Carvalho [2] notes that over a consecutive 10-year period, the cumulative rainfall of 50 mm required to initiate seeds germination, and the agricultural year can occur anytime between September and November. This variability significantly impacts the duration of the favorable autumn period and the productive potential of forage. Similarly, combined rainfall during March and April can range from 40 to 325 mm, influencing the length of the spring season. Notably, April rainfall plays a critical role in the growth and development of pastures.

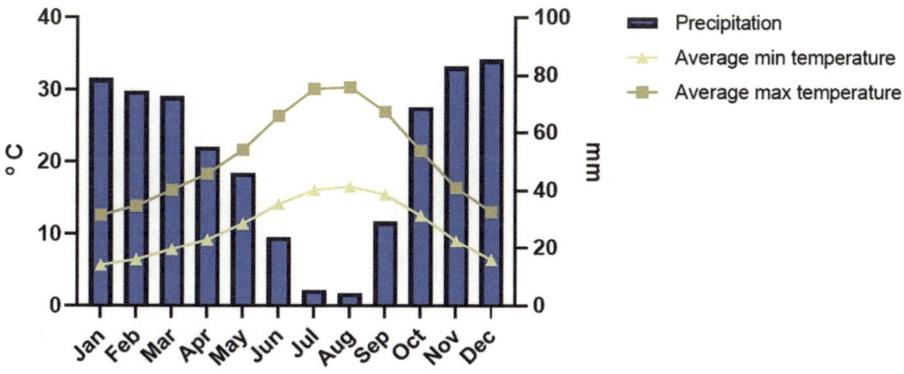

Figure 6.
Average monthly precipitation and average minimum and average maximum monthly temperature for Évora, between 1871 and 2007. Data collected from the Portuguese Institute for Sea and Atmosphere [31].

Prolonged droughts are common in the Mediterranean region and have become more frequent in recent years due to climate change. As shown in **Figure 6**, between 1871 and 2007, the average monthly temperature was 19.6°C, with a maximum of 24.7°C and a minimum of 16.6°C, for the region of Évora, Portugal [31]. Notably, the average maximum temperature has shown an upward trend, increasing as the years progress toward the present. The irregularities of the Mediterranean climate, exacerbated by climate change, have made dryland pasture production in this region increasingly unpredictable, resulting in severely constrained periods for biomass growth.

According to Ferreira [33], the Mediterranean climate has two unfavorable seasons for plant production: the hot, dry summer and the wet, cold winter in the interior region of Alentejo. Therefore, in a Mediterranean climate, the production of pastures and fodder is very limited by the climatic conditions, in addition to the relief conditions and the physical and chemical characteristics of the soil. It is the only climate in which the maximum annual temperatures coincide with the minimum levels of precipitation. The total annual rainfall is often identical to that of temperate climates, but the concentration of rainfall in the autumn/winter season and the lack of water in the hot seasons cause serious inconveniences for agriculture.

4.2 Dry-land pastures

Over the last 35 years, the area of permanent pastures has increased in Portugal, although the majority are natural dry lands without any type of improvement. Furthermore, only 1% of pastures in Portugal is irrigated [7]. The germination and development of pasture plants depends on the characteristics of the soil and the weather conditions. The production of dryland pasture depends on rainfall and temperature, presenting considerable variability between years and within the same year. Normally, in the Mediterranean region, pasture plants begin to germinate and develop with the onset of the rains in autumn, provided there are adequate temperatures and radiation. For general germination occurrence, 50 mm of rainfall is required [2]. During the autumn, there can be a slight peak in production, which added to the scarce winter production, is responsible for producing around 15–35% of the total annual production, however, is not observed when the first significant rains do not occur until later in the year [34].

With the onset of winter and the accompanying drop in temperatures, pasture production ceases entirely. Despite the presence of water in the soil, the extremely low temperatures prevent any green matter from being produced. In spring, however, the combination of adequate water, optimal temperatures, and sufficient sunlight creates ideal conditions for pasture plant growth. During this period, dryland pastures reach their peak production, achieving values of approximately 50 to 120 kg of dry matter (DM)/ha/day. This peak accounts for roughly 65–85% of the total annual pasture production [34].

By the end of May, when pastures generally begin to dry out, there is a very rapid growth of grass, with an increase in dry matter content and a marked decrease in digestibility, energy, and protein [35]. In summer, the pasture dries out completely, leaving the remaining biomass, which, although abundant, has a low food value. To make these agricultural systems economically profitable, the land factor is used, leading to extensive production systems and, if intensification is chosen, irrigation is generally used as a way of increasing productivity.

The permanent pastures currently found in the South-Montado region exhibit very low DM production [23]. According to Carvalho [2], addressing soil acidity and mitigating the toxic effects of aluminum and manganese, along with the application of phosphorus, can lead to a fivefold increase in productivity. Carvalho [2] also emphasizes that dryland pastures must remain the foundation of these systems, as they represent the most cost-effective feed source, and the region lacks sufficient water resources to irrigate the entire area. Enhancing pasture productivity can be readily achieved by correcting soil acidity and applying phosphorus.

Belo et al. [23] refer to studies conducted to support their findings: (a) one study reported DM production values of 800 kg/ha/year in Montado pastures; (b) another study noted that DM production rarely exceeds 1500 kg/ha/year; and (c) a third study observed DM production of 695 kg/ha in autumn/winter and 2014 kg/ha in spring. However, Efe Serrano [36] reports a significantly higher DM production of around 3000 kg/ha/year. In a recent study, Serrano et al. [37] compared pasture DM production under and outside the canopy. They found average DM production values of 437, 1232, 1804, 2751, and 2363 kg/ha under the canopy and 425, 1868, 2987, 3582, and 6191 kg/ha outside the canopy for December, March, April, May, and June, respectively.

Although rainfed pasture is the most cost-effective feed option for ruminants, its nutritional value often falls short of meeting the animals' production requirements, necessitating costly supplementation. To address these deficiencies, farmers frequently need to cultivate and store fodder supplements or acquire additional external feed resources.

The first step to improve the nutritional value of pasture, is to improve soil fertility. Since the main problem with the soils in the Montado region is acidity and manganese toxicity, this requires the application of magnesium-rich dolomitic limestone [38]. In addition, the recommended process for restoring pastures in the Mediterranean region is to increase soil fertility by applying phosphate fertilizers [39]. On the other hand, the soil under the tree canopy is more fertile than the soil outside the canopy [24], which shows the importance of trees for the grass growth and the production systems.

In a study conducted in the Montado ecosystem, Serrano et al. [37] observed that mean crude protein (CP) levels under the tree canopy varied across months: 22.9% in December, 22.4% in February, 15.9% in March, 11.2% in May, and 8% in June. Outside the canopy, CP values followed a similar seasonal pattern, starting at 21.3%

in December and declining to 6.3% in June, with intermediate values of 19.8% in February, 13.5% in March, and 9.8% in May.

Figures 7 and **8** show the percentages of CP and neutral detergent fiber (NDF), respectively, of pasture samples collected throughout the pasture's vegetative cycle, in plots subject to four different treatments (UC - without application of dolomitic limestone and continuous grazing - 1 AUE/ha; UD - without application of dolomitic limestone and deferred grazing - 2 AUE/ha; TD - with application of dolomitic limestone and deferred grazing - 2 AUE/ha; TC - with application of dolomitic limestone and continuous grazing - 1 AUE/ha).

As can be seen in **Figure** 7, throughout the pasture's vegetative cycle, CP values were always above the threshold required for ruminants (above 7%). Even at the end of May, when the pasture was already dry, CP values were close to 10% in all treatments. The percentage of NDF tended to be higher in the TC treatment (**Figure 8**), and lower in the UC treatment, where no corrective was applied.

Besides protecting animals from the weather and heat, in agro-sylvo-pastoral systems, trees also act as soil protectors and contribute to increasing soil fertility. Moreno et al. [40] found that the organic matter content under the canopy of holm oaks was around twice as high as that found beyond the canopy. It is very common in the Montado to see plants with greater nutritional value under the canopy in autumn, winter, and spring, as there is greater soil fertility than those that appear outside the canopy, where *Rumex bucephalophorus* L. appears above all. Usually, this plant is regarded as a bioindicator of soil acidity and Mn toxicity.

Pastures are typically established on low-fertility soils, which are often shallow, making it impractical to cultivate more profitable crops. Moreover, the floristic composition of natural pastures in these areas is often compromised by poor grazing management practices. Improving pastures involves implementing sustainable modifications to their natural floristic composition to enhance both productivity and quality.

The improvement of pastures, Mediterranean or otherwise, thus consists of altering or 'unbalancing' bad herbaceous associations by introducing or making species of better forage value prevail [36]. The recommended process for restoring pastures in the Mediterranean region is to increase soil fertility by applying phosphate fertilizers [39]. As a way of improving productivity in extensive livestock production systems,

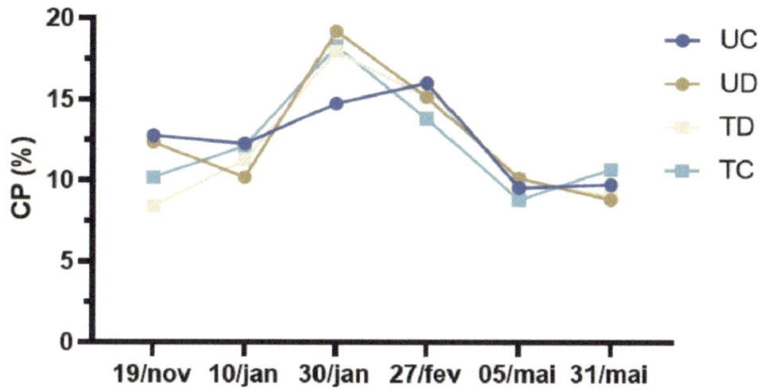

Figure 7.
Percentage of crude protein (CP) in four treatments: Authors' unpublished data. UC – without application of dolomitic limestone and continuous grazing; UD – without application of dolomitic limestone and deferred grazing; TD – with application of dolomitic limestone and deferred grazing; TC – with application of dolomitic limestone and continuous grazing.

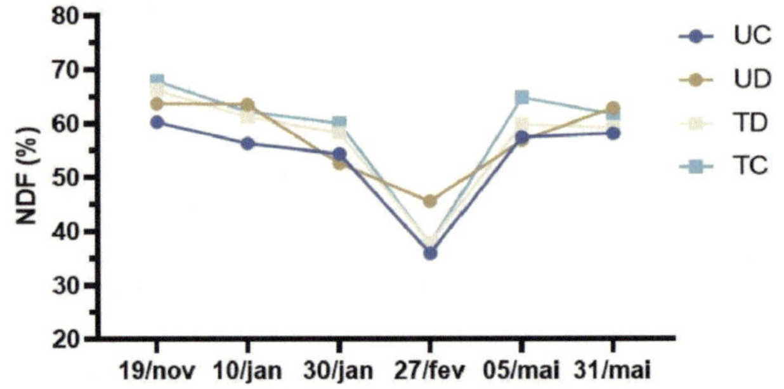

Figure 8.
Percentage of neutral detergent fiber (NDF) in four treatments: Authors' unpublished data. UC – without application of dolomitic limestone and continuous grazing; UD – without application of dolomitic limestone and deferred grazing; TD – with application of dolomitic limestone and deferred grazing; TC – with application of dolomitic limestone and continuous grazing.

the application of phosphate fertilizers and/or the application of limestone correctives and the introduction of biodiverse pastures are also common practices [25].

The structure and composition of the plant communities that make up pastures are generally affected by grazing [18, 41] and, above all, by selective grazing [42]. Selective grazing occurs when the animal load is low regarding the green mass produced [43]. Faria [44] notes that the shift from sheep to cattle on farms in the Iberian Peninsula brought changes to animal management practices, such as the number of grazing days and animal rotation, as well as alterations in grass structure and pasture floristic composition. Conversely, Zhu et al. [45], in a study conducted in China, found that grazing by cattle, sheep, and goats did not influence the species richness of pasture plants; however, it did significantly reduce plant biomass and increase variability in plant heights. According to Voisin and Lecomte [18] in a pasture sown with *Poa pratensis* and white clover, the percentages of both vary throughout the year of sowing and subsequent years, depending on the grazing time of each plot and can be as high as 80% clover in grazing systems with one cut per week, down to just 1% clover in systems with only one cut every 12 weeks.

The sustainable intensification of these systems involves, firstly, improving permanent pastures, of which around 90% are considered poor (IEEA 2013 cited by [2]). In a study comparing the productivity of natural pastures with biodiverse, legume-rich pastures in the Montado, Simões et al. [46] observed that the increase in DM production, which in some cases doubled, together with a higher proportion of nutritionally valuable species, allowed the grazing animal load to be tripled.

Following the experimental work referred above, in which the effect of four treatments was evaluated (UC; UD; TD; TC), at the end of spring (end of May) the botanical species belonging to the *Poaceae* and *Fabaceae* families were identified (**Table 1**); of the total of 32 different botanical species identified, 20 belong to the Poaceae family and 12 to the Fabaceae family. The plot where the most plants were identified was the one subject to continuous grazing without application of soil amendment (UC). The plot with the fewest botanical species identified was the one subject to continuous grazing and application of dolomitic limestone (TC). In this last plot, the fewest botanical species belong to the *Fabaceae* family. The application of dolomitic limestone led to greater development of grasses during the autumn/winter, which were not consumed because the stocking rate was relatively low, thus preventing the penetration of solar radiation

Botanical species	UC	UD	TD	TC
Poaceae **family**				
Vulpia geniculata (L.) Link	x	x	x	x
Agrostis pourretii Willd.	x	x	x	x
Hordeum murinum (Link) Arcang.	x	x	x	x
Lolium perenne L.	x	x	x	x
Lolium rigidum L.	x	x	x	
Avena barbata (Tab.Morais) Romero Zarco	x		x	x
Briza maxima L.	x			x
Briza minor L.	x			x
Gaudinia fragilis (L.) P. Beauv.	x	x	x	x
Aira caryophyllea L.	x			
Phalaris arundinacea L.	x		x	x
Cynodon dactylon (L.) Pers.		x	x	x
Taeniatherum caput-medusae (L.) Nevski			x	x
Agrostis stolonifera L.			x	x
Lamarckia aurea (L.) Moench	x			
Cynosurus echinatus L.	x			
Vulpia bromoides (L.) S.F.Gray	x			
Agrostis castellana Boiss. & Reut.				x
Holcus lanatus L.				x
Avena sterilis L.	x	x	x	x
Fabaceae **family**				
Trifolium subterraneum L.	x	x		
Trifolium glomeratum L.	x	x	x	x
Trifolium campestre Schreb.	x	x	x	x
Trifolium dubium Sibth.	x	x	x	x
Trifolium repens L.	x	x	x	
Trifolium scabrum L.	x			
Trifolium resupinatum L.	x	x	x	
Biserrula pelecinus L.	x		x	
Medicago polymorpha L.	x			x
Lathyrus angulatus L.	x			x
Lotus subbiflorus Lag.	x	x	x	
Trifolium angustifolium L.	x	x	x	x

Authors' unpublished data. UC – without application of dolomitic limestone and continuous grazing; UD – without application of dolomitic limestone and deferred grazing; TD – with application of dolomitic limestone and deferred grazing; TC – with application of dolomitic limestone and continuous grazing. "x" marks the presence of the botanical species.

Table 1.
Botanical species identified in dry-land pastures under Montado ecosystem.

into the soil, essential for the germination of legumes. In addition, most of these species are prostrate and therefore did not receive enough light to develop.

4.3 Achieving sustainability

Montado plays an important role in carbon sequestration, especially because they are made up of long-lived cork and holm oak trees, which promote carbon storage over very long periods [24]. On the other hand, pastures under Montado cover also play a fundamental role in carbon sequestration. According to Pinto-Correia et al. [5], this system has great potential for accumulating carbon in its arboreal and herbaceous biomass and in the soil. The same authors stated that healthy cork oak forests with reasonable tree cover can sequester between less than 1 and more than 3 tonnes of carbon per hectare every year. Overall, carbon sequestration in the Montado is estimated at 6.7 tonnes of CO_2/ha/year [3].

Balanced management of the Montado also contributes to preventing rural fires. Nowadays this is of great importance in Portugal, as in recent years fires in the centre and north of the country have not only destroyed forests and property but have also killed hundreds of people. As reported by Oliveira et al. [47] agroforestry systems, by reducing the amount of fuel, promoting the formation of horizontal and vertical discontinuities in the vegetation and forming clearings in the landscape, can be a means of controlling the number and severity of fires.

The spatial and temporal heterogeneity of the Montado promotes a substantial wealth of ecological niches [24]. We can find more than 135 species of vascular plants per 0.1 ha in the Montado [48], many of them rare or with protected status such as *Narcissus fernandesii*, *B. cavanillesii*, *Armeria pinifolia*, *Centaurea coutinhoi*, *Halimium verticillatum*, *Ruscus aculleatus* and *Narcissus bulbocodium* [24].

The diversity of tree cover in the Montado allows for the development of herbaceous and shrubby strata dominated by Mediterranean flora with a high expression of angiosperm species that are pollinated by insects, particularly bees [5]. This is why, cork oak forests are so important for the beekeeping sector, with around 2200 beekeepers in the Alentejo and Algarve regions (southern Portugal) [49]. The value of honey amounts to more than 20 million euros per year, particularly speciality types (monofloral, Protected Designation or Organic Production) [5].

Mushrooms are also present in the Montado ecosystems with great diversity, namely edible mushrooms with economic value, such as *Amanita caesarea* (Caesar's mushroom), *Amanita ponderosa* (heavy amidella), many Boletes species (e.g., *B. aereus* and *B. aestivalis*), *Cantharellus cibarius* (golden chanterelle), *Craterellus cornucopoides* (horn of plenty), and desert truffles: *Terfezia arenaria*, *Terfezia fanfani*, and *Terfezia Leptoderma* [5].

More than 28 species of fauna with protected status are also found in the Montado and, in addition, the cork oak and holm oak forests are home to around a hundred other animal species, which appear in the annexes to the EU Habitats and Birds Directives [24]. The Montado provides habitat, food and shelter for many species, including many bird species, which are strongly linked to this ecosystem, such as *Elanus caeruleus* (black-winged kite), *Circaetus gallicus* (short-toed snake eagle) or *Hieraaetus pennatus* (booted eagle) [24]. This ecosystem is also of great importance to winter migratory birds such as *Vanellus vanellus* (Northern lapwing) and *Grus grus* (common crane).

According to Pinto-Correia et al. [5], of the 71 species of terrestrial mammals (both flying and non-flying) reported in mainland Portugal, over 95% are found in the Montado ecosystem. However, their presence may vary, ranging from occasional and fragmented occurrences to widespread distributions with high abundances. Among the species occasionally present is the critically endangered Iberian lynx (*Lynx pardinus*).

The shrubby areas and rugged slopes of the Montado also support a diverse array of fauna, including rabbits, hares, lizards, snakes, and wild house mice. These species play a critical role in the food chain, serving as prey for carnivores such as the common genet (*Genetta genetta*), least weasel (*Mustela nivalis*), red fox (Vulpes vulpes), and European wildcat (Felis silvestris) [50].

The sustainability of this unique ecosystem relies on pastures composed of self-reseeding annual grassland species. The longevity and persistence of these pastures are closely tied to the soil seed bank, which ensures the renewal of the pasture's production cycle with the first autumn rains [5]. This natural forage base, complemented by the fruits of holm oak (acorns) and cork oak (lande), along with occasional supplementation from pruned branches, underpins a low-cost production system. When paired with appropriate stocking rates and effective management, this approach greatly reduces costs compared to intensive or poorly managed systems with excessive stocking densities, which are heavily dependent on preserved feeds such as hay, straw, haylage, silage, and concentrates.

5. Conclusions

The Montado (in Portugal) or Dehesa (in Spain), recognized by the European Environment Agency as a High-Value Natural Agricultural System, represents a vital component of biological and cultural heritage. It was included on Portugal's national list of potential UNESCO World Heritage Sites in 2016 [24].

In addition to its economic contributions—such as the production of cork, firewood, livestock (beef, mutton, pork, and goat), mushrooms, herbs, and honey—the Montado plays a critical role in providing essential ecosystem services. These include water cycle regulation, carbon sequestration, erosion control, support for high biodiversity, recreational and leisure opportunities, and reinforcement of local cultural identity [5].

This is an ecosystem designed for multifunctional agricultural and forestry production (agro-sylvo-pastoral system) because it is made up of various subsystems and integrated and interdependent production systems typical of Mediterranean environments. Montados are understood as multifunctional production systems, i.e. systems that, in the process of producing wood, cork or fruit, give rise to other goods and services that society has come to appreciate (biodiversity, carbon sequestration, hunting, environmental protection, and many others).

The Montado is a man-made ecological ecosystem designed around the sustainable use of natural resources, complemented by extensive agricultural activities. When managed in accordance with the principles necessary to maintain balance across its various subsystems, it has proven to be environmentally friendly. Despite the diverse goods and services generated by the Montado, farmers remain heavily reliant on income from cork production and livestock farming.

The Montado plays a crucial role in soil conservation, protection, and enhancement, making it an essential tool in combating desertification—an issue of particular significance in the Mediterranean region. Here, forest loss, demographic pressure, and climate change amplify the risk of desertification. Due to their immense socio-economic and ecological value, oak forests are critical in forming forest barriers against desertification. These forests are not only vital for preserving ecosystems but also serve as the backbone of life on Earth [24].

As an agro-sylvo-pastoral ecosystem, the Montado is of immense importance in the Alentejo region and throughout Portugal, covering a significant area. Its preservation and improvement are essential for supporting local populations, maintaining biodiversity, and generating economic wealth. By sustaining the Montado, we ensure the settlement of rural communities, the conservation of unique ecosystems, and the production of high-quality goods that are unparalleled in their value.

Acknowledgements

MED (https://doi.org/10.54499/UIDB/05183/2020; https://doi.org/10.54499/UIDP/05183/2020) and CHANGE (https://doi.org/10.54499/LA/P/0121/2020).

Funding

This work was funded by the SUMO Project: Montado Sustainability (PRR-C05-i03-I-000066), Investment supported by the PRR - Recovery and Resilience Plan and by the European Funds Next Generation EU and by National Funds through the FCT - Foundation for Science and Technology under the Project UIDB/05183.

Conflict of interest

The authors declare no conflict of interest.

Author details

Emanuel Carreira[1], Flávio Silva[1,2,3*], João Serrano[1], Ana Geraldo[1], Paulo Infante[4], Rui Charneca[1], Cristina Conceição[1] and Alfredo Pereira[1]

1 MED—Mediterranean Institute for Agriculture, Environment and Development, and CHANGE—Global Change and Sustainability Institute, Institute for Advanced Studies and Research, University of Évora, Évora, Portugal

2 Department of Animal Science, Veterinary and Animal Research Centre and Al4AnimalS, University of Trás-os-Montes e Alto Douro, Vila Real, Portugal

3 Center for Research and Development in Agrifood Systems and Sustainability, Polytechnic Institute of Viana do Castelo - Agrarian School of Ponte de Lima, Refóios do Lima, Portugal

4 Research Center in Mathematics and Applications, University of Évora, Évora, Portugal

*Address all correspondence to: fsilva@uevora.pt

References

[1] Harris DR, Fuller DQ. Agriculture: Definition and overview. In: Smith C, editor. Encyclopedia of Global Archaeology. New York: Springer; 2014. pp. 104-113

[2] Carvalho M. A seca e a intensificação sustentável dos sistemas agro-pecuários: desafios e soluções. Dossier secas e cheias: como gerir? Indústria e ambiente 110 Maio/Junho, Portugal. 2018

[3] Potes JM. O Montado no Portugal Mediterrânico. Lisboa, Portugal: Edições Colibri; 2011. 193 p

[4] APORMOR News. Montermor-O-Novo Capital Nacional Da Pecuária Extensiva. In: Reportagem da EXPOMOR – A Maior Feira de Pecuária Extensiva. Portugal: APORMOR News; 2024

[5] Pinto-Correia T, Ribeiro N, Potes J. Livro verde dos Montados. In: Edição: ICAAM - Instituto de Ciências Agrárias e Ambientais Mediterrânicas. Portugal: Universidade de Évora; 2013

[6] Canva. Available from: https://www.canva.com/ [Accessed: January 5, 2025]

[7] INE – Instituto Nacional de Estatística. Retrato da Agricultura Nacional – 2023. Portugal: INE – Instituto Nacional de Estatística; 2024

[8] Beetz AE, Rinehart L. Rotational grazing. In: Encyclopedic Dictionary of Landscape and Urban Planning. Berlin/Heidelberg, Germany: Springer; 2010. p. 857

[9] Crespo DG. Em tempos de crise qual o papel das pastagens e forragens no desenvolvimento da agricultura. Agrotec: Revista técnico – científica agrícola. 2011;**1**:18-21

[10] Norman HC, Wilmot MG, Thomas DT, Barrett-Lennard EG, Masters DG. Sheep production, plant growth and nutritive value of a saltbush-based pasture system subject to rotational grazing or set stocking. Small Ruminant Research. 2010;**91**:103-109

[11] Teague R, Provenza F, Norton B, Steffens T, Barnes M, Kothmann M, et al. Benefits of multi-paddock grazing management on rangelands: Limitations of experimental grazing research and knowledge gaps. In: Schroder HG, editor. Grass Lands: Ecology, Management and Restoration. Hauppauge, NY, USA: Nova Science Publishers, Inc.; 2008. pp. 41-80. ISBN 978-1-60692-023-7

[12] Morand-Fehr P, Fedele V, Decandia M, Frileux Y. Influence of farming and feeding systems on composition and quality of goat and sheep milk. Small Ruminant Research. 2007;**68**(1-2):20-34. DOI: 10.1016/j.smallrumres.2006.09.019

[13] Reis DG, Calafate LC. Os montados—Conciliação entre Homem e Natureza. Captar. 2013;**4**:13-26

[14] Grigoli A, Todaro M, Di Miceli G, Genna V, Tornambè G, Alicata ML, et al. Effects of continuous and rotational grazing of different forage species on ewe milk production. Small Ruminant Research. 2012;**106**:S29-S36

[15] Holechek JL. Considerations concerning grazing systems. Rangelands. 1983;**5**:208-211

[16] Teague WR, Dowhower SL. Patch dynamics under rotational and continuous grazing management in large, heterogeneous paddocks. Journal of Arid Environments. 2003;**53**:211-229

[17] Barcella M, Filipponi F, Assini S. A simple model to support grazing management by direct field observation. Agriculture, Ecosystems and Environment. 2016;**234**:107-117

[18] Voisin A, Lecomte A. La Vaca y la Hierba: Cómo Obtener Buenos Rendimientos del Ganado Vacuno. Madrid, Spain: Editorial Tecnos; 1968

[19] Barriga PAB. Produção de Novilhos em Pastagem de Inverno Com Diferentes Intensidades de Pastejo e Adubação Nitrogenada em Integração Lavoura-Pecuária. Tese de Doutoramento [thesis]. Pato Branco, Brazil: Universidade Tecnológica Federal do Paraná, Câmpus; 2019

[20] Hussain Z, Kemp PD, Horne DJ, Jaya IKD. Pasture production under densely planted young willow and poplar in a silvopastoral system. Agroforestry Systems. 2009;**76**:351-362

[21] Zhang N, Wang M, Wang N. Precision agriculture—A worldwide overview. Computers and Electronics in Agriculture. 2002;**36**:113-132

[22] Serrano J, Carreira E, Shahidian S, de Carvalho M, Marques da Silva J, Paniagua LL, et al. Impact of deferred versus continuous sheep grazing on soil compaction in the Mediterranean Montado ecosystem. AgriEngineering. 2023;**5**:761-776. DOI: 10.3390/agriengineering5020047

[23] Belo CC, Coelho I, Rolo J, Reis P. Sistemas agroflorestais em Portugal continental. Parte II: Montados, condições de uso do solo e evolução. Revista de Ciências Agrárias. 2014;**37**:122-130

[24] Batista T, Mascarenhas JM, Mendes P. Montado's ecosystem functions and services: The case study of Alentejo Central-Portugal. Problems Landscape Ecology. 2017;**XLIV**:15-27

[25] Serrano J. Utilização eficiente dos recursos no Montado: Sensores próximos e detecção remota no apoio à tomada de decisão. Perspetiva. 2019;**2020**:18-25

[26] David TS, Pinto CA, Nadezhdina N, Kurz-Besson C, Henriques MO, Quilhó T, et al. Root functioning, tree water use and hydraulic redistribution in Quercus suber trees: A modeling approach based on root sap flow. Forest Ecology and Management. 2013;**307**:136-146

[27] Serrano J, Peça J, Silva JM, Shahidian S, Carvalho M. Phosphorus dynamics in permanent pastures: Differential fertilizing and the animal effect. Nutrient Cycling in Agroecosystems. 2011;**90**(1):63-74

[28] Feio M. Clima e Agricultura. In: Exigências Climáticas das Principais Culturas e Potencialidade Agrícolas do Nosso Clima. Lisboa, Portugal: Ministério da Agricultura, Pescas e Alimentação; 1991. ISBN 972-9175-25-X

[29] Moreira N. Agronomia das Forragens e Pastagens. Vila Real, Portugal: Sector Editorial; Universidade de Trás-os-Montes e Alto Douro; 2002. 183 p

[30] Fonseca A. O Montado no Alentejo: (Séculos XV a XVIII). Lisboa, Portugal: Edições Colibri; 2004. ISBN 972-772-505-8

[31] IPMA. Portuguese Institute of the Sea and Atmosphere. 2024. Available from: https://www.ipma.pt/ [Accessed: December 2, 2024]

[32] Lopes M, Monteiro AC, Ribeiro I, Sá E, Martins H, Coutinho M, et al. Alterações Climáticas e Gestão da Água em Portugal. Revista Brasileira de Geografia Física. 2012;**6**:1333-1357

[33] Ferreira DB. Evolução da paisagem de montado no Alentejo interior ao longo do século XX: Dinâmica e incidências ambientais. Finisterra. 2001;**36**:179-193

[34] Crespo D. O Papel das Pastagens Biodiversas Ricas em Leguminosas na Reabilitação da Agricultura Alentejana e na Qualidade dos seus Produtos Tradicionais. Évora: Ruraltec, Universidade de Évora; 2006

[35] Pedroso RJR. Desmame faseado em bovinos de carne (problemática dos anos difíceis) [thesis]. Évora: Universidade de Évora; 2006. 90 p

[36] Efe Serrano J. Pastures in Alentejo: Technical Basis for Characterization, Grazing and Improvement. Évora, Portugal: Universidade de Évora—ICAM; Gráfica Eborense; 2006. pp. 165-178

[37] Serrano J, Shahidian S, Da Silva JM, Sales-Baptista E, De Oliveira IF, De Castro JL, et al. Tree influence on soil and pasture: Contribution of proximal sensing to pasture productivity and quality estimation in montado ecosystems. International Journal of Remote Sensing. 2017;**39**:4801-4829

[38] Carvalho M, Goss MJ, Teixeira D. Manganese toxicity in Portuguese Cambisols derived from granitic rocks: Causes, limitations of soil analyses and possible solutions Toxicidade de manganês nos Cambissolos derivados de rochas graníticas em Portugal: Causas, limitações das análises de solo. Revista de Ciências Agrárias. 2015;**38**:518-527

[39] Daniels MB, Delaune P, Moore PA, Mauromoustakos A, Chapman SL, Langston JM. Soil phosphorus variability in pastures: Implications for sampling and environmental management strategies. Journal of Environmental Quality. 2001;**30**:2157-2165

[40] Moreno G, Obrador JJ, García A. Impact of evergreen oaks on soil fertility and crop production in intercropped dehesas. Agriculture, Ecosystems and Environment. 2007;**119**:270-280

[41] Miao F, Guo Z, Xue R, Wang X, Shen Y. Effects of grazing and precipitation on herbage biomass, herbage nutritive value, and yak performance in an alpine meadow on the Qinghai-Tibetan plateau. PLoS One. 2015;**10**:e0127275

[42] Fabian Y, Sandau N, Bruggisser OT, Kehrli P, Aebi A, Rohr RP, et al. Diversity protects plant communities against generalist molluscan herbivores. Ecology and Evolution. 2012;**2**:2460-2473

[43] Ribatski EO, dos Santos FS, Neto FN. Revisão sobre utilizações de RPAs no cenário agrícola. Brazilian Journal of Technology. 2018;**1**:313-333

[44] Faria N. Predicting agronomical and ecological effects of shifting from sheep to cattle grazing in highly dynamic Mediterranean dry grasslands. Land Degradation and Development. 2019;**30**:300-314

[45] Zhu H, Wang D, Wang L, Bai Y, Fang J, Liu J. The effects of large herbivore grazing on meadow steppe plant and insect diversity. Journal of Applied Ecology. 2012;**49**:1075-1083

[46] Simões NM, Pereira MS, Quintas A, Barradas AM, Vicente C, Belo CC, et al. Annual legume species for improvement of pastures. In: Ramalho Ribeiro JMC, Horta AEM, Mosconi C, e Rosati A, editors. Animal Products from the Mediterranean área. EAAP. Vol. 119. Santarém, Portugal: Wageningen Academic Publishers; 2006. pp. 43-49

[47] Oliveira F, Moreno G, López L, Cunha M. Origem, distribuição e funções

dos sistemas agro-florestais. Pastagens e Forragens. 2007;**28**:93-115

[48] Díaz-Villa MD, Marañón T, Arroyo J, Garrido B. Soil seed bank and floristic diversity in a forest-grassland mosaic in southern Spain. Journal of Vegetation Science. 2003;**14**(5):701-709

[49] Gabinete de Planeamento, Políticas e Administração Geral. Análise Sectorial da Apicultura. Portugal: Gabinete de Planeamento, Políticas e Administração Geral; 2020

[50] Regato-Pajares P, Jiménez-Caballero S, Castejón M, Elena-Rosselló R. Recent landscape evolution in Dehesa woodlands of western Spain. In: Recent Dynamics of the Mediterranean Vegetation and Landscape. England: Wiley; 2004. pp. 57-72

Chapter 12

Extensive Farming Systems

Livia Vidu and Dănuț Nicolae Enea

Abstract

Animal husbandry is an ancient occupation that formed the basis of the evolution of the human species. The consumption of animal products has contributed to the development of healthy humans, considering that foods such as milk and eggs are considered complete, providing all the essential amino acids that the human body cannot synthesise. The most interesting extensive animal farming is that related to transhumance, in the case of sheep herds, where the economic and social benefits left their mark on the evolution of rural human societies. This nomadic system led to the spread of the most valuable animal breeds and the development of milk, meat and wool processing techniques, with influences up to the postmodern period.

Keywords: animal, soil, farm, milk, meat

1. Introduction

Currently, animal husbandry is viewed as an inherent and integral aspect of domestic, family-run farms and intensive livestock farms. Discontinuing this practice would have specific implications for human nutrition. Nevertheless, a cursory examination of the historical trajectory reveals that the current state of affairs is the culmination of millennia of collective endeavour, gradually, depending on the species and the region, being influenced both by the economic interest that the species in question has had, but also by its ease of growth.

The first domesticated animal for human consumption was the sheep, the dog was domesticated before the sheep [1], but for other purposes (guarding, defence, hunting) about 14,000–12,000 years ago.

Sheep were domesticated 11,000 years ago [2]. They were domesticated mainly for their meat and wool. Along with sheep, goats were domesticated at the same time. Their ancestors were represented by Asiatic mouflon (*Ovis orientalis*) and Bezoar ibex (*Capra aegagrus*). The domestication occurred in Eastern Anatolia and the Iranian Zagros Mountains [3, 4].

As with sheep and goats, there is archaeological and genomic evidence to suggest that the cattle (*Bos taurus*), that are bred today originated from the domesticated *Bos primigenius primigenius*, which was realised in Fertile Crescent during the Neolithic, 10,000 years ago [5–8]. Further, after approximately 1500 years, a second phase of domestication was successfully completed, in the Indus Valley from *Bos primigenius namadicus*, where from the separation from the taurine branch, zebu cattle were born [9]. A third hypothesis concerning the domestication of the cattle is also put forward.

According to this one, the third domestication took place in Africa, 8000–9000 years ago, which gave rise to the divergent African taurine cattle. In order to substantiate this hypothesis, there is archaeological evidence, which is corroborated by a comparative osteological analysis of ancient wild and domestic cattle from Europe, Asia and Africa [10–14].

With regard to the swine, it can be stated that the domestic pig is a domesticated form of the Eurasian wild boar (*Sus scrofa*) [15]. The earliest evidence of pig domestication is thought to date back approximately 9000 years to the Near East. It is hypothesised that this process may have occurred on multiple occasions, with local wild boar populations being the source of the initial domestication. It is curious to note that the DNA of modern European pigs is derived from European wild boar, rather than Near Eastern animals. This has prompted researchers to hypothesise that pigs may have been domesticated independently in Europe. The results of DNA analysis indicate that the majority of Near Eastern ancestors of European domestic pigs became extinct approximately 3000 years ago as a consequence of gradual interbreeding with European wild boars. Regardless of their origin, we can state with certainty that their flexibility and adaptability have allowed them to survive and prosper in a multitude of habitats, even in areas where humans have introduced new conditions. They have proven to be a valuable source of high-quality meat for humans.

Another species for which there are still questions as to the origin and purpose of its domestication is the horse. The practice of horse breeding, potentially for the purpose of food production, may have commenced in a limited capacity on the European steppes during the early Eneolithic period, circa 5000 BC. This was a period during which horses were first incorporated alongside cattle and sheep [16, 17]. Indirect evidence that horses were domesticated comes from the circular and semicircular corrals that have been found in the Eneolithic settlements of Botai [18, 19]. Textual and artistic evidence of domesticating horses' dates back only to the late third millennium BC [20].

Animal foods are regarded as a rich source of nutrients, particularly protein, due to their unique composition [21]. Additionally, foods of animal origin offer a diverse range of micronutrients that are often challenging to obtain in sufficient quantities from plant-based sources alone [22], such as zinc, iron, vitamin B12, selenium and phosphorus [23, 24]. For example, for an average adult just 100 g of cooked lean beef will provide half of the daily needs of protein, a large quantity of selenium, niacin, iron and zinc, despite contributing only 10% of the daily energy and fat intake [25]. Animal protein foods are a rich source of essential vitamins and minerals that are often lacking in the diets of many people worldwide [25–27]. Consumption of cow's milk compared to beverages of plant origin has been shown to be correlated with increased height in childhood [28]. The consumption of milk by children has been linked to an increase in lean mass, a reduction in body fat [29] and an enhancement in the body's overall protein balance [30], this is of paramount importance in light of the fact that, in the present era, obesity is becoming increasingly prevalent at an early age [28, 31]. The consumption of protein-rich, animal-based diets have been demonstrated to promote satiety, reduce the tendency to engage in emotional eating or food-related reward behaviours, and enhance the overall quality of an individual's diet [32, 33]. Concurrently, a higher consumption of foods of animal origin over a 20-year period has been linked to the preservation of muscle mass and functional performance in older adults [34].

2. Materials and methods

The present manuscript was conceived in the context of the contemporary drive to achieve superior outcomes in a more expeditious manner across a range of disciplines, including animal breeding. The objective was to demonstrate that satisfactory yields can be achieved with minimal human intervention and that there are many meadows and agricultural lands that can be exploited. The objective of this paper is to demonstrate that animals reared extensively can achieve high-quality production. In order to achieve the set goal, an analysis was conducted of the methods of domestication of the main species of economic interest and the ways in which they can be reared extensively. To illustrate the significance of this subject, the advantages of consuming foods of animal origin were also examined.

In order to achieve the aforementioned objectives, an exhaustive bibliographical study of the extant studies on this topic has been conducted.

3. Definition of the extensive animal husbandry system

Extensive animal husbandry requires minimal effort from the farmer, relying mostly on natural or semi-natural grasslands [35]. Given the relatively low challenges of this livestock system, the worldwide spread is quite large, reaching 25% of the earth's surface, from very arid areas to regions with abundant vegetation and good soil quality. This system is also related to the traditions and culture of some peoples.

We must also mention that quite large areas are represented by lands of poor value for agriculture, which can only be exploited by raising animals, especially sheep and goats that can easily reach the most difficult places. According to statistics, approximately 61 million ha are represented by permanent meadows on the surface of Europe. In the USA and Canada, the areas are much larger, with the addition of improved or artificial pastures.

In general, this rearing system is recognised by the fact that it requires low consumption of energy, materials and labour, and the performances achieved by the animals are average.

Growth systems animal have evolved, at the present time, they are classified into extensive, semi-intensive and intensive-industrial, depending on the quality of biological material exploited, the degree of technicality used in exploitation, the composition of the rations used in feeding animal, the age at which they reach the slaughter weight, the size and degree of specialisation of the farms and the rhythmicity of production throughout the year.

4. Peculiarities of growing different species of animals in extensive system

4.1 Breeding of sheep and goats in extensive system

Amongst all domestic animal species, the extensive system is best suited to sheep and goats, due to the morpho-physiological peculiarities of these species.

Shepherding is an old occupation, which by its attributions led to the early days of agriculture. Although it is a hard job, it is practiced all over the globe, primarily to ensure the daily living of the family.

The breeding of sheep in an extensive system, in addition to the economic importance, also has social and scientific importance through the dissemination of the biological material over an area as wide as possible.

Based on the primary products obtained from sheep and goats, real industries have developed in the poorer areas of the world that have provided jobs to local communities (processing of milk into cheeses and fresh products, processing of meat into traditional dishes and products of controlled origin, textile industry, especially carpets based on obtained wool), leather goods industry (processing of raw hides).

A distinctive feature of the extensive sheep farming system refers to the phenomenon of transhumance, that is, the movement of flocks of sheep, in the winter in lowland areas with a milder climate and in the winter in the mountainous regions where the vegetation is rich. Together with these movements of shepherds between the mountains and the plains, an occupation, moreover, attested during the Roman Empire, certain breeds of sheep resistant to long journeys, with the structure of the wool fibre to allow water to drain from the precipitations (also developed Turcana from Romania) or drought-resistant, with fat deposition on the tail (Awassi breed from Israel).

Transhumance has experienced limitations over time, currently taking place in limited areas. Certainly, this system does not require investment in shelters, the feeding is mainly done with volume fodder from the pasture and vegetable residues from agriculture (**Figure 1**).

There are areas of the globe, such as Australia, where the breeding of Merino sheep for wool production is carried out exclusively in an extensive system.

Sheep farming in an extensive system faces major challenges related to the limitations imposed by environmental protection legislation, the possibility of fully or partially using plant resources, the resistance of sheep breeds, but also their lower productivity.

Figure 1.
Sheep grazing (Source: https://www.123rf.com/stock-photo/sheep_farm.html).

In this context, the extensive production system for sheep and goats in the immediate period will suffer major influences from precision livestock farming technologies, which will be introduced into traditional systems as a necessity to cover the shortage of labour in agriculture.

Thus, we will witness the use of remote sensing systems, sensor-technologies that provide information about the health of the animals, the degree of rumination, the monitoring of reproductive activity, but also about the state of the pasture, the height of the grass (different types of thorns). These modern means will contribute to the transformation of pastoral systems for raising small ruminants, with important advantages for farmers. Monitoring herds, respectively each animal through precise livestock farming will contribute to making optimal management decisions.

The basis of feeding dairy cattle during the grazing season is green fodder. It consists of grass.

Grass represents herbaceous plants (annual, perennial), more commonly grasses on meadows.

Green grass is a broader concept, which includes both grass from permanent and temporary meadows, as well as that from the culture of fodder plants (annual or perennial), respectively a mixture of plants from the grass and leguminous families (annual and perennial), as well as different occasional resources.

The grass on the permanent (natural) meadows consists of herbaceous vegetation, with spontaneous growth, often used by grazing. It is composed of plants belonging to different families, especially grasses (50–90%), legumes (0–40%) and other botanical families that have a low weight, which makes the nutritional value variable.

As for the grass on the temporary meadows, it is more valuable, as it comes from mixtures, often optimised of grasses and legumes (60/40%), differentiated by geographical areas.

In general, the nutritional value of grass from permanent or temporary pastures depends on two essential factors, namely:

- The botanical composition where three groups of plants can be distinguished—grasses (they lignify earlier, being richer in energy), legumes (they are richer in protein, energy and calcium) and other botanical families that have a nutritional value, yield and lower digestibility;

- Vegetation stage at use (palatability and corresponding nutritional value have grasses in the sprouting stage and legumes in budding).

In practice, four types of meadows can be found according to the botanical composition: rich in grasses (over 70%); balanced (50–70%—grasses); rich in legumes (50–70%) and rich in other plants (over 50% other plants or plants + legumes). In permanent meadows, the botanical composition should be 65–75% grasses, 20–25% legumes and 5–10% other families, and in temporary meadows 60% grasses and 40% legumes [35].

4.2 Cattle breeding in extensive system

The world agricultural economy is particularly influenced by cattle production; and here we are talking about milk-producing cows, animals for meat, but also about buffaloes with their products. It is known that they have a significant contribution to the total production, thus they contribute more than 96% to the total production of milk, with almost 30% of the meat production and more than 90% of the total skins used in specialised industries.

The extensive system of rearing dairy cows is, in fact, a subsistence one, used especially in traditional households, where most of the time the herds of animals are very small.

We must note a particularly important fact related to the synthesis of milk. If, in beef production, only 20% of the nutritional value of the ingested feed is transformed into weight gain, in the case of milk, the transformation of the nutritional value of the feed into a useful product is, on average, 40%.

An important feature of this system is the feeding of dairy cows, which is based on bulk fodder grazed directly in the summer, and medium-quality hay and other vegetable by-products are used in the winter. The administration of concentrated feed is done especially during the winter, but the quantities are limited. We must also note the housing conditions, which are simple, rustic, without important interior arrangements. Regarding the provision of welfare norms in such conditions, it can be stated that they are provided at a minimal level. Also, the milking technologies used are simple, with a low degree of mechanisation, or even hand milking.

As far as raising dairy cows on organic farms is concerned, this is the superior form of extensive farming, where special attention is paid to both arable and pasture soils. The organic dairy farm is correlated with the fodder resources, ensuring a maximum load of 170 kg N/ha/year, respectively the equivalent of two cows per hectare. Also, this system requires the raising of dairy cows in a free system, with permanent access to the paddock and the use of seasonally differentiated fodder, in the winter from the stock and in the summer on the pasture. In the ration structure, the percentage of silage is limited to 50% and concentrates to 30%.

The biological material used in organic farms consists of hardy animals in good health from local, less improved breeds that also have good productive and biological longevity.

The chemical composition of the milk obtained in the organic system is clearly superior to that obtained in the conventional system, especially in terms of the content of essential fatty acids and vitamins.

Traditional systems of raising beef production impose several conditions that lead to obtaining direct benefits, namely: the existence of cattle breeds specialised for meat production, good and medium-quality pastures and a market for cattle carcasses (**Figure 2**).

The traditional breeding of beef cattle requires that the calves grow alongside the mother cows, utilising both the pasture and the mother's milk until the age of 6–8 months.

Animals raised on pasture in an extensive system reach average weights of 450–600 kg at the age of 22–24 months, with a slaughter yield of over 65%. Average daily gains recorded when reared on pasture are between 900 and 1400 g per day, depending on the quality of the pasture.

Where climatic conditions become unfavourable due to the extension of drought periods, concentrate supplements are used, as well as hay from farm stores.

Breeding beef cattle in an extensive system is an advantageous economic activity because the animals remain on the pasture for long periods of the year (those that are resistant to extreme climatic conditions can also remain during the heatwave, but also during the cold periods), housing is done in simple constructions, from local materials, reproduction is natural, using harem mounting, and activities that require human intervention are very few.

Figure 2.
Cattle from the Aberdeen Angus breed bred in an extensive system (Source: developed by the author Livia Vidu).

In such farms, special attention is paid to the state of the pasture, in the sense that the exaggerated use of some pasture plots can lead to the disappearance of some plants from the floristic composition. That is why grazing is done for specific periods, depending on the number of animals and pedo-climatic conditions. Pastures must be maintained by overseeding, manure spreading and removal of woody vegetation.

Buffaloes are animals that lend themselves excellently to growing in an extensive or organic system. They are animals with a robust constitution, biological longevity of up to 30 years, and a productive one of 20 years, they have resistance to diseases (cases of tuberculosis or foot-and-mouth disease are very rare). As far as feeding is concerned, they make good use of coarse forages and grassy vegetation on the banks of rivers.

Buffaloes are used for traction, especially in tropical and subtropical agriculture, where they participate in approximately 50% of agricultural work and transport on swampy and steep terrain.

In buffaloes, the thermoregulatory mechanism is weaker (the physiological response of buffaloes to climatic conditions through thermolysis and thermogenesis is imperfect), which makes buffaloes react inappropriately to heat and cold. The skin of buffaloes is very thick, with a specific structure (lacking glands sweating, developed sebaceous glands) and pigmented in black, which requires that the pastures have nearby swamps or running water, and where there are no special places for dipping animals, which in very hot weather face itchy skin (**Figure 3**).

Buffaloes have a high capacity to utilise pasture, especially vegetation with low nutritional value, on acid soils.

Between the two species, respectively taurine and buffalo, there are also some differences regarding the morpho-histology of the digestive system. Thus, buffaloes, compared to cows, present the following particularities: greater capacity to ingest fodder by 5–10%; the length of the digestive tube reaches up to 20 times the length of the animal's head; the volume of the rumen and the network in the gastric complex is very large (88%); the papillae of the rumen are more developed in the anterior part

Figure 3.
Immersion of buffaloes in water during summer (Source: developed by the author Livia Vidu).

and more often in the posterior part; ruminal movements are slower, implying a lower exit speed of the digested mass in the rumen.

When evaluating the food requirements of buffaloes, the chemical composition of milk is of particular importance, which presents very high values (17.5% SU, 7.9% fat, 4.9% lactose and 4.0% protein).

The exploitation of buffalo milk obtained in an extensive system is mostly done through traditional preparations.

The processing of buffalo milk ensures different assortments of products specific to each region (acidified products, fresh products—cream, ghee-butter, cheeses). The classification of the types of buffalo milk cheeses is done as follows:

- Soft cheeses (water content over 45%)—Karish, Mish and Damiati (in Egypt); Madhar (Iraq), Mozzarella (Italy); Alghab (Syria) and Vlădeasa (Romania);

- Semi-hard cheeses (water content 40–45%)—Beyaz peymeri (Turkey);

- Hard cheeses (water content below 40%)—Brăila (Romania); Rahss (Egypt); White brine (Bulgaria); Akkavi (Syria) [36, 37].

Regarding the fattening of young buffaloes in the extensive system, average daily gains between 400 and 700 g were obtained, with capitalisation at 28 months at weights between 450 and 500 kg. Regarding organoleptic properties, the meat of the youth is tasty, tender and juicy, and the colour is darker due to the higher iron content.

4.3 Pig breeding in extensive system

Throughout the historical stages, the methods of raising pigs have seen a continuous evolution, from capturing sows with piglets and rearing in pens to automated technologies that monitor all technological parameters with the help of sensors. Thus, for many rural communities, the most common practice is that at the beginning of spring breeding sows and young pigs are left free in forests, marshy areas and swamps. The feeding of these animals is done with herbs, fruits and tree seeds, without any human intervention. As far as reproduction is concerned, mating is done freely with boars from the herd or with wild boars. The animals shelter themselves from the extreme weather, surviving without the intervention of man, who towards the end of autumn collects the pigs (pigs, sows and piglets born in the meantime), brings them to the shelter, and fattens them for a short period of time, in order to slaughter them. This system is still practiced in many countries, especially in the Balkan area (**Figure 4**).

The extensive system of growth of pigs has been applied since domestication, is characterised by the maintenance of pigs in simple and cheap shelters, without inside arrangements, provided with paddocks, without possibilities to regulate the internal temperature, with watering and feeding in exchange and with the manual evacuation of droppings. Pigs from local breeds are exploited, late and with low performance, but resistant to environmental, feeding and exploitation conditions. Pigs from improved breeds or their crosses with local breeds can also be used, but in this case, the productive potential of the animals is only partially exploited.

This system of growth is applied in some households traditionally, where 1–2 pigs are fattened per year, for own consumption. The best results are obtained when seasonal breeding is applied, with sows mounted in the fall for farrowing in the spring, with piglets weaned in late March and early April. The feeding of the fattened pigs in the extensive system is based on local fodder such as grass harvested by grazing directly from the pasture, green mass administered at the barter or in the paddock, parcels of beet, nettles, alfalfa, clover, sweet potatoes, potatoes downgraded from human consumption, beets, beets, kitchen scraps and occasionally acorns, whey, buttermilk, etc. Small amounts of concentrated feed are added to them.

The economic efficiency of this breeding system is reduced, due to the long period of time in which the slaughter weight is achieved (12–16 months), due to the high consumption of fodder to obtain the growth and the lower quality of the carcass.

Figure 4.
Bazna breed—extensive breeding system (Source: developed by the author Livia Vidu).

In general, the animals are slaughtered at the beginning or during the winter, the pigs reaching weights of 120–130 kg or even more.

The modern world today is divided in terms of pig farming, from the classical conventional systems with total confinement of the animals to the alternative systems, with variants of open-air farming, silvopastoral farming and organic farming.

Studies carried out over time have revealed that more than 60% of pork consumers in France consider it important that animals have access to the outdoors, and 44% of them look with concern at the surface provided to each animal.

It is also interesting that the modern man is willing to pay more for pork obtained in a pasture system compared to the classic closed system. Thus, the study shows that Americans are willing to pay 2.02 USD more for this meat.

At the same time, we notice that the incidence of some diseases, such as respiratory and locomotor diseases, in pigs raised in an extensive system, respectively in the open air, is lower.

5. Conclusions

Extensive farming systems represent the easiest means of raising animals, regardless of the geographical area and the financial power of the farmer. They can also provide secure sources of income for isolated rural areas of the world.

Regarding the future of extensive farming systems, we believe that they will turn into systems partially dependent on precision livestock farming.

Acknowledgements

Financial support for the publication of this chapter is provided by the University of Agronomic Sciences and Veterinary Medicine from Bucharest.

Author details

Livia Vidu* and Dănuț Nicolae Enea
University of Agronomic Sciences and Veterinary Medicine from Bucharest, Bucharest, Romania

*Address all correspondence to: liviavidu@gmail.com

References

[1] Turnbull PF, Reed CA. The terminal pleistocene fauna of palegawra cave, a Zardian occupation site in northeastern Iraq. Fieldiana Anthropology. 1974;**63**(3):81-146

[2] Hartung J. A brief history of animal production. In: Animal Shelter. Academic Wageningen; 2013. pp. 21-34

[3] Rezaei HR. Phylogénie moleculaire du Genre Ovis (Mouton et Mouflons); Implications pour la conservation du genre et pour l'origine de l'espèce domestique [PhD thesis]. Univ. Joseph Fourier, Grenoble I; 2007

[4] Naderi S et al. The domestication process of goats inferred from large-scale mitochondrial DNA analysis of wild and domestic individuals. Proceedings of the National Academy of Sciences of the United States of America. 2008;**105**:17659-17664

[5] Pitt D, Sevane N, Nicolazzi EL, et al. Cattle domestication: Two or three events? Evolutionary Applications. 2019;**12**:123-136. DOI: 10.1111/eva.12674

[6] Bruford MW, Bradley DG, Luikart G. DNA markers reveal the complexity of animal domestication. Nature Reviews Genetics. 2003;**4**(11):900-910. DOI: 10.1038/nrg1203

[7] Ajmone-Marsan P, Garcia JF, Lenstra JA. On the origin of cattle: How cattle became and colonized the world. Evolutionary Anthropology: Issues, News, and Reviews. 2010;**19**(4):148-157. DOI: 10.1002/evan.20267

[8] MacHugh DE, Larson G, Orlando L. Taming the past: Ancient DNA and the study of animal domestication. Annual Review of Animal Biosciences. 2017;**5**:329-351. DOI: 10.1146/annurev-animal-022516-022747

[9] Loftus RT, MacHugh DE, Bradley DG, Sharp PM, Cunningham P. Evidence for two independent domestications of cattle. Proceedings of the National Academy of Sciences of the United States of America. 1994;**91**:2757-2761. DOI: 10.1073/pnas.91.7.2757

[10] Applegate A, Gautier A, Duncan S. Northern mounds of the Late Neolithic Nabta ceremonial complex. In: Wendorf F, Schild R, Associates, editors. Settlement of the Egyptian Sahara. Volume 1: The Archeology of Nabta Playa. New York, NY: Kluwer; 2001. pp. 468-488. DOI: 10.1007/978-1-4615-0653-9

[11] Grigson C. An African origin for African cattle? Some archaeological evidence. African Archaeological Review. 1991;**9**:119-144. DOI: 10.1007/BF01117218

[12] Grigson C. Bos africanus (Brehm)? Notes on the Archaeozoology of Native African Cattle. In: Blench RM, MacDonald KC, editors. The Origins and Development of African Animals. Archaeology, Genetics, Linguistics and Ethnography. London, UK: UCL Press; 2000. pp. 38-60

[13] Stock F, Gifford-Gonzalez D. African cattle genetics and domestication. African Archaeological Review. 2013;**30**:51-72. DOI: 10.1007/s10437-013-9131-6

[14] Bradley DG, MacHugh DE, Cunningham P, Loftus RT. Mitochondrial diversity and the origins of African and European cattle. Proceedings of the National Academy of Sciences.

1996;**93**(10):5131-5135. DOI: 10.1073/pnas.93.10.5131

[15] Giuffra E, Kijas JM, Amarger V, Carlborg O, Jeon JT, Andersson L. Origin of the domestic pig: Independent domestication and subsequent introgression. Genetics. 2000;**154**(4):1785-1791. DOI: 10.1093/genetics/154.4.1785

[16] Brown D, Anthony D. Bit wear, riding and the Botai site of Kazakstan. Journal of Archaeological Science. 1998;**25**(4):331-347

[17] Anthony DW, Brown DR. Eneolithic horse exploitation in the Eurasian steppes: Diet, ritual and riding. Antiquity. 2000;**74**(283):75-86

[18] Olsen SL. Early domestication of horses in the Eurasian steppe. In: Documenting Domestication: New Genetic and Archaeological Paradigms. 2006. pp. 245-269

[19] Stiff AR, Capo RC, Gardiner JB, Olsen SL, Rosenmeier MF. Geochemical evidence of possible horse domestication at the copper age botai settlement of Krasnyi Yar, Kazakhstan. In: At the Annual Meeting in Philadelphia. 2006. Available from: http://gsa.confex.com/gsa/2006AM/finalprogram/abstract_116407.htm

[20] Levine MA. The domestication and early history of the horse. In: The Domestic Horse: Its Behavioral Origins, Development, and Management. 2005. pp. 5-22

[21] Sheffield S, Fiorotto ML, Davis TA. The nutritional importance of animal foods in a healthy diet. Frontiers in Nutrition. 2024;**11**:1424912

[22] Murphy SP, Allen LH. The nutritional importance of foods of animal origin. Journal of Nutrition. 2003;**133**(11):3932S-3935S

[23] US Department of Agriculture (USDA). Laboratory Methods and Application of Food Composition. 2020. Available from: https://www.ars.usda.gov/northeast-area/beltsville-md-bhnrc/beltsville-human-nutrition-research-center/methods-and-application-of-food-composition-laboratory/mafcl-site-pages/sr28-page-reports/ [Accessed: August 11, 2024]

[24] Phillips SM, Fulgoni VL, Heaney RP, Nicklas TA, Slavin JL, Weaver CM. Commonly consumed protein foods contribute to nutrient intake, diet quality, and nutrient adequacy. The American Journal of Clinical Nutrition. 2015;**101**:1346S-1352S. DOI: 10.3945/ajcn.114.084079

[25] Klurfeld DM. Research gaps in assessing the relationship between meat and health. Meat Science. 2015;**109**:86-95. DOI: 10.1016/j.meatsci.2015.05.022

[26] Neumann C, Harris DM, Rogers LM. Contribution of animal foods to improving diet quality and function in children in the developing world. Nutrition Research. 2002;**22**:193-220. DOI: 10.1016/S0271-5317(01)00374-8

[27] Adesogan AT, Havelaar AH, McKune SL, Eilittä M, Dahl GE. Food of animal origin: Sustainability problem or malnutrition and sustainability solution? Perspective matters. Global Food Security. 2020;**25**:100325. DOI: 10.1016/j.gfs.2019.100325

[28] Fryar CD, Carroll MD, Affu J. Prevalence of overweight, obesity, and severe obesity among children and adolescents aged 2-19 years: United States, 1963-1965 to 2017-2018. In: NCHS Health E-Stats. 2020. Available from:

https://www.cdc.gov/nchs/data/hestat/obesity-child-17-18/overweight-obesity-child-H.pdf [Accessed: August 11, 2024]

[29] Karagounis LG, Volterman KA, Breuillé D, Offord EA, Emady-Azar S, Moore D. Protein intake at breakfast promotes positive whole-body protein balance in a dose-dependent manner in healthy children: A randomized trial. The Journal of Nutrition. 2019;**148**:729-737. DOI: 10.1093/jn/nxy026

[30] Kang K, Sotunde OF, Weiler HA. Effects of milk and milk product consumption on growth among children and adolescents aged 6 to 18 years: A meta-analysis of randomized controlled trials. Advances in Nutrition. 2019;**10**:250-261. DOI: 10.1093/advances/nmy081

[31] World Health Organization (WHO). Obesity and Overweight. 2024. Available from: https://www.who.int/news-room/fact-sheets/detail/obesity-and-overweight [Accessed: August 11, 2024]

[32] Paddon-Jones D, Leidy H. Dietary protein and muscle in older people. Current Opinion in Clinical Nutrition and Metabolic Care. 2014;**17**:5-11. DOI: 10.1097/MCO.0000000000000011

[33] Leidy HJ, Ortinau LC, Douglas SM, Hoertel HA. Beneficial effects of a high-protein breakfast on appetitive, hormonal, and neural signals controlling the regulation of energy intake in "breakfast-skipping" overweight/obese late adolescent girls. The American Journal of Clinical Nutrition. 2013;**97**:677-688. DOI: 10.3945/ajcn.112.053116

[34] Hruby A, Sahni S, Bolster D, Jacques PF. Protein intake and functional integrity in aging: Follow-up from the Framingham Heart Study. The Journals of Gerontology. Series A, Biological Sciences and Medical Sciences. 2020;**75**:123-130. DOI: 10.1093/gerona/gly201

[35] Georgescu G, Vidu L. Monograph on the Growth of Buffaloes Worldwide and in Romania. Bucharest: Ceres Publishing House; 2008

[36] Borghese A. Buffalo Production and Research. Rome, Italy: Instituto Sperimentale per la Zootecnia; 2005

[37] Borghese A. Buffalo Livestock and Products. Rome, Italy: Instituto Sperimentale per la Zootecnia; 2013

Section 8

Adaptation to Changing Environment

Chapter 13

Sheep Migration Patterns and Behavioral Adaptations in Indian Grazing Lands: Understanding Stress Factors and Environmental Adaptation

Vinod Bhateshwar, Basant Kumar Bhinchhar, Hitesh Muwal and Paramveer Palriya

Abstract

Sheep play a crucial role in India's pastoral economy, navigating vast grazing lands through complex migration patterns shaped by diverse environmental stressors such as temperature extremes, forage variability, and predation risks. GPS tracking and behavioral observations reveal that these animals employ adaptive strategies such as social cohesion, herd dynamics, and individual learning to mitigate these challenges and optimize resource use. Sheep display flexible foraging behaviors and selective grazing, adjusting to fluctuating conditions. Stress markers, like cortisol levels, further highlight the impact of these environmental factors on sheep well-being and resilience. This study integrates ecological and ethological perspectives to explore how environmental stressors influence migration patterns and behavioral adaptations among sheep populations in India. The findings emphasize the importance of understanding these dynamics for sustainable pastoral management and conservation efforts. By uncovering the mechanisms of stress resilience and adaptation, the research offers valuable insights into the ecological functioning of grazing ecosystems. It informs strategies to protect sheep welfare and pastoral livelihoods in the face of environmental change.

Keywords: behavioral adaptations, grazing lands, India, sheep migration, stress

1. Introduction

Sheep migration patterns and behavioral adaptations play a crucial role in their survival and productivity within Indian grazing lands [1]. India's diverse topography, ranging from the Himalayan foothills to the vast plains and coastal regions, offers a variety of habitats for sheep populations to thrive [2]. Understanding the intricacies of their migration patterns and behavioral adaptations is essential for comprehending their responses to stress factors and environmental changes. Sheep, primarily raised for wool and meat production, exhibit complex migration behaviors influenced by factors such as food availability, climatic variations, and human activities [3]. In India, where a significant portion of sheep husbandry occurs in extensive grazing systems, the movement of flocks across different landscapes is a common sight [4]. These migration patterns not only are driven by the search for adequate forage but also serve as a means to regulate exposure to environmental stressors.

The behavioral adaptations of sheep further enhance their resilience to the challenges posed by their environment. From flock dynamics to individual responses, these adaptations reflect a fine-tuned balance between survival instincts and learned behaviors [5]. For instance, sheep exhibit hierarchical structures within their flocks, where dominant individuals often lead the group to preferred grazing areas while subordinate members follow [6]. Such social dynamics help in efficient foraging and protection against predators. Moreover, sheep display remarkable physiological adaptations to cope with environmental stressors [7]. Their ability to conserve water in arid regions, tolerance to temperature fluctuations, and capacity to digest a wide range of vegetation highlights their evolutionary flexibility [8]. Additionally, sheep demonstrate behavioral plasticity, adjusting their grazing strategies in response to changing environmental conditions. For example, they may exhibit crepuscular grazing behavior, feeding during the cooler hours of dawn and dusk to avoid midday heat stress [9]. However, despite these adaptations, sheep populations in India face numerous stress factors that challenge their survival and productivity. Anthropogenic activities such as habitat degradation, competition for resources with other livestock species, and encroachment of grazing lands further exacerbate these challenges [10]. Understanding the interplay between migration patterns, behavioral adaptations, and stress factors is crucial for devising sustainable management strategies to ensure the welfare of sheep populations and the ecosystems they inhabit.

In this chapter, we aim to explore the intricate relationship between sheep migration patterns, behavioral adaptations, and environmental stress factors in Indian grazing lands. By synthesizing existing literature and empirical evidence, we seek to provide insights into the mechanisms driving sheep behavior and their responses to environmental changes. Ultimately, this understanding will contribute to the development of informed conservation and management practices that promote the resilience and sustainability of sheep populations in India's diverse landscapes.

2. Methodology

This chapter examines methodologies for studying sheep migration patterns and behavioral adaptations in Indian grazing lands using online data collection methods. Researchers utilize online surveys and virtual interviews with shepherds and farmers *via* digital platforms to gather qualitative insights on migration patterns and grazing practices. Remote sensing and satellite imagery, accessed through tools like Google Earth Engine, monitor grazing areas, and environmental conditions climate data from sources such as NASA's Earth data and NOAA are used to correlate environmental factors with sheep behavior, while social media and this approach offers a comprehensive framework for understanding stress factors and adaptive strategies in sheep, contributing to sustainable grazing practices and improved animal welfare.

2.1 Sheep migration patterns in India

Sheep migration patterns in India can vary depending on climate, geography, and human activity. Generally, migrations often follow seasonal patterns in regions where sheep herding is common. Here is an overview:

2.1.1 Himalayan region

Sheep herding is a traditional practice in the Himalayan region, particularly in states like Himachal Pradesh, Uttarakhand, and Jammu & Kashmir. During the summer months, shepherds move their flocks to high-altitude grazing lands as the lower regions become too hot [11]. This migration pattern, known as transhumance, allows sheep to graze on fresh vegetation in the alpine meadows. The migration pattern of sheep in the Himalayan region can vary depending on factors such as climate, availability of pasture, and traditional herding practices of the local communities [12]. However, there are some general trends observed:

2.1.1.1 Seasonal migration

Sheep in the Himalayan region often engage in seasonal migration, moving to higher altitudes during the summer months and descending to lower elevations during the winter [13]. This migration pattern is influenced by the availability of fresh pasture at higher elevations during the warmer months and the need to avoid harsh winter conditions at lower altitudes.

2.1.1.2 Altitudinal migration

In mountainous regions like the Himalayas, sheep may migrate vertically, moving between different altitudinal zones in search of suitable grazing areas [14]. They may ascend to alpine meadows and high pastures during the summer and descend to lower slopes or valleys in the winter.

Transhumance: Transhumance is a traditional practice of seasonal movement of livestock between fixed summer and winter pastures [15]. In the Himalayan region, transhumance is common among pastoral communities that follow a seminomadic lifestyle. These communities may move their sheep along established migration routes, taking advantage of the available resources at different times of the year [11].

2.1.1.3 Herding practices

The migration pattern of sheep in the Himalayan region is often guided by herders who know the local terrain and grazing areas [16]. Herders may lead their flocks along well-established routes, which may have been used for generations.

2.1.1.4 Environmental factors

Environmental factors such as snowfall, rainfall, and the condition of grazing lands can also influence the migration pattern of sheep [17]. Unpredictable weather patterns or natural disasters like avalanches may alter migration routes or timing [18].

2.1.1.5 Conservation concerns

Human activities, including infrastructure development, deforestation, and overgrazing, can disrupt traditional migration patterns and impact the availability of grazing lands for sheep [19]. Conservation efforts often aim to preserve these migratory routes and maintain the ecological balance of the Himalayan region [20]. Overall, the migration pattern of sheep in the Himalayan region is a complex interplay of ecological, cultural, and economic factors, shaped by centuries of coexistence between humans and livestock in these rugged landscapes.

2.1.2 Rajasthan and Gujarat

In the western desert regions of Rajasthan and parts of Gujarat, sheep herding communities, such as the Rabari and Bharvad tribes, practice nomadic pastoralism. Their migration patterns are influenced by the availability of water and forage. During the dry season, they move to areas where water sources are more reliable and vegetation is still available [21]. The migration pattern of sheep in Rajasthan and Gujarat is primarily influenced by the availability of grazing lands, water sources, and seasonal climate changes. In these regions, especially in arid and semiarid areas, sheep herding is a traditional occupation, and many communities depend on it for their livelihood [22]. During the dry season, which typically lasts from October to June, sheep herders move their flocks to areas with better grazing opportunities and access to water [23]. They often travel long distances in search of suitable pastures and water sources. This movement is essential to ensure that the sheep have enough food and water to sustain themselves during the harsh conditions of the dry season.

In Rajasthan, for example (**Figure 1**), sheep herders often migrate to regions such as the Thar Desert and the Aravalli Range during the dry season [24]. These areas

Figure 1.
Sheep herders migrate to regions such as the Aravalli range during the dry season. Source: Developed by the author Dr. Vinod Bhateshwar.

offer some vegetation even during the dry months, and herders may rely on wells or other water sources to provide for their flocks. Similarly, in Gujarat, sheep herders migrate to regions such as the Rann of Kutch and the Gir forest during the dry season. These areas have grasslands and scrub vegetation that can support grazing, and herders may also utilize water sources such as rivers or artificial ponds. During the monsoon season, which typically lasts from July to September, the migration pattern may reverse as herders return to their home villages or settlements. The availability of rainwater and the regeneration of pastures during this time provided sufficient resources for the sheep closer to their home bases [25]. Overall, the migration pattern of sheep in Rajasthan and Gujarat is a cyclical process that allows herders to optimize the use of available resources and ensure the well-being of their flocks throughout the year.

2.1.3 Central India

In states like Maharashtra, Madhya Pradesh, and Chhattisgarh, where there are vast grasslands and agricultural areas, sheep herding is also practiced. Here, migrations may not be as pronounced as in the Himalayas or desert regions, but shepherds may still move their flocks in search of better grazing lands, especially during the monsoon season when grasslands are more abundant. Sheep migration patterns in Central India can vary depending on factors such as seasonal changes, availability of grazing areas, and human activity [26]. However, there is not a standardized migration pattern for sheep in Central India as there might be in regions where transhumance (seasonal movement of livestock) is more common.

In general, sheep in Central India may move in search of suitable grazing grounds and water sources, especially during dry seasons when pastures become scarce. They might migrate to higher elevations during hotter months to escape the heat and descend to lower areas during colder months for better forage [3]. Local herding practices, land use patterns, and environmental conditions play significant roles in shaping these migration patterns. Additionally, factors like agricultural activities, urbanization, and the presence of predators can also influence the movement of sheep herds in the region. The dynamic nature of sheep migration in Central India, it would be beneficial to consult with local shepherds, livestock experts, or research studies for more specific information about particular areas and seasons.

2.1.4 Coastal regions

Coastal areas such as parts of Karnataka, Kerala, and Tamil Nadu also have sheep herding communities. These communities may not have extensive migration patterns like those in the Himalayas or deserts, but they may move their flocks within coastal areas based on seasonal changes in vegetation and weather patterns.

2.1.5 Human influence

In recent years, urbanization, land development, and changes in land use have impacted traditional sheep migration routes in many parts of India [27]. As a result, some migration patterns may have become more restricted or altered.

Human influence can have significant effects on sheep migration in India. Here are some factors to consider:

2.1.5.1 Habitat loss and fragmentation

Human activities such as agriculture, urbanization, and infrastructure development can lead to the loss and fragmentation of natural habitats for sheep [28]. This can disrupt traditional migration routes and force sheep to alter their movements or inhabit suboptimal habitats.

2.1.5.2 Fencing and barriers

The construction of fences, roads, and other barriers can impede the natural movement of sheep herds [28]. These barriers can disrupt migration patterns, isolate populations, and lead to conflicts with humans when sheep attempt to navigate around them.

2.1.5.3 Land use changes

Changes in land use patterns, such as the conversion of grasslands to croplands or urban areas, can reduce available grazing areas for sheep [29]. This can force them to migrate longer distances in search of suitable forage, leading to increased competition with other wildlife or domestic livestock.

2.1.5.4 Human-wildlife conflict

As human populations expand into sheep habitats, conflicts between humans and sheep can arise [30]. This may result in retaliatory killings of sheep by farmers or shepherds, especially if sheep damage crops or compete with livestock for resources.

2.1.5.5 Climate change

While not directly caused by humans, climate change, largely driven by human activities, can significantly impact sheep migration patterns. Changes in temperature, precipitation patterns, and the availability of water and forage can alter the timing and routes of sheep migrations, leading to challenges in adaptation for both sheep and the communities that depend on them [31].

2.1.5.6 Overgrazing and resource depletion

Unregulated grazing by domestic livestock, including sheep, can lead to overgrazing and depletion of vegetation in certain areas [32]. This can disrupt natural ecosystems, reduce biodiversity, and exacerbate soil erosion and desertification, affecting both sheep migration patterns and the overall health of the landscape [33].

Addressing these issues requires a holistic approach that considers the needs of both humans and wildlife. It involves sustainable land management practices, conservation efforts to preserve critical habitats, community-based natural resource management initiatives, and policies that promote coexistence between humans and wildlife. By mitigating human impacts and promoting sustainable practices, we can help ensure the long-term viability of sheep migrations in India and the ecosystems they inhabit.

2.2 Behavioral adaptations

Migration can significantly impact the behavioral adaptations of sheep in India, especially considering the diverse environments and cultural practices across the country. Here are some potential effects:

2.2.1 Forced migration and stress

In regions where forced migration is due to environmental factors like drought or flooding, sheep may experience heightened stress levels. This can lead to changes in their behavior, such as increased aggression or fearfulness [4]. Forced migration and stress can have significant effects on sheep behavior, as they do in many other animal species. Sheep are social animals, and forced migration can disrupt their social structures [6]. When groups are forcibly moved or combined with other groups, it can lead to conflicts as they establish new hierarchies and territories. Stress can exacerbate these conflicts, leading to aggressive behaviors within the flock [34].

Sheep may become more aggressive or fearful when exposed to stressors associated with forced migration (**Figure 2**). This can manifest as increased aggression toward other sheep or humans, or as heightened vigilance and fearfulness in response to perceived threats. Stress can affect sheep's appetite and feeding behavior. Forced migration may disrupt their access to familiar grazing areas or food sources, leading to reduced feeding and weight loss [4]. Additionally, stressed sheep may exhibit decreased interest in feeding or spend less time grazing due to

Figure 2.
The flock of sheep was exposed to stressors associated with long-distance forced migration. Source: Developed by the author Dr. Vinod Bhateshwar.

heightened anxiety or discomfort. Stress can impact reproductive behaviors in sheep, including mating and maternal care [35]. Forced migration may disrupt breeding cycles or cause ewes to exhibit decreased receptivity to rams. In pregnant ewes, stress can increase the risk of pregnancy loss or result in reduced maternal care behaviors toward offspring [36]. Forced migration can disrupt sheep's natural movement patterns and migration routes. Stressed sheep may exhibit erratic or aimless movement behaviors as they attempt to cope with the stress of the migration process [37]. This can result in disorganized flock movement and difficulty in navigating unfamiliar terrain.

Stress weakens the immune system, making sheep more susceptible to diseases and infections [38]. Forced migration can expose sheep to new pathogens or environmental stressors, further compromising their health and increasing the risk of disease outbreaks within the flock [31]. The effects of forced migration and stress on sheep behavior may persist beyond the immediate migration period. Chronic stress can lead to long-term changes in behavior, including altered stress responses, increased fearfulness, and reduced resilience to future stressors [39].

Overall, forced migration and stress can have multifaceted effects on sheep behavior, influencing social dynamics, feeding and grazing behavior, reproductive patterns, movement patterns, susceptibility to disease, and long-term behavioral outcomes. Proper management practices and measures to minimize stress can help mitigate these effects and promote the welfare of migrating sheep populations.

2.2.1.1 Adaptation to climate

Sheep migration can have both positive and negative effects on adaptation to climate, depending on various factors such as the specific characteristics of the migration, the local ecosystem, and management practices [40]. Sheep that migrate to higher altitudes during the summer to escape the heat in the plains may exhibit behavioral adaptations such as seeking cooler areas during the day and grazing at night [41]. They may also develop a higher tolerance for cold temperatures. Sheep migration can lead to more efficient utilization of pasture resources, allowing areas to recover and regenerate, which can enhance ecosystem resilience to climate variations [42]. Migration may expose sheep to a wider variety of forage types, which can promote better nutrition and resilience to environmental changes. Through migration, sheep populations may encounter diverse environmental conditions, which can drive natural selection for traits that enhance adaptation to local climates [43].

However, uncontrolled or poorly managed migration can lead to overgrazing in certain areas, exacerbating soil erosion, reducing vegetation cover, and degrading habitats [33]. This can hinder ecosystem adaptation to climate change. Migration can facilitate the spread of diseases between sheep populations, especially if they move between areas with different disease pressures [44]. Climate change may also alter disease dynamics, potentially increasing the risk of transmission. In some cases, sheep migration routes may overlap with habitats of native wildlife. This can lead to competition for resources and disrupt natural ecosystems, affecting the ability of native species to adapt to changing climates.

In management strategies implementing rotational or managed grazing practices can help mitigate the negative impacts of overgrazing associated with migration. Regular monitoring of sheep health, combined with appropriate disease prevention and control measures, can help minimize the spread of diseases during migration [45]. Restoration efforts in areas impacted by overgrazing can enhance ecosystem resilience and support adaptation to climate change. Engaging with local communities, landowners, and stakeholders to develop collaborative management plans for sheep migration can help balance the needs of both human communities and natural ecosystems [46]. The various factors related to sheep migration and their impact on climate adaptation, highlighting management practices, environmental conditions, and interactions between sheep and local ecosystems are summarized in **Table 1**.

Overall, sheep migration's impact on climate adaptation can vary widely depending on management practices, environmental conditions, and the interactions between sheep and local ecosystems. Sustainable management strategies that promote ecosystem health and resilience are crucial for ensuring that sheep migration contributes positively to adaptation efforts.

S. No.	Factor	Description	Impact on climate adaptation
1.	Grazing management	Rotational grazing, stocking rates, grazing duration, and rest periods	Promotes plant regrowth, soil health, and biodiversity, enhancing ecosystem resilience
2.	Water management	Availability and management of water sources for sheep	Ensures sustainable water use, supports plant and animal life, and reduces overgrazing near water sources
3.	Soil health	Soil composition, erosion control, and nutrient cycling	Maintains soil fertility and structure, prevents erosion, and enhances carbon sequestration
4.	Vegetation types	Diversity of plant species, forage quality, and seasonal availability	Supports diverse diets for sheep, reduces reliance on single forage type, and improves ecosystem stability
5.	Climate conditions	Temperature, precipitation patterns, and extreme weather events	Influences migration timing and routes, affects forage availability, and requires adaptive management
6.	Ecosystem interactions	Relationships between sheep, predators, other herbivores, and plant communities	Balances grazing pressure, promotes natural predator-prey dynamics, and supports biodiversity
7.	Habitat fragmentation	Impact of human activities, land use changes, and barriers to migration	Limits migration routes, reduces access to forage and water, and necessitates connectivity conservation
8.	Traditional knowledge	Indigenous and local knowledge of migration patterns, grazing practices, and environmental indicators	Enhances adaptive management, integrates sustainable practices, supports cultural heritage
9.	Technological integration	Use of GPS tracking, remote sensing, and IoT devices for monitoring sheep and environmental conditions	Provides real-time data for informed decision-making, improves efficiency, reduces overgrazing
10.	Policy and regulation	Government policies, subsidies, and regulations related to grazing and land use	Shapes management practices, encourages sustainable practices, mitigates negative environmental impacts

Source: Developed by the authors Dr. Vinod Bhateshwar and Basant Kumar Bhinchhar.

Table 1.
Factors affecting sheep migration and their impact on climate adaptation.

2.2.1.2 Social dynamics

The effect of migration on sheep social dynamics can vary depending on several factors, including the species of sheep, the type of migration, and environmental conditions. Migration can impact the cohesion of sheep groups. During migration, sheep often move in herds, which can strengthen social bonds among individuals within the group [47]. Migration may influence the establishment or reinforcement of social hierarchies within sheep herds. Dominant individuals may assert their leadership during migration, guiding the group and making decisions about when and where to move. Migration can be a stressful experience for sheep, especially if they encounter unfamiliar environments or predators along the migration route [48]. This stress can lead to increased aggression within the group as individuals compete for resources or jostle for positions within the herd hierarchy.

Migration may impact the communication patterns of sheep. During migration, sheep may use vocalizations, body language, and scent marking to maintain contact with group members, navigate their environment, and coordinate movement [49]. Migration provides opportunities for social learning among sheep. Younger individuals may learn migration routes and behaviors from older, more experienced group members, contributing to the transmission of knowledge within the herd [50]. Migration allows sheep to access different grazing areas and seasonal resources. The movement of sheep herds across landscapes can influence vegetation dynamics and ecosystem processes, shaping the availability and distribution of resources for both sheep and other species [51]. Migration can facilitate genetic mixing among sheep populations. As herds move between different areas, they may encounter individuals from other groups or populations, leading to gene flow and genetic diversity within the overall sheep population [52]. Overall, migration plays a significant role in shaping the social dynamics of sheep herds, impacting their behavior, communication, and interactions with both their environment and other individuals within the group.

2.2.1.3 Feeding behavior

Sheep migration can significantly impact their feeding behavior due to various factors such as the availability of forage, competition with other animals, environmental conditions, and the physiological state of the sheep themselves [41]. During migration, sheep may encounter different grazing areas with varying vegetation types and densities. This can influence their feeding behavior as they adapt to the available forage. They may spend more time searching for suitable food sources or adjust their diet based on what is accessible in each location [34]. In areas where multiple herds or species of animals migrate together, there can be increased competition for food resources. Sheep may need to alter their feeding behavior to compete effectively with other grazers, such as by grazing at different times of the day or selecting less-preferred forage if preferred options are scarce (**Figure 3**) [51]. Sheep have specific nutritional requirements based on factors like age, reproductive status, and stage of production. Migration may affect the availability of forage that meets these nutritional needs, leading to changes in feeding behavior as sheep prioritize obtaining essential nutrients [53].

Sheep are social animals and often exhibit flocking behavior, especially during migration [6]. Social dynamics within the flock can influence feeding behavior, with dominant individuals accessing preferred forage more readily or influencing the movement patterns of the entire group. Environmental factors such as weather, terrain, and water availability can impact sheep migration routes and grazing patterns.

Figure 3.
During migration grazing at different times of the day or selecting less-preferred forage when sheep-preferred options are scarce. Source: Developed by the author Dr. Vinod Bhateshwar.

Extreme weather conditions or challenging terrain may necessitate changes in feeding behavior as sheep adapt to the environment to meet their nutritional requirements [54]. During migration, sheep may encounter increased predation risk, especially if they move through areas inhabited by predators. This can influence their feeding behavior, causing them to be more vigilant or selective in their foraging to minimize the risk of predation. The distance and duration of migration can affect sheep feeding behavior. Long-distance migrations may require sheep to prioritize energy conservation, leading to adjustments in feeding patterns, while shorter migrations may allow for more consistent grazing behavior [55]. The factors influencing sheep migration and their impact on feeding behavior are listed in **Table 2**.

Overall, the effects of sheep migration on feeding behavior are complex and multifaceted, influenced by a combination of ecological, social, and physiological factors. Understanding these dynamics is important for managing grazing lands and ensuring the welfare and productivity of sheep populations.

2.2.1.4 Predator avoidance

Migration can expose sheep to different predators along their routes. This may lead to changes in vigilance behavior, with sheep becoming more cautious or developing strategies to avoid predation, such as traveling in larger groups or utilizing terrain features for protection [48]. When sheep migrate in large groups, there is a dilution effect where individual sheep have a lower risk of predation because predators are less likely to focus on any one individual. Predators may find it difficult to single out and attack prey amidst a large moving flock [56]. Sheep tend to be more vigilant during

S. No.	Factor	Description	Impact on feeding behavior
1.	Seasonal changes	Variations in temperature, precipitation, and forage availability across seasons	Alters the availability and quality of forage, influencing grazing patterns and diet diversity
2.	Grazing pressure	Intensity and duration of grazing in specific areas	This can lead to overgrazing, depletion of preferred forage species, and shifts to less desirable forage
3.	Water availability	Proximity and accessibility of water sources	Affects grazing locations and movement patterns, with sheep staying closer to water sources
4.	Habitat fragmentation	Barriers such as fences, roads, and human settlements	Limits access to diverse grazing areas, reducing diet variety and potentially overgrazing certain areas
5.	Forage diversity	The presence of multiple plant species and types	Promotes a varied diet, allowing sheep to balance nutrient intake and avoid toxic plants
6.	Topography	Terrain features such as hills, valleys, and plains	Influences grazing efficiency and energy expenditure, with sheep preferring certain terrains for feeding
7.	Predator presence	Risk of predation from wild animals	This may cause sheep to avoid certain areas or change grazing times to safer periods
8.	Human activities	Agricultural practices, land use changes, and human disturbances	Can disrupt grazing patterns, force migrations, and alter feeding behavior
9.	Social structure	Flock dynamics, leadership, and social interactions within the herd	Determines movement patterns and grazing site selection, with dominant sheep influencing flock behavior
10.	Disease and parasite load	Presence of diseases and parasitic infestations	Can reduce grazing efficiency and alter feeding behavior due to health impacts

Source: Developed by the authors Dr. Vinod Bhateshwar and Basant Kumar Bhinchhar.

Table 2.
Various factors affect sheep migration and their impact on feeding behavior.

migration, as they are exposed to a higher risk of predation. This heightened vigilance can make it easier for them to detect predators and take evasive action.

Migration often involves herding behavior, where sheep stick together closely. This group behavior can serve as a form of defense against predators, as larger groups are more intimidating and can collectively defend against attacks [57]. Migration may lead sheep to areas with better natural defenses against predators, such as higher ground or areas with dense vegetation. This can reduce their vulnerability to predation during migration periods. While migration offers some benefits in terms of predator avoidance, it also exposes sheep to different risks [58]. For example, during migration, sheep may have to navigate unfamiliar terrain, which can increase their vulnerability to predators who are more familiar with the area [48]. The movement of large groups of sheep during migration can attract predators, leading to dynamic predator-prey interactions. Predators may actively follow migrating herds, which can influence the timing and route of migration for the sheep [59]. Overall, the effect of sheep migration on predator avoidance is multifaceted and depends on various factors such as the behavior of both predators and prey, the landscape, and the presence of other environmental factors.

2.2.1.5 Human interaction

Migration routes may intersect with human settlements or infrastructure, exposing sheep to various human activities. This could lead to habituation or avoidance behaviors in response to human presence, impacting their grazing patterns and movement. Human activities such as urbanization, agriculture, and infrastructure development can fragment natural habitats, disrupting traditional migration routes for sheep [19]. This fragmentation may force sheep to alter their routes or navigate through unfamiliar territories, which can increase their vulnerability to predation and decrease access to essential resources like food and water [60]. Physical barriers like fences, highways, and other human-made structures can obstruct sheep migration paths [61]. These barriers can disrupt the natural flow of migration and may require sheep to deviate from their usual routes or expend additional energy to navigate around them. These barriers can sometimes lead to injuries or fatalities among migrating sheep. Human presence, including recreational activities such as hiking, camping, and off-road vehicle use, can disturb sheep and cause them to avoid areas of high human activity. This disturbance can lead to changes in migration patterns as sheep seek quieter and less disturbed areas, potentially altering their routes and timing of movement [62]. Human activities can indirectly influence predation risk for migrating sheep. For example, the presence of human settlements or livestock can attract predators to areas that sheep frequent during migration, increasing the likelihood of predation events [63]. In response, sheep may alter their migration routes to avoid areas with heightened predation risk, leading to changes in their overall migration patterns. While not directly related to human interaction, human-induced climate change can have significant impacts on sheep migration routes by altering habitat conditions and resource availability [64]. Changes in temperature, precipitation patterns, and vegetation dynamics can influence the timing and distribution of food and water resources along migration routes, prompting sheep to modify their movements in response to these environmental changes [65]. Overall, human interaction can disrupt the natural behavior of sheep and alter their migration routes, which can have implications for their survival, population dynamics, and ecosystem functioning. Conservation efforts aimed at minimizing human-induced disturbances and preserving critical habitat corridors are essential for maintaining healthy and resilient sheep populations.

2.2.1.6 Reproductive patterns

Migration can significantly impact the reproductive patterns of sheep due to various factors related to changes in the environment, nutrition, stress levels, and social dynamics and may affect the timing of mating seasons and birthing patterns [26]. Sheep often exhibit seasonal breeding patterns, influenced by factors like day length and temperature. Migration may alter the timing or availability of breeding resources, affecting the synchronization of estrus cycles among ewes [66]. Migration can lead to changes in available forage and water sources. Inadequate nutrition can negatively impact reproductive health, leading to decreased fertility rates, lower conception rates, and increased embryonic loss [67]. The physical exertion of migration, especially over long distances or challenging terrain, can induce stress in sheep. Chronic stress can disrupt hormonal balances, affecting reproductive hormone secretion and leading to irregular estrus cycles or anestrus [68]. Migration often involves the mixing of different sheep populations, introducing new social hierarchies and dynamics. Dominance hierarchies within groups can influence mating behavior and access to mates, potentially affecting reproductive success. Migration routes may expose sheep to increased predation risk, causing heightened stress levels. Fear of predation can trigger physiological responses that impact reproductive hormones and behavior [48].

Migration may expose sheep to varying climates and weather conditions. Extreme temperatures or weather events can induce stress and impact reproductive function, particularly during critical stages such as mating or pregnancy. Migration can increase exposure to parasites such as internal worms, which can affect sheep's health and reproductive performance. Parasite infestations can lead to nutrient deficiencies, anemia, and overall poor reproductive outcomes. Migration can facilitate genetic exchange between different sheep populations. This genetic mixing may introduce traits related to reproductive performance, such as fertility or fecundity, which can influence future breeding patterns [69]. Overall, the effects of migration on sheep reproductive patterns are complex and multifaceted, depending on various environmental, physiological, and social factors. Management practices that mitigate stress, ensure adequate nutrition, and minimize exposure to disease and predators are essential for maintaining optimal reproductive health in migrating sheep populations. Here is a comprehensive **Table 3** summarizing the various factors that influence sheep migration and their impacts on reproductive patterns.

S. No.	Factor	Description	Impact on reproductive patterns
1.	Seasonal breeding patterns	Influenced by day length and temperature changes	Migration may alter the timing or availability of breeding resources, affecting the synchronization of estrus cycles among ewes
2.	Nutritional status	Availability and quality of forage impacting body condition and energy reserves	Inadequate nutrition can lead to decreased fertility, lower conception rates, and increased embryonic loss
3.	Stress levels	Physical and environmental stress from migration and habitat changes	Chronic stress can disrupt hormonal balances, causing irregular estrus cycles or anestrus
4.	Social dynamics	Changes in flock structure, dominance hierarchies, and mating opportunities	Altered social dynamics can influence mating behavior and reproductive success
5.	Water and forage availability	Changes in available resources due to migration	Affects nutrition intake and health, potentially impacting fertility and overall reproductive performance
6.	Predation risk	Increased stress levels due to exposure to predators	Fear of predation can affect reproductive hormone secretion and behavior, potentially reducing fertility rates
7.	Climate and weather conditions	Exposure to varying climates and extreme weather events during migration	Extreme temperatures or weather events can stress sheep and impact reproductive function
8.	Parasite infestations	Exposure to parasites such as internal worms	Parasite infections can lead to nutrient deficiencies, anemia, and poor reproductive outcomes
9.	Genetic exchange	Mixing of different sheep populations during migration	Introduces genetic traits related to reproductive performance, influencing future breeding patterns

Source: Developed by the authors Dr. Vinod Bhateshwar and Basant Kumar Bhinchhar.

Table 3.
Various factors impact reproductive patterns in sheep.

2.2.1.7 Cultural influences

Sheep migration can have various effects on cultural influences, depending on the context and the specific cultural practices of the communities involved. In many pastoralist cultures, sheep migration is a deeply ingrained tradition. The movement of sheep herds from one grazing area to another may be a central aspect of cultural identity and heritage [70]. This tradition can be passed down through generations, shaping cultural practices, rituals, and oral traditions. Sheep migration can significantly influence economic activities in pastoralist societies. The timing and routes of migration can impact trading patterns, market dynamics, and economic relationships with neighboring communities [71]. Economic exchanges related to sheep farming, such as wool and meat production, can also contribute to cultural exchanges and interactions. Sheep migration often involves communal cooperation among pastoralist communities. The coordination required for successful migrations fosters social cohesion and strengthens bonds within the community. Cultural values such as cooperation, reciprocity, and solidarity are reinforced through collective efforts to manage and execute the migration process. Pastoralist cultures develop extensive knowledge about the land, weather patterns, and ecosystems through generations of sheep migration [11].

This ecological knowledge forms an integral part of cultural identity and influences decision-making processes related to resource management and adaptation to environmental changes. Sheep migration routes often intersect with trade routes and pathways of cultural exchange. As pastoralist communities move their herds, they may encounter other groups with different cultural practices, languages, and belief systems [72]. These encounters can facilitate cultural exchange, leading to the adoption of new customs, technologies, and ideas. Sheep migration can sometimes lead to conflicts, especially in areas where resources are scarce or contested [73]. Conflicts over grazing rights, water access, or land ownership can arise between pastoralist communities or with sedentary agriculturalists.

However, these conflicts also provide opportunities for cultural negotiation, conflict resolution, and the establishment of customary laws and agreements. In the face of modernization and external pressures, sheep migration serves as a means of preserving traditional cultural practices and knowledge systems [74]. At the same time, pastoralist cultures may also adapt their migration practices in response to changing socio-economic conditions, environmental challenges, and government policies. Overall, sheep migration plays a multifaceted role in shaping cultural influences, contributing to the resilience, adaptation, and diversity of pastoralist societies around the world.

Overall, the effects of migration on the behavioral adaptations of sheep in India are multifaceted and influenced by a variety of factors including environmental conditions, social dynamics, and cultural practices. Understanding these dynamics is crucial for effective sheep management and conservation efforts in diverse landscapes.

2.3 Environmental stress factors

Sheep migrations can be stressful events for the animals involved, primarily due to changes in their environment, social dynamics, and physical exertion. Here are some common stressors encountered by sheep during migration:

2.3.1 Environmental changes

Sheep migrating often encounter changes in terrain, weather conditions, and vegetation. Extreme weather events such as storms, heavy rain, or heat waves can cause stress by disrupting their usual grazing patterns and shelter options [75]. Sheep migrate in search of food, water, and suitable grazing grounds. Environmental changes such as droughts, floods, or changes in vegetation patterns can alter the availability and distribution of these resources. If the usual migration routes become disrupted or if resources become scarce, sheep may experience stress due to inadequate nutrition or dehydration [1]. Sheep are sensitive to temperature changes. Extreme heat or cold can induce stress, affecting their ability to migrate effectively. Climate change, leading to more frequent and intense heatwaves or cold spells, can disrupt migration patterns and increase stress levels among sheep [3]. Changes in environmental conditions can also influence predator-prey dynamics. For instance, alterations in vegetation cover or water availability may affect the distribution and abundance of predators. Sheep may experience heightened stress if they perceive increased predation risk during migration.

Environmental changes can facilitate the spread of infectious diseases among sheep populations [76]. Crowding during migration, compromised immune responses due to stress, and altered environmental conditions conducive to pathogen survival can increase the likelihood of disease transmission, leading to additional stress among sheep [77]. Sheep are social animals that rely on group cohesion during migration. Environmental changes that disrupt social structures or lead to increased competition for resources within the group can induce stress [77]. For example, if resource availability decreases, dominant individuals may monopolize access to food and water, causing stress among subordinate sheep. Environmental changes can also cause psychological stress among sheep [36]. Disorientation, confusion, and anxiety may result from sudden alterations in familiar landscapes or disruptions to established migration patterns. Additionally, noise pollution from human activities or natural disasters such as storms or wildfires can further exacerbate stress levels [78].

To mitigate the impact of environmental changes on sheep migration stress, conservation efforts should focus on preserving and restoring natural habitats, implementing sustainable land management practices, and monitoring and managing wildlife populations. Additionally, measures to enhance resilience in sheep populations, such as providing supplementary feeding during periods of resource scarcity or establishing wildlife corridors to facilitate safe migration, can help alleviate stress and promote population health.

2.3.1.1 Predator threats

Environmental predator threats can significantly impact the stress levels of sheep. When sheep perceive a predator threat, their bodies release stress hormones like cortisol [48]. Elevated cortisol levels indicate heightened stress, which can have negative impacts on their health and well-being [79]. Sheep may exhibit altered behavior in response to predator threats. They may become more vigilant, spending more time scanning their surroundings for potential threats. This heightened vigilance can lead to decreased feeding and resting behaviors, which can further impact their health and productivity [48]. Prolonged exposure to predator threats can disrupt reproductive patterns in sheep [48]. Stress hormones can interfere with reproductive hormones, leading to decreased fertility and reproductive success [80]. Chronic stress resulting from persistent predator threats can weaken the immune system of sheep, making them more susceptible to diseases and infections. This can lead to higher mortality rates and reduced overall herd health [48].

Stress can also hinder the growth and development of lambs [81]. Pregnant ewes experiencing stress due to predator threats may produce smaller and weaker offspring, affecting the overall productivity of the flock [48]. Continual exposure to predator threats can result in long-term psychological stress in sheep. This can manifest in various ways, including decreased resilience to future stressors, altered social behavior, and even symptoms of anxiety or depression [82].

To mitigate these effects, sheep farmers often implement predator control measures such as using guardian animals, installing fences, or employing deterrents like lights or sound devices to reduce the risk of predator attacks. Additionally, providing a safe and secure environment for sheep, such as well-designed shelters or pasture management strategies, can help minimize stress levels and promote overall welfare.

2.3.1.2 Social hierarchy

The influence of environmental social hierarchy on sheep migration is an interesting aspect of animal behavior and ecology [6]. Sheep, like many other social animals, exhibit a hierarchical social structure within their groups. This hierarchy often determines access to resources such as food, water, and mates, as well as social interactions within the group [83]. In the context of migration, the social hierarchy can play a significant role in shaping the behavior of individual sheep and the dynamics of the entire flock. Here are some ways in which environmental social hierarchy may impact sheep migration, in sheep groups, dominant individuals often emerge as leaders, guiding the movements of the flock [41]. These leaders may have a greater influence on migration decisions, determining when and where the flock moves based on their knowledge of the environment and available resources. Sheep lower in the social hierarchy may follow dominant individuals to ensure access to resources during migration [84]. Dominant individuals may have priority access to preferred foraging areas or water sources, and subordinate sheep may benefit from following them to gain access to these resources.

Dominant individuals within the flock may also provide a sense of security and protection during migration [85]. Subordinate sheep may feel safer following dominant individuals, especially in unfamiliar or potentially dangerous environments. The social hierarchy can contribute to the cohesion of the flock during migration. Subordinate sheep may be more likely to stay with the group and follow dominant individuals, maintaining the integrity of the flock as it moves across the landscape [41]. In some cases, the social hierarchy within the flock may lead to competition and conflict during migration. Dominant individuals may compete for leadership roles, leading to challenges or aggressive interactions within the group [6]. This competition could influence the direction or pace of migration. Sheep may learn from each other during migration, with dominant individuals playing a key role in transmitting knowledge about the landscape and optimal migration routes. This social learning can help the flock adapt to changes in the environment or respond to challenges encountered during migration [6].

Overall, the environmental social hierarchy within sheep groups can have significant effects on migration behavior, shaping the decisions of individual sheep and the collective movement of the flock across the landscape. Understanding these dynamics is important for conservation efforts and the management of sheep populations in natural habitats.

2.3.1.3 Physical exertion

Physical exertion can have various effects on sheep migration, depending on factors such as the duration, intensity, and timing of the exertion, as well as environmental conditions and the health and fitness of the sheep involved [86]. Here are some potential effects: Physical exertion can influence the speed at which sheep migrate. If the exertion is strenuous, such as climbing steep terrain or traversing long distances without rest, it may slow down the migration process as the sheep need to conserve energy [87].

Exertion increases the energy expenditure of sheep. They need to consume more food and water to replenish the expended energy, which can affect their grazing patterns and may lead them to seek out rest stops more frequently during migration. Prolonged or excessive exertion without adequate rest can lead to fatigue and physical strain in sheep [88]. This can make them more susceptible to injuries, illnesses, or conditions like heat stress or exhaustion, particularly if they are already in poor health or if environmental conditions are harsh. Sheep may exhibit changes in behavior in response to physical exertion during migration. They may become more restless, vocalize more, or display signs of stress or agitation. Alternatively, they may become more focused and determined as they navigate challenging terrain or obstacles.

Physical exertion can affect the cohesion and dynamics of the sheep group [89]. Fatigue or injury in some individuals may slow down the entire group, while stronger or more resilient individuals may lead the way or provide support to weaker members. The presence of physical exertion may influence the choice of migration routes. Sheep may prefer paths that offer easier terrain or access to resources like food, water, or shelter to mitigate the effects of exertion and conserve energy [4]. Prolonged or intense physical exertion during migration could potentially impact the reproductive success of sheep, particularly if it leads to stress, malnutrition, or injuries that affect fertility or the ability to care for offspring [39].

Overall, while physical exertion is a natural part of sheep migration, its effects can be complex and varied, influenced by a range of factors that interact with each other and with the intrinsic characteristics of the sheep population and their environment.

2.3.1.4 Food and water availability

Competition for resources within the flock or encountering scarcity can cause stress and nutritional deficiencies [90]. Sheep often migrate in response to seasonal changes in vegetation and water availability. During the dry season, they may move to areas where water is more abundant and vegetation is still relatively lush [91]. Conversely, during the wet season, they might migrate to higher elevations where vegetation is more abundant and water is plentiful. Sheep are selective grazers and prefer areas with high-quality forage [92]. They will migrate to areas where food resources are abundant and of good quality. If food availability decreases in one area due to overgrazing or environmental changes, sheep may move to other areas where forage is more plentiful. Access to water is crucial for sheep survival, especially in arid and semiarid regions (**Figure 4**). Sheep will often follow migration routes that lead to reliable water sources such as rivers, streams, or seasonal waterholes [93]. Migration routes may change depending on the availability of water along the way.

Sheep may also migrate to areas where they can avoid predators. This could mean moving to higher elevations where visibility is better or to areas with natural barriers that make it more difficult for predators to approach [94]. Human activities such as

Figure 4.
Water scarcity for sheep in arid and semiarid regions in India. Source: Developed by the author Dr. Vinod Bhateshwar.

agriculture, urbanization, and fencing can impact the availability of food and water for sheep. In some cases, sheep may be forced to alter their migration routes or adapt to changes in their environment due to human intervention [95].

Overall, food and water availability are primary drivers of sheep migration routes, and these routes can vary widely depending on environmental conditions, seasonal changes, and human activities in the area. Understanding these factors is crucial for conservation efforts and sustainable management of sheep populations.

2.3.1.5 Human interference

Human interference can significantly impact sheep migration routes in various ways. Construction of roads, fences, and buildings can obstruct traditional migration paths, forcing sheep to detour or change their routes altogether [96]. These barriers can disrupt the natural flow of migration and isolate populations, leading to reduced genetic diversity and potential habitat fragmentation. Human activities such as agriculture, urbanization, and industrial development can alter the landscape and disrupt habitats along migration routes [97]. Conversion of natural habitats into farmland or urban areas can reduce available grazing areas for sheep and increase the likelihood of conflicts with human settlements. Pollution, deforestation, overgrazing by livestock, and resource extraction activities can degrade the quality of habitats along migration routes [98]. Loss of vegetation cover, water sources, and shelter can negatively impact sheep populations, making it difficult for them to find food and rest during their migrations. Increased human presence along migration routes can disturb sheep behavior and cause them to avoid certain areas or alter their movement patterns [99].

Activities such as recreational hiking, camping, hunting, and livestock grazing can create disturbances, noise, and potential conflicts with migrating sheep.

Overall, human interference can disrupt the natural processes that govern sheep migration, leading to habitat loss, fragmentation, and increased vulnerability to environmental stressors. Conservation efforts aimed at mitigating human impacts, preserving key habitats, and maintaining connectivity along migration routes are essential for ensuring the long-term survival of sheep populations.

2.3.1.6 Separation from lambs

Separation from lambs during migration can cause significant distress and anxiety among both ewes and their offspring [48]. This separation disrupts the strong bond between the mother and her lamb, leading to various behavioral and physiological stress responses [100]. For ewes, the separation from their lambs can result in increased vocalizations, restlessness, and attempts to reunite with their young [101]. This anxiety stems from their maternal instinct to protect and care for their lambs, and being unable to do so can cause significant stress.

Lambs, on the other hand, also exhibit signs of distress when separated from their mothers. They may bleat frequently, exhibit restlessness, and show decreased activity as they attempt to find their mothers [102]. The lack of maternal care can lead to increased vulnerability, reduced feeding, and overall impaired well-being. In addition to behavioral changes, both ewes and lambs can experience physiological stress responses [103]. Elevated levels of cortisol, a stress hormone, can be detected in their systems, indicating a state of heightened stress [104]. This can affect their immune function, growth, and overall health, making them more susceptible to diseases and other health issues.

Efforts to minimize separation during migration are crucial for maintaining the welfare of both ewes and lambs [105]. Ensuring that they remain together can help reduce stress, promote normal behavior, and support their overall health and well-being during the challenging migration process.

2.3.1.7 Unfamiliar territory

Sheep are creatures of habit and may experience stress when navigating unfamiliar terrain or encountering new obstacles along migration routes [106]. This behavior is rooted in their instincts and social structure. Sheep rely heavily on familiar surroundings and established pathways to feel secure. When faced with new environments or unexpected barriers, they can become anxious or disoriented [88]. This stress response can affect their overall well-being and movement efficiency.

To mitigate these issues, shepherds and farmers often take steps to gradually introduce sheep to new areas, using techniques such as leading with experienced animals or creating temporary enclosures that allow sheep to acclimate at their own pace. Understanding and managing the stress responses of sheep in unfamiliar territory is crucial for maintaining their health and ensuring smooth migrations or transitions between pastures.

2.3.1.8 Transportation

Transportation can indeed be a highly stressful experience for domesticated sheep due to several factors. When sheep are transported, they often face confinement in unfamiliar and sometimes cramped conditions, which can induce anxiety and stress [107]. The noise from the vehicle and external environment, along with the movement and vibrations, adds to their discomfort. If not handled properly, this stress can negatively impact their health and welfare. Proper handling during transportation is crucial to minimize stress, including ensuring adequate space, ventilation, and gentle handling practices [108]. Additionally, reducing the duration of transportation and providing rest periods can help mitigate some of the stress associated with the journey. By addressing these factors and implementing best practices, the welfare of sheep during transportation can be significantly improved, reducing the stress and associated negative impacts on their health and productivity.

2.3.1.9 Health issues

Migration can significantly impact the health of sheep, often exacerbating existing health issues or making them more susceptible to new diseases [109]. The act of migration itself can be highly stressful for sheep. This stress can weaken their immune system, making them more vulnerable to diseases. Stress factors include changes in the environment, handling during transportation, and disruption of their regular routines. Long-distance travel can lead to exhaustion. Fatigue reduces the animals' ability to fend off illnesses and can exacerbate existing health problems, such as respiratory conditions or joint issues [88]. When sheep are moved to new environments, they may encounter unfamiliar pathogens to which they have no immunity [110]. This exposure can lead to outbreaks of diseases such as foot and mouth disease, bluetongue, or other infectious diseases prevalent in the new location [111].

Migration often involves changes in diet and water sources. Inadequate nutrition or changes in diet can affect sheep's health, leading to issues like malnutrition or digestive problems [112]. Moving to a different climate can expose sheep to weather extremes they are not accustomed to, such as extreme heat or cold, which can exacerbate conditions like pneumonia or heat stress [3]. Migration can increase exposure to parasites, both internal and external, such as worms, ticks, and lice. These parasites can thrive in different environmental conditions, leading to increased infestation and subsequent health issues [113]. In conclusion, the process of migration can present significant health challenges for sheep, necessitating careful management and preventive measures to mitigate these risks.

Addressing these stressors requires careful planning and management practices to ensure the welfare of migrating sheep populations. Providing access to adequate food, water, and shelter, and minimizing disturbances can help reduce stress and promote a successful migration.

Sheep, like many other animals, experience a range of physiological responses to cope with stressors during migration. These responses involve changes in hormone levels, heart rate variability, and immune function, which help the animals adapt to the physical and psychological challenges they encounter.

2.4 Physiological responses to stress

Sheep, like many other animals, experience a range of physiological responses to cope with stressors during migration. These responses involve changes in hormone levels, heart rate variability, and immune function, which help the animals adapt to the physical and psychological challenges they encounter.

2.4.1 Hormone levels

Hormones play a crucial role in the physiological responses to stress in sheep, just as they do in other mammals. When sheep encounter stressors, whether they are environmental, social, or physiological, their bodies initiate a cascade of hormonal responses to help them cope. Here is how hormones contribute to these responses:

2.4.1.1 Cortisol

Often referred to as the stress hormone, cortisol is released by the adrenal glands in response to stress. In sheep, cortisol levels rise in situations of stress, such as during handling, transportation, or exposure to predators [114]. Cortisol helps mobilize energy reserves by increasing blood glucose levels, ensuring that the sheep have the energy needed to respond to the stressor [39]. It also suppresses nonessential functions such as growth, reproduction, and immune responses temporarily to prioritize survival.

2.4.1.1.1 Adrenaline (epinephrine)

Another hormone released by the adrenal glands, adrenaline is part of the immediate stress response known as the fight-or-flight response [115]. In sheep, as in other animals, adrenaline triggers rapid physiological changes, including increased heart rate, rapid breathing, and heightened alertness [88]. These changes prepare the sheep to either confront the stressor or flee from it.

2.4.1.1.2 Thyroid hormones

Thyroid hormones, such as thyroxine and triiodothyronine, also play a role in the stress response [116]. They help regulate metabolism, which can be altered during stressful situations. Thyroid hormones influence energy expenditure and thermoregulation, ensuring that the sheep can adapt to the demands imposed by the stressor [103].

2.4.1.1.3 Vasopressin

This hormone, also known as antidiuretic hormone (ADH), is released by the pituitary gland in response to stress. Vasopressin helps regulate water balance in the body by promoting water reabsorption in the kidneys [117]. During stress, vasopressin levels increase, reducing urine output and helping maintain blood pressure and hydration levels [118].

2.4.1.1.4 Oxytocin

While often associated with social bonding and reproduction, oxytocin also plays a role in stress responses. In sheep, oxytocin can have calming effects and help mitigate the negative impacts of stress [119]. It promotes social bonding and affiliative behaviors, which can reduce stress levels and improve overall well-being [120].

These hormones work together to enable sheep to cope with various stressors they encounter in their environment. However, prolonged or chronic stress can have detrimental effects on sheep health and welfare, disrupting hormone balance and leading to various physiological and behavioral issues. Therefore, sheep farmers and caretakers need to minimize stressors and provide appropriate management practices to ensure the well-being of their animals.

2.4.2 Heart rate variability (HRV)

Heart rate variability (HRV) refers to the variation in time intervals between heartbeats, influenced by the autonomic nervous system (ANS). It is often used as a measure of physiological stress in various organisms, including sheep. In sheep, HRV can be indicative of their response to stressors [121]. When faced with stress, whether it is due to environmental factors, handling, or social dynamics within a flock, sheep typically exhibit changes in their autonomic nervous system activity, which can affect HRV.

Research has shown that acute stressors can lead to a decrease in HRV in sheep. For example, exposure to novel environments, handling procedures, or sudden loud noises can trigger a stress response, leading to a decrease in HRV [122]. Conversely, when sheep are in a relaxed or non-stressful state, their HRV tends to be higher. Monitoring HRV in sheep can provide valuable insights into their welfare and management. By assessing HRV responses, farmers and researchers can identify stressors in the environment or management practices that maybe negatively impacting the animals. This information can then be used to implement strategies to reduce stress and improve the overall well-being of the flock. In summary, HRV serves as a useful tool for assessing the physiological responses of sheep to stressors, helping to inform management practices aimed at promoting animal welfare.

2.4.2.1 Immune function

The immune system plays a crucial role in the physiological responses to stress in sheep, as it does in all mammals [123]. Stress can come from various sources in a sheep's environment, such as changes in diet, temperature, social interactions, or predator threats. Here is how the immune function intersects with stress responses in sheep:

2.4.2.1.1 Impact on immune function

Stress triggers the release of stress hormones like cortisol, which can suppress the immune system's function. In sheep, as in humans, prolonged stress can lead to decreased immune cell activity, making them more susceptible to infections and diseases [124].

2.4.2.1.2 Susceptibility to infections

When sheep are stressed, their immune system's ability to respond to pathogens may be compromised [125]. This can increase the risk of infections such as bacterial pneumonia or parasitic infestations like gastrointestinal nematodes.

2.4.2.1.3 Wound healing

Stress can impair wound healing in sheep by reducing the immune response at the site of injury [126]. This can prolong recovery times and increase the risk of secondary infections.

2.4.2.1.4 Vaccination response

Stress can also affect the efficacy of vaccinations in sheep. High levels of stress hormones can dampen the immune response to vaccines, leading to reduced protection against infectious diseases [127].

2.4.2.1.5 Nutritional immunity

Stress can influence nutrient utilization and absorption in sheep, which in turn affects their immune function [128]. Proper nutrition is essential for maintaining a robust immune system and mitigating the negative effects of stress.

2.4.2.1.6 Behavioral changes

Stress can induce changes in sheep behavior, such as increased aggression or reduced social interactions, which may further impact their susceptibility to diseases. Social stress, for example, can disrupt the hierarchy within a flock and lead to increased disease transmission [129].

2.4.2.1.7 Management practices

Understanding the interaction between stress and immune function is crucial for implementing effective management practices in sheep farming. Minimizing stressors, providing adequate nutrition, and ensuring proper vaccination protocols can help support the immune health of sheep and improve overall flock productivity [130].

In summary, the immune function in sheep is closely linked to their physiological responses to stress. Minimizing stressors and promoting overall welfare is essential for maintaining a healthy immune system and reducing the risk of diseases in sheep populations.

2.4.3 Behavioral and metabolic adjustments

Sheep, like all animals, experience stress in various forms, and their bodies respond with a complex interplay of behavioral and metabolic adjustments. Here is how sheep typically respond:

2.4.3.1 Behavioral responses

2.4.3.1.1 Flight or fight

When faced with a stressor, sheep may exhibit a flight response by trying to escape the situation or a fight response by standing their ground and showing defensive behaviors [131].

2.4.3.1.2 Grazing and resting patterns

Stress can disrupt normal feeding and resting behaviors in sheep. They may reduce their food intake and spend less time resting, which can affect their overall health and productivity [132].

2.4.3.1.3 Social interactions

Stress can alter social interactions among sheep, leading to changes in dominance hierarchies, aggression levels, and affiliative behaviors within the flock.

2.4.3.1.4 Metabolic responses

2.4.3.1.4.1 Activation of the hypothalamic: Pituitary: Adrenal (HPA) axis

In response to stress, the HPA axis is activated, leading to the release of cortisol from the adrenal glands [133]. Cortisol helps mobilize energy reserves by increasing glucose levels in the blood.

2.4.3.1.4.2 Energy metabolism

Stress can lead to alterations in energy metabolism in sheep. Glucose is redirected from nonessential tissues to organs vital for the stress response, such as the brain and muscles [134].

2.4.3.1.4.3 Immune function

Prolonged stress can suppress the immune system in sheep, making them more susceptible to diseases and infections [48].

2.4.3.1.4.4 Gastrointestinal function

Stress can disrupt gastrointestinal function, leading to reduced feed intake, altered nutrient absorption, and digestive disorders in sheep [135].

2.4.3.1.5 Long-term adaptations

2.4.3.1.5.1 Habituation

Sheep may become habituated to chronic stressors over time, leading to reduced physiological responses to repeated stress exposure [136].

2.4.3.1.5.2 Epigenetic changes

Prolonged or severe stress can induce epigenetic changes in sheep, altering gene expression patterns that may persist across generations [137].

2.4.3.1.5.3 Health and productivity

Chronic stress can have detrimental effects on the health and productivity of sheep, leading to reduced growth rates, reproductive performance, and wool quality [39].

Understanding the behavioral and metabolic adjustments in physiological responses to stress is crucial for implementing effective management strategies to minimize stress and promote the well-being of sheep in farming and husbandry practices.

In summary, sheep cope with the stressors of migration through a complex interplay of hormonal changes, autonomic nervous system adjustments, and immune function modulation. These physiological responses are critical for maintaining energy balance, protecting against infections, and ensuring survival during the demanding migratory period.

2.5 Genetic adaptations

Genetic adaptations in sheep populations reflect millennia of evolutionary pressure to thrive in diverse environments. These adaptations enable them to migrate over long distances and survive in various climates and terrains. Here are some key areas of research and findings:

2.5.1 Wool characteristics

Sheep have been selectively bred for wool traits that offer protection against harsh climates. Different breeds exhibit variations in wool type, density, and color, all of which contribute to their adaptability [138]. For instance, Merino sheep, originating from Spain, have fine wool suited for cold environments, while breeds like the Barbados Blackbelly have coarser wool adapted to warmer climates [139].

2.5.2 Heat tolerance

Sheep in arid and hot regions have developed mechanisms to cope with heat stress. Genetic studies have identified genes associated with heat tolerance, such as those involved in thermoregulation and sweat gland activity [140]. Understanding these genetic factors can inform breeding programs aimed at developing heat-tolerant sheep breeds.

2.5.3 Foraging behavior

Sheep exhibit diverse foraging behaviors influenced by their genetic makeup and environmental conditions. Research has shown that certain genes are associated with grazing preferences, browsing tendencies, and the ability to thrive on different types of vegetation [141]. These genetic adaptations allow sheep to exploit varied food sources in their habitats, enhancing their survival prospects.

2.5.4 Disease resistance

Sheep populations in different regions have evolved resistance to local pathogens and parasites. Genetic studies have identified genes involved in immune response and disease resistance, offering insights into natural defense mechanisms. Breeding for disease resistance can help reduce the reliance on chemical interventions and enhance the overall health of sheep populations [142].

2.5.5 Hoof structure

Sheep inhabiting mountainous regions have adapted to rugged terrain through changes in hoof morphology. Genetic research has revealed variations in genes related to hoof development and strength, which contribute to agility and stability in rocky landscapes. Understanding these genetic adaptations can inform breeding strategies aimed at improving hoof health and durability.

2.5.6 Migration patterns

Genetic analyses have provided insights into the historical migration patterns of sheep populations. By studying genetic markers, researchers can trace the ancestry and dispersal routes of different breeds, shedding light on past human migrations and trade routes. This information is valuable for conservation efforts and maintaining genetic diversity within sheep populations.

Overall, genetic adaptations in sheep populations underscore the dynamic interplay between natural selection, environmental pressures, and human intervention. Continued research into sheep genomics holds promise for understanding and harnessing genetic diversity to enhance the resilience and productivity of these important livestock species.

2.6 Human impacts on migration

Human activities can significantly influence sheep migration behavior and adaptation through various means such as land use changes, infrastructure development, and grazing regulations. Here is an examination of each of these factors:

2.6.1 Land-use changes

2.6.1.1 Urbanization

The expansion of urban areas can encroach upon traditional sheep grazing lands, reducing available habitat and forcing migrations to alternative areas.

2.6.1.2 Agricultural expansion

Controlling natural habitats into agricultural lands can fragment migration routes, disrupt access to seasonal forage, and lead to conflicts between livestock and farmers.

2.6.1.3 Deforestation

Deforestation alters the landscape, reducing the availability of shelter and forage for migrating sheep. It can also increase soil erosion, affecting the quality of grazing lands.

2.6.2 Infrastructure development

2.6.2.1 Roads and highways

Construction of roads and highways can fragment migration corridors, creating barriers that disrupt traditional migration patterns. Sheep may be reluctant to cross roads due to the increased risk of predation and collisions with vehicles.

2.6.2.2 Fences

Fencing associated with infrastructure development can impede the movement of sheep, obstructing migration routes and leading to population isolation. Moreover, fences can increase the risk of entanglement and injury for migrating animals.

2.6.2.3 Dams and reservoirs

The construction of dams and reservoirs can flood grazing lands and disrupt water sources, altering migration routes and limiting access to essential resources for sheep.

2.6.3 Grazing regulations

2.6.3.1 Overgrazing

Poorly regulated grazing practices can lead to overgrazing, degrading vegetation, and reducing forage availability along migration routes. This can force sheep to travel longer distances in search of suitable feeding grounds, increasing energy expenditure and vulnerability to predation.

2.6.3.2 Seasonal grazing restrictions

Grazing regulations restricting access to certain areas during critical times of the year can disrupt traditional migration patterns, forcing sheep to alter their routes or remain in less suitable habitats.

2.6.3.3 Predator control

Human interventions aimed at predator control, such as culling or fencing, can indirectly influence sheep migration behavior by altering predator-prey dynamics along migration routes.

Overall, human activities have the potential to significantly impact sheep migration behavior and adaptation by altering habitat availability, fragmenting migration corridors, and disrupting access to essential resources. Sustainable land management practices and conservation efforts are essential to mitigate these impacts and ensure the long-term viability of sheep populations and their migratory behaviors.

2.7 Future research directions

Understanding sheep migration patterns and behavioral adaptations in Indian grazing lands requires a multifaceted approach that combines traditional ecological studies with cutting-edge technology and interdisciplinary collaboration. Here are some future research directions that could enhance our understanding:

2.7.1 GPS tracking and remote sensing

Implementing GPS tracking devices on individual sheep can provide high-resolution data on movement patterns, grazing behavior, and habitat preferences. Combining this with remote sensing techniques such as satellite imagery and drones can offer a comprehensive view of landscape dynamics and resource availability.

2.7.2 Social network analysis

Investigate the social structure within sheep herds to understand how social interactions influence migration decisions and resource utilization. Social network analysis can reveal leadership dynamics, communication patterns, and information transfer among individuals within the flock.

2.7.3 Genomic studies

Explore the genetic basis of migratory behavior and adaptation to different environments in Indian sheep breeds. Comparative genomics can identify genetic markers associated with traits such as migratory tendency, foraging efficiency, and tolerance to environmental stressors.

2.7.4 Ethnographic studies

Collaborate with local communities and traditional shepherds to document Indigenous knowledge and practices related to sheep migration. Ethnographic research can provide insights into historical migration routes, cultural significance of grazing lands, and traditional management strategies.

2.7.5 Climate change impacts

Investigate how climate change is affecting sheep migration patterns and grazing behavior in Indian landscapes. Long-term monitoring coupled with climate modeling can elucidate the drivers of change and potential adaptation strategies for both sheep and pastoral communities.

2.7.6 Ecosystem services assessment

Quantify the ecosystem services provided by migratory sheep grazing, including vegetation management, soil nutrient cycling, and biodiversity conservation. Understanding the ecological benefits of traditional grazing practices can inform sustainable land management policies.

2.7.7 Agent-based modeling

Develop agent-based models to simulate sheep movement and decision-making processes in response to changing environmental conditions and human activities. These models can help predict future migration patterns and assess the effectiveness of management interventions.

2.7.8 Interdisciplinary collaboration

Foster collaboration between ecologists, geneticists, climatologists, social scientists, and pastoral communities to integrate diverse perspectives and methodologies. Interdisciplinary research frameworks can generate holistic insights into the complex dynamics of sheep migration systems.

By pursuing these research directions, we can deepen our understanding of sheep migration patterns and behavioral adaptations in Indian grazing lands, ultimately supporting the conservation of biodiversity, sustainable land use practices, and the livelihoods of pastoral communities.

3. Conclusion

The study of sheep migration patterns, behavioral adaptations, and their responses to environmental stress factors in Indian grazing lands reveals a fascinating interplay between instincts and environmental influences. Through centuries of evolution, sheep have developed sophisticated strategies for survival in diverse habitats, demonstrating remarkable resilience to challenges such as food scarcity, climatic variability, and human disturbances. However, the increasing pressures of anthropogenic activities pose significant threats to sheep populations and the ecosystems they inhabit. Sustainable management practices that prioritize habitat conservation, promote responsible grazing practices, and mitigate human-induced stressors are imperative for safeguarding the welfare of sheep and preserving the integrity of grazing lands. By fostering a deeper understanding of the complex dynamics shaping sheep behavior, we can work toward harmonizing human activities with the needs of these invaluable livestock species and the ecosystems they support.

Conflicts of interest

The authors disclose that they have no conflicting interests.

Author details

Vinod Bhateshwar[1*], Basant Kumar Bhinchhar[1], Hitesh Muwal[2] and Paramveer Palriya[3]

1 Department of Livestock Production Management, Sri Karan Narendra Agriculture University, Jobner, Jaipur, Rajasthan, India

2 Department of Animal Production, Maharana Pratap University of Agriculture and Technology. Udaipur, Rajasthan, India

3 Department of Animal Husbandry and Dairying, Sam Higginbottom University of Agriculture Technology and Sciences, Prayagraj, Uttar Pradesh, India

*Address all correspondence to: vinod.lpm@sknau.ac.in

References

[1] Naqvi SMK, De K, Gowane GR. Sheep production system in arid and semi-arid regions of India. Annals of Arid Zone. 2013;**52**:3-4

[2] Rawat GS, Adhikari BS. Ecology and Management of Grassland Habitats in India. ENVIS Bulletin: Wildlife & Protected Areas. Dehradun, India: Wildlife Institute of India; 2015;**17**:1-240

[3] Sejian V, Samal L, Soren NM, Bagath M, Krishnan G, Vidya MK, et al. Adaptation strategies to counter climate change impact on sheep. In: Sheep Production Adapting to Climate Change. Singapore: Springer Nature Pte Ltd; 2017. pp. 413-430

[4] Kilgour RJ, Waterhouse T, Dwyer CM, Ivanov ID. Farming systems for sheep production and their effect on welfare. The Welfare of Sheep. Springer Science+Business Media B.V.; 2008;**6**:213-265

[5] Beaver BV, Höglund D. Efficient Livestock Handling: The Practical Application of Animal Welfare and Behavioral Science. London, UK: Academic Press; 2015:1-230. eBook ISBN: 9780124172876

[6] Fisher A, Matthews L. The social behaviour of sheep. Social Behaviour in Farm Animals. 2001;**6**:211-245

[7] Sejian V, Bhatta R, Gaughan JB, Dunshea FR, Lacetera N. Adaptation of animals to heat stress. Animal. 2018;**12**(s2):s431-s444

[8] Nair MR, Sejian V, Silpa MV, Fonseca VF, de Melo Costa CC, Devaraj C, et al. Goat as the ideal climate-resilient animal model in tropical environment: Revisiting advantages over other livestock species. International Journal of Biometeorology. 2021;**65**:2229-2240

[9] Guthmann A, Onyango MB, Iannarilli F, Packer C. Livestock activity shifts large herbivore temporal distributions to their crepuscular edges. Journal of Animal Ecology. 2024;**93**(2):231-245

[10] Ikhuoso OA, Adegbeye MJ, Elghandour MM, Mellado M, Al-Dobaib SN, Salem AZ. Climate change and agriculture: The competition for limited resources amidst crop farmers-livestock herding conflict in Nigeria-a review. Journal of Cleaner Production. 2020;**272**:123104

[11] Bhasin V. Pastoralists of Himalayas. Journal of Human Ecology. 2011;**33**(3):147-177

[12] Singh R, Sharma RK, Babu S, Bhatnagar YV. Traditional ecological knowledge and contemporary changes in the agro-pastoral system of upper Spiti landscape, Indian trans-Himalayas. Pastoralism. 2020;**10**:15

[13] Sharma G, Tambe S, Rawat GS, Arrawatia ML. Yak herding and associated transboundary issues in the Sikkim Himalaya, India. In: Yak Move Transboundary Challenges and Opportunities for Yak Raising in a Changing Hindu Kush Himalayan Region. Kathmandu, Nepal: International Centre for Integrated Mountain Development; 2016. pp. 93-112

[14] Ahmad S, Mir N, Bhat S, Singh J. High altitude pasturelands of Kashmir Himalaya: Current status, issues and future strategies in a changing climatic scenario. Current Journal

of Applied Science and Technology. 2018;**27**:1-10

[15] Aryal S, Maraseni T, Cockfield G, de Bruyn LL. Transhumance, livestock mobility and mutual benefits between crop and livestock production. Sustainable Agriculture Reviews 31: Biocontrol. Cham: Springer; 2018;**31**:25-39

[16] Dev I, Singh V, Misri B. Socio-economic profile of migratory graziers and participatory appraisal of forage production and utilization of an alpine pasture in north-west Himalaya. ENVIS Bulletin: Himalayan Ecology and Development; 2003;**11**(2):52-96

[17] Mysterud A, Iversen C, Austrheim G. Effects of density, season and weather on use of an altitudinal gradient by sheep. Applied Animal Behaviour Science. 2007;**108**(1-2):104-113

[18] White KS, Hood E, Wolken GJ, Peitzsch EH, Bühler Y, Wikstrom Jones K, et al. Snow avalanches are a primary climate-linked driver of mountain ungulate populations. Communications Biology. 2024;**7**(1):423

[19] Wassie SB. Natural resource degradation tendencies in Ethiopia: A review. Environmental Systems Research. 2020;**9**(1):1-29

[20] Das AP. Conservation efforts for east Himalayan biodiversity and need for the establishment of corridors. In: Ghosh C, Das AP, editors. Recent Studies in Biodiversity and Traditional Knowledge in India. Kolkata: Sarat Book House; 2011. pp. 329-346

[21] Putfarken D, Dengler J, Lehmann S, Härdtle W. Site use of grazing cattle and sheep in a large-scale pasture landscape: A GPS/GIS assessment. Applied Animal Behaviour Science. 2008;**111**(1-2):54-67

[22] Misra AK, Subramanyam KV, Babu MV, Reddy TY, Shivarudrappa B, Ramakrishna YS. Improving the livelihood of landless and marginal farmers through sheep rearing in rainfed agro-ecosystem of India. Livestock Research for Rural Development. 2006;**18**(5)

[23] Louhaichi M, Chand K, Misra A, Gaur MK, Ashutosh S, Johnson DE, et al. Increasing Resilience of Livestock Migration in the Arid Areas of India: A Case Study of Livestock Mobility in the State of Rajasthan, India. Lebanon: International Center for Agricultural Research in the Dry Areas (ICARDA); 2015

[24] Singh V. Environmental migration as planned livelihood among the rebaris of Western Rajasthan, India. Global Environment. 2012;**5**(9):50-73

[25] Treydte AC, Schmiedgen A, Berhane G, Tarekegn KD. Rangeland forage availability and management in times of drought–A case study of pastoralists in Afar, Ethiopia. Journal of Arid Environments. 2017;**139**:67-75

[26] Shinde AK, Sejian V. Sheep husbandry under changing climate scenario in India: An overview. Indian Journal Animal Sciences. 2013;**83**(10):998-1008

[27] Namgail T, Bhatnagar YV, Mishra C, Bagchi S. Pastoral nomads of the Indian Changthang: Production system, landuse and socioeconomic changes. Human Ecology. 2007;**35**:497-504

[28] Gregory A, Spence E, Beier P, Garding E. Toward best management practices for ecological corridors. Land. 2021;**10**(2):140

[29] Trubins R. Land-use change in southern Sweden: Before and

after decoupling. Land Use Policy. 2013;**33**:161-169

[30] Papouchis CM, Singer FJ, Sloan WB. Responses of desert bighorn sheep to increased human recreation. The Journal of Wildlife Management. 2001;**65**(3):573-582

[31] Naqvi SM, De K, Kumar D, Sahoo A. Mitigation of climatic change effect on sheep farming under arid environment. Abiotic stress management for resilient. Agriculture. Singapore: Springer; 2017. pp. 455-474. DOI: 10.1007/978-981-10-5744-1_22

[32] Morris K, Reich P. Understanding the relationship between livestock grazing and wetland condition. In: Arthur Rylah Institute for Environmental Research Technical Report Series. No. 252. Heidelberg, Victoria: Department of Environment and Primary Industry; 2013. p. 252

[33] Wassie SB. Natural resource degradation tendencies in Ethiopia: A review. Environmental Systems Research. 2020;**1**:1-29

[34] Hinch GN. Understanding the natural behaviour of sheep. In: Advances in Sheep Welfare. Sawston, UK: Woodhead Publishing Series in Food Science, Technology and Nutrition; 2017. pp. 1-15. DOI: 10.1016/B978-0-08-100718-1.00001-7

[35] Van Wettere WH, Kind KL, Gatford KL, Swinbourne AM, Leu ST, Hayman PT, et al. Review of the impact of heat stress on reproductive performance of sheep. Journal of Animal Science and Biotechnology. 2021;**12**:1-8

[36] Dwyer C. Reproductive management (including impacts of prenatal stress on offspring development). In: Advances in Sheep Welfare. Sawston, UK: Woodhead Publishing Series in Food Science, Technology and Nutrition; 2017. pp. 131-152. DOI: 10.1016/B978-0-08-100718-1.00007-8

[37] Webster J. Animal Welfare: Limping towards Eden: A Practical Approach to Redressing the Problem of our Dominion over the Animals. John Wiley & Sons; 2008

[38] Colditz IG. Effects of the immune system on metabolism: Implications for production and disease resistance in livestock. Livestock Production Science. 2002;**75**(3):257-268. DOI: 10.1016/S0301-6226(01)00320-7

[39] Tufekci H, Sejian V. Stress factors and their effects on productivity in sheep. Animals. 2023;**13**(17):2769

[40] Hoffmann I. Adaptation to climate change–exploring the potential of locally adapted breeds. Animal. 2013;7(s2):346-362

[41] Squires VR. Ecology and behaviour of domestic sheep (*Ovis aries*): A review. Mammal Review. 1975;**5**(2):35-57

[42] Hiernaux P, Dardel C, Kergoat L, Mougin E. Desertification, adaptation and resilience in the Sahel: Lessons from long term monitoring of agro-ecosystems. In: The End of Desertification? Disputing Environmental Change in the Drylands. Berlin, Heidelberg: Springer-Verlag; 2016. pp. 147-178

[43] George W, Zoltan B, Andras J, Szilvia K. Impacts of climate change on sheep genetic diversity: A review. Natural Resources and Sustainable Development. 2020;**10**(2):240-261

[44] Altizer S, Bartel R, Han BA. Animal migration and infectious disease risk. Science. 2011;**331**(6015):296-302

[45] Kiss IZ, Green DM, Kao RR. The network of sheep movements within Great Britain: Network properties and their implications for infectious disease spread. Journal of the Royal Society Interface. 2006;**3**(10):669-677

[46] Gann GD, McDonald T, Walder B, Aronson J, Nelson CR, Jonson J, et al. International principles and standards for the practice of ecological restoration. Restoration Ecology. 2019;**27**(S1):S1-S46

[47] Festa-Bianchet M. The social system of bighorn sheep: Grouping patterns, kinship and female dominance rank. Animal Behaviour. 1991;**42**(1):71-82

[48] Dwyer CM. How has the risk of predation shaped the behavioural responses of sheep to fear and distress? Animal Welfare. 2004;**13**(3):269-281

[49] Ward A, Webster M. Sociality: The Behaviour of Group-Living Animals. Berlin, Germany: Springer; 2016. p. 407

[50] Berger DJ, German DW, John C, Hart R, Stephenson TR, Avgar T. Seeing is be-leaving: Perception informs migratory decisions in Sierra Nevada bighorn sheep (Ovis canadensis sierrae). Frontiers in Ecology and Evolution. 2022;**10**:742275

[51] di Virgilio A, Morales JM. Towards evenly distributed grazing patterns: Including social context in sheep management strategies. London, UK: Peer J Publishing; 2016;**4**:e2152

[52] Buchalski MR, Navarro AY, Boyce WM, Vickers TW, Tobler MW, Nordstrom LA, et al. Genetic population structure of peninsular bighorn sheep (Ovis canadensis nelsoni) indicates substantial gene flow across US–Mexico border. Biological Conservation. 2015;**184**:218-228

[53] Salem HB, Smith T. Feeding strategies to increase small ruminant production in dry environments. Small Ruminant Research. 2008;**77**(2-3):174-194

[54] Joy A, Dunshea FR, Leury BJ, Clarke IJ, DiGiacomo K, Chauhan SS. Resilience of small ruminants to climate change and increased environmental temperature: A review. Animals. 2020;**10**(5):867

[55] Liddell CL. Behavioural response of sheep to forage resources and parasite risk in an extensive grazing system [doctoral dissertation]; University of Bristol. 2021

[56] Curio E. The Ethology of Predation. Berlin, Germany: Springer Science & Business Media; 1976;**7**:1-250

[57] Huntingford FA, Turner AK, Dowine LM. Animal Conflict. Chapman & Hall Animal Behaviour Series. Dordrecht: Springer; 2013. pp. 1-448. DOI: 10.1007/978-94-009-3145-9

[58] Spitz DB. Does Migration Matter? Causes and Consequences of Migratory Behavior in Sierra Nevada Bighorn Sheep. Graduate Student Theses, Dissertations, & Professional Papers. University of Montana: ProQuest LLC; 2015. pp. 1-166, 10793. Available from: https://scholarworks.umt.edu/etd/10793

[59] Lia-Rognli E, Nummedal SA. Sheep and predator interactions: An investigation into the behavioural patterns of sheep during attacks and the feasibility of predictive modelling using GPS data [Master's thesis]; NTNU. 2023

[60] Fuller A, Mitchell D, Maloney SK, Hetem RS. Towards a mechanistic understanding of the responses of large terrestrial mammals to heat and aridity

associated with climate change. Climate Change Responses. 2016;**3**:1-9

[61] Flesch AD, Epps CW, Cain JW III, Clark M, Krausman PR, Morgart JR. Potential effects of the United States-Mexico border fence on wildlife. Conservation Biology. 2010;**24**(1):171-181

[62] Drent RH, Eichhorn G, Flagstad A, Van der Graaf AJ, Litvin KE, Stahl J. Migratory connectivity in Arctic geese: Spring stopovers are the weak links in meeting targets for breeding. Journal of Ornithology. 2007;**148**:501-514

[63] Braczkowski A, Fattebert J, Schenk R, O'Bryan C, Biggs D, Maron M. Evidence for increasing human-wildlife conflict despite a financial compensation scheme on the edge of a Ugandan National Park. Conservation Science and Practice. 2020;**2**(12):e309

[64] Abrahms B, Carter NH, Clark-Wolf TJ, Gaynor KM, Johansson E, McInturff A, et al. Climate change as a global amplifier of human–wildlife conflict. Nature Climate Change. 2023;**13**(3):224-234

[65] Bett B, Kiunga P, Gachohi J, Sindato C, Mbotha D, Robinson T, et al. Effects of climate change on the occurrence and distribution of livestock diseases. Preventive Veterinary Medicine. 2017;**137**:119-129

[66] Ahmad Pampori Z, Ahmad Sheikh A, Aarif O, Hasin D, Ahmad BI. Physiology of reproductive seasonality in sheep–an update. Biological Rhythm Research. 2020;**51**(4):586-598

[67] Joshi P. Nutrition and reproduction in sheep. Food & Agribusiness Management (FABM). 2022;**3**(2):48-52

[68] Naqvi SM, Kumar D, Paul RK, Sejian V. Environmental stresses and livestock reproduction. Environmental Stress and Amelioration in Livestock Production. Berlin, Heidelberg: Springer; 2012. pp. 97-128. DOI: 10.1007/978-3-642-29205-7_5

[69] Gootwine E. Genetics and breeding of sheep and goats. In: Animal Agriculture. Elsevier, Academic Press; 2020. pp. 183-198. DOI: 10.1016/B978-0-12-817052-6.00010-0

[70] Belliggiano A, Bindi L, Ievoli C. Walking along the sheeptrack... rural tourism, ecomuseums, and bio-cultural heritage. Sustainability. 2021;**13**(16):8870

[71] Turner MD, Williams TO. Livestock market dynamics and local vulnerabilities in the Sahel. World Development. 2022;**30**(4):683-705

[72] Dyson-Hudson R, Dyson-Hudson N. Nomadic pastoralism. Annual Review of Anthropology. 1980;**9**(1):15-61

[73] Steimann B. Conflicting strategies for contested resources: Pastoralists' responses to uncertainty in post-socialist rural Kyrgyzstan. In: Pastoral Practices in High Asia: Agency of 'development' Effected by Modernisation, Resettlement and Transformation. Dordrecht: Springer; 2012. pp. 145-160

[74] Sharifian A, Fernandez-Llamazares A, Wario HT, Molnar Z, Cabeza M. Dynamics of pastoral traditional ecological knowledge: A global state-of-the-art review. Ecology and Society. 2022;**27**(1):14

[75] Masters DG, Blache D, Lockwood AL, Maloney SK, Norman HC, Refshauge G, et al. Shelter and shade for grazing sheep: Implications for animal welfare and production and for landscape health. Animal Production Science. 2023;**63**(7):623-644

[76] Manjunathareddy GB, Sajjanar B, Sejian V. Impact of climate change on sheep disease occurrences and its management. Sheep Production Adapting to Climate Change. 2017:197-207

[77] Estevez I, Andersen IL, Naevdal E. Group size, density and social dynamics in farm animals. Applied Animal Behaviour Science. 2007;**103**(3-4):185-204

[78] Cooke SJ, Galassi DM, Gillanders BM, Landsman SJ, Hammerschlag N, Gallagher AJ, et al. Consequences of "natural" disasters on aquatic life and habitats. Environmental Reviews. 2022;**31**(1):122-140

[79] Cockram MS. A review of behavioural and physiological responses of sheep to stressors to identify potential behavioural signs of distress. Animal Welfare. 2004;**13**(3):283-291

[80] Ramya S, Poornima P, Jananisri A, Geofferina IP, Bavyataa V, Divya M, et al. Role of hormones and the potential impact of multiple stresses on infertility. Stress. 2023;**3**(2):454-474

[81] Brien FD, Cloete SW, Fogarty NM, Greeff JC, Hebart ML, Hiendleder S, et al. A review of the genetic and epigenetic factors affecting lamb survival. Animal Production Science. 2014;**54**(6):667-693

[82] Basile F, Capaccia C, Zampini D, Biagetti T, Diverio S, Guelfi G. Omics insights into animal resilience and stress factors. Animals. 2020;**11**(1):47

[83] Ramseyer A, Boissy A, Thierry B, Dumont B. Individual and social determinants of spontaneous group movements in cattle and sheep. Animal. 2009;**3**(9):1319-1326

[84] Rohwer S. The social significance of avian winter plumage variability. Evolution. Oxford: Oxford University Press; 1975;**29**(4):593-610

[85] Pandey LN, Gyawali R. Constraints and potential of goat and sheep production under transhumance management system in the high mountainous regions of Nepal. In: Conference: Research and Development Strategies for Goat Enterprises in Nepal. Kathmandu, Nepal, Kanibahal, Lalitpur: Siddartha Printing Press; 2013. pp. 92-101

[86] Van Dijk JG, Matson KD. Ecological immunology through the lens of exercise immunology: New perspective on the links between physical activity and immune function and disease susceptibility in wild animals. Integrative and Comparative Biology. 2016;**56**(2):290-303

[87] Valdez R, Krausman PR. Mountain Sheep of North America. Tucson: University of Arizona Press; 1999. pp. 1-353

[88] Gregory NG. Physiology and Behaviour of Animal Suffering. Oxford, UK, Ames, Iowa: Wiley-Blackwell Science; 2008. pp. 1-268. DOI: 10.1002/9780470752494

[89] Michelena P, Gautrais J, Gerard JF, Bon R, Deneubourg JL. Social cohesion in groups of sheep: Effect of activity level, sex composition and group size. Applied Animal Behaviour Science. 2008;**112**(1-2):81-93

[90] Misra AK. Climate change and challenges of water and food security. International Journal of Sustainable Built Environment. 2014;**3**(1):153-165

[91] Gionfriddo JP. Summer Habitat Use by Mountain Sheep. Tucson: The

University of Arizona Press; 1984. pp. 1-37. Available from: http://hdl.handle.net/10150/275221

[92] Behnke RH, Fernandez- Gimenez ME, Turner MD, Stammler F. Pastoral migration: Mobile systems of livestock husbandry. In: Milner-Gulland EJ, Fryxell J, Sinclair ARE, editors. Animal Migration–A Synthesis. Oxford: Oxford University Press; 2011. pp. 144-171

[93] Wolanski E, Gereta E, Borner M, Mduma S. Water, migration and the Serengeti ecosystem. American Scientist. 1999;**87**:526-533

[94] DeCesare NJ, Pletscher DH. Movements, connectivity, and resource selection of Rocky Mountain bighorn sheep. Journal of Mammalogy. 2006;**87**(3):531-538

[95] Serranito B, Cavalazzi M, Vidal P, Taurisson-Mouret D, Ciani E, Bal M, et al. Local adaptations of Mediterranean sheep and goats through an integrative approach. Scientific Reports. 2021;**11**(1):21363

[96] Wilcove DS. No Way Home: The Decline of the World's Great Animal Migrations. Wasbington, London: Island Press; 2008. pp. 1-123

[97] Said MY, Ogutu JO, Kifugo SC, Makui O, Reid RS, de Leeuw J. Effects of extreme land fragmentation on wildlife and livestock population abundance and distribution. Journal for Nature Conservation. 2016;**34**:151-164

[98] Kumar R, Kumar A, Saikia P. Deforestation and forests degradation impacts on the environment. In: Environmental Degradation: Challenges and Strategies for Mitigation. Cham: Springer International Publishing; 2022. pp. 19-46

[99] Frid A, Dill L. Human-caused disturbance stimuli as a form of predation risk. Conservation Ecology. 2002;**6**(1):11

[100] Karakus F. Weaning stress in lambs. Journal of International Scientific Publications: Agriculture & Food. 2014;**2**:165-170

[101] Mora-Medina P, Mota-Rojas D, Arch-Tirado E, Orozco-Gregorio H. Animal Welfare in Lambs: Ewe-Lamb Separation. Large Animal Review. 2015. pp. 39-44

[102] Dodd CL, Pitchford WS, Edwards JEH, Hazel SJ. Measures of behavioural reactivity and their relationships with production traits in sheep: A review. Applied Animal Behaviour Science. 2012;**140**(1-2):1-15

[103] Endris M, Feki E. Review on effect of stress on animal productivity and response of animal to stressors. Journal of Animal Veterinary Advances. 2021;**20**(1):1-14

[104] Cook CJ. Oxytocin and prolactin suppress cortisol responses to acute stress in both lactating and non-lactating sheep. Journal of Dairy Research. 1997;**64**(3):327-339

[105] Savage K. Sheep-Work: Food Empires and Pastoral Resistance in the Intermountain West. [Master Thesis]. Salt Lake City, UT: The University of Utah; 2010. pp. 1-56

[106] Hoare B. Animal Migration: Remarkable Journeys in the Wild. Berkeley: University of California Press; 2009. pp. 1-176

[107] McKenna E. Livestock: Food, Fiber, and Friends. Athens, Georgia: University of Georgia Press; 2018. pp. 1-53

[108] Bhatt N, Singh NP, Mishra AK, Kandpal D, Jamwal S. A detailed review of transportation stress in livestock and its management techniques. International Journal of Livestock Research. 2021;**11**(1):30-41

[109] Caminade C, McIntyre KM, Jones AE. Impact of recent and future climate change on vector-borne diseases. Annals of the New York Academy of Sciences. 2019;**1436**(1):157-173

[110] Broom DM, Kirkden RD. Welfare, stress, behaviour and pathophysiology. In: Veterinary Pathophysiology. Ames, Iowa: Blackwell; 2004. pp. 337-369

[111] Bianchi RM, Panziera W, Faccin TC, Almeida GLD, Cargnelutti JF, Flores EF, et al. Clinical, pathological and epidemiological aspects of outbreaks of bluetongue disease in sheep in the central region of Rio Grande do Sul. Pesquisa Veterinária Brasileira. 2017;**37**:1443-1452

[112] Morgante M. Digestive disturbances and metabolic-nutritional disorders. In: Dairy Sheep Nutrition. Wallingford UK: CABI Publishing; 2004. pp. 165-190

[113] Yadav N, Upadhyay RK. Global effect of climate change on seasonal cycles, vector population and rising challenges of communicable diseases: A review. Journal of Atmospheric Science Research. 2023;**6**(1):21-59

[114] Broom DM. Transport stress in cattle and sheep with details of physiological, ethological and other indicators. Deutsche Tierarztliche Wochenschrift. 2003;**110**(3):83-88

[115] Tort L, Teles M. The endocrine response to stress-a comparative view. In: Basic and Clinical Endocrinology Up-to-date. London, UK: IntechOpen; 2011. pp. 263-286

[116] Todini L. Thyroid hormones in small ruminants: Effects of endogenous, environmental and nutritional factors. Animal. 2007;**1**(7):997-1008

[117] Jackson EK. Vasopressin and other agents affecting the renal conservation of water. In: Goodman & Gilmans. The Pharmaological Basis of Therapeutics. 11th ed. New York: McGraw Hill; 2006. pp. 771-788

[118] Sun Q, Dimopoulos G, Nguyen DN, Tu Z, Nagy N, Hoang AD, et al. Low-dose vasopressin in the treatment of septic shock in sheep. American Journal of Respiratory and Critical Care Medicine. 2003;**168**(4):481-486

[119] Muhammad M, Stokes JE, Manning L. Positive aspects of welfare in sheep: Current debates and future opportunities. Animals. 2022;**12**(23):3265

[120] Walker SC, McGlone FP. The social brain: Neurobiological basis of affiliative behaviours and psychological well-being. Neuropeptides. 2013;**47**(6):379-393

[121] Sutherland MA, Worth GM, Dowling SK, Lowe GL, Cave VM, Stewart M. Evaluation of infrared thermography as a non-invasive method of measuring the autonomic nervous response in sheep. PLoS One. 2020;**15**(5):e0233558

[122] Gilbert-Gregory S. Sensory and perception. Introduction to Animal Behavior and Veterinary Behavioral Medicine. Hoboken, New Jersey: John Wiley & Sons, Inc; 2024. pp. 65-89

[123] Xiao J, Guo W, Han Z, Xu Y, Xing Y, Phillips CJ, et al. The effects of housing on growth, immune function and antioxidant status of young female lambs in cold conditions. Animals. 2024;**14**(3):518

[124] Bradford BJ, Contreras GA. Adipose tissue inflammation: Linking physiological stressors to disease susceptibility. Annual Review of Animal Biosciences. 2024;**12**(1):261-281

[125] Li B, Wu K, Duan G, Yin W, Lei M, Yan Y, et al. Folic acid and taurine alleviate the impairment of redox status, immunity, rumen microbial composition and fermentation of lambs under heat stress. Animals. 2024;**14**(7):998

[126] de Jong IE, Wells RG. In utero extrahepatic bile duct damage and repair: Implications for biliary atresia. Society for Pediatric and Developmental Pathology. 2024. DOI: 10.1177/10935266241247479

[127] Hua KXQ, Peng Y, Sun Y, He R, Luo R, Jin H. The role of hormones in the regulation of lactogenic immunity in porcine and bovine species. Domestic Animal Endocrinology. 2024;**88**:106851

[128] Zhang R, Wei M, Zhou J, Yang Z, Xiao M, Du L, et al. Effects of organic trace minerals chelated with oligosaccharides on growth performance, blood parameters, slaughter performance and meat quality in sheep. Frontiers in Veterinary Science. 2024;**11**:1366314

[129] Oldham L. Affective and Cognitive Components of Agonistic Behaviour in Pigs: Implications for Animal Welfare and Social Behaviour. Edinburgh, UK: The University of Edinburgh; 2024. pp. 1-257

[130] Hassan LM, Omer EA. Review Sudan's Sheep Production: Limitations and Prospects. IntechOpen; 2024. pp. 1-19. DOI: 10.5772/intechopen.114158

[131] Houpt KA. Domestic Animal Behavior for Veterinarians and Animal Scientists. 7th edition. Hoboken, New Jersey, United States: Wiley-Blackwell; 2024. pp. 1-496

[132] Ibrahim A, Ngadiyono N, Maulana H, Atmoko BA. Effect of shearing on thermo-physiological, behavior, and productivity traits of two Indonesian local sheep breeds. Tropical Animal Science Journal. 2024;**47**(1):42-52

[133] Davies KL, Miles J, Camm EJ, Smith DJ, Barker P, Taylor K, et al. Prenatal cortisol exposure impairs adrenal function but not glucose metabolism in adult sheep. Journal of Endocrinology. 2024;**260**:e230326

[134] Saroha V. Developmental programming of the cell stress response and metabolic inflammation in liver and adipose tissue in an ovine model [doctoral dissertation]; University of Nottingham. 2017

[135] Hyder I, Ravi Kanth Reddy P, Raju J, Manjari P, Srinivasa Prasad C, Aswani Kumar K, et al. Alteration in rumen functions and diet digestibility during heat stress in sheep. Sheep Production Adapting to Climate Change. Singapore: Springer; 2017. pp. 235-265

[136] Destrez A, Deiss V, Leterrier C, Calandreau L, Boissy A. Repeated exposure to positive events induces optimistic-like judgment and enhances fearfulness in chronically stressed sheep. Applied Animal Behaviour Science. 2014;**154**:30-38

[137] Brien FD, Cloete SWP, Fogarty NM, Greeff JC, Hebart ML, Hiendleder S, et al. A review of the genetic and epigenetic factors affecting lamb survival. Animal Production Science. 2014;**54**(6):667-693

[138] Leite JHGM, Da Silva RG, Asensio LAB, de Sousa JER, da

Silva WST, da Silva WE, et al. Coat color and morphological hair traits influence on the mechanisms related to the heat tolerance in hair sheep. International Journal of Biometeorology. 2020;**64**:2185-2194

[139] Gowane GR, Gadekar YP, Prakash V, Kadam V, Chopra A, Prince LLL. Climate change impact on sheep production: Growth, milk, wool, and meat. Sheep Production Adapting to Climate Change. Singapore: Springer; 2017. pp. 31-69

[140] Pantoja MHDA. Thermoregulation responses in sheep: A cellular approach [doctoral dissertation]; Universidade de Sao Paulo. 2022

[141] Skarpe C, Hester AJ. Plant traits, browsing and gazing herbivores, and vegetation dynamics. Ecological Studies. 2008;**195**:217

[142] Rahmann G, Seip H. Alternative management strategies to prevent and control endo-parasite diseases in sheep and goat farming systems–A review of the recent scientific knowledge. Land Bauforschung Volkenrode. 2007;**2**:75-88

Chapter 14

Resilience, Sustainability, and the Role of Livestock in Rural Food Systems: A Case Study from Ethiopia

Asrat Guja Amejo

Abstract

This chapter examines the resilience and sustainability of smallholder farming systems in Sub-Saharan Africa (SSA), with a particular emphasis on Ethiopia. Historically, traditional agricultural systems in this region have maintained a symbiotic balance between crop cultivation and livestock rearing, fostering sustainable ecosystems. However, contemporary farming practices have disrupted this equilibrium, leading to a compartmentalized approach that threatens the sustainability of these systems. This chapter underscores the critical integration of crop and livestock systems as a strategy for optimizing resource utilization, enhancing productivity, and improving livelihoods. It assesses the influence of spatial factors such as topography, elevation, aspect, weather extremes, and market proximity on resource distribution and agricultural practices. Furthermore, it explores the socio-economic and cultural contexts of smallholder agriculture, highlighting the role of livestock in diversifying income and providing buffers against economic and environmental shocks. The complex dynamics of mixed crop-livestock systems are examined, demonstrating their efficacy in maximizing resource use, promoting sustainability, and bolstering food security. The chapter concludes by addressing the challenges and opportunities for enhancing the resilience and sustainability of smallholder farming through adaptive management practices, supportive policy frameworks, and targeted research and development interventions, advocating for a holistic approach to sustainable agricultural development in SSA.

Keywords: sustainable agriculture, smallholder farming systems, crop-livestock integration, climate adaptation, socio-economic resilience, cultural significance

1. Introduction

Livestock are integral to Ethiopian agriculture, particularly within smallholder farming systems, where they contribute significantly to household income, nutritional diversity, and cultural practices. This study examines the complex dynamics of livestock rearing and crop production, emphasizing their synergistic roles in bolstering resilience and sustainability within Ethiopia, with a comparative analysis of rural food systems across Sub-Saharan Africa (SSA). Here, resilience denotes the system's capacity to absorb disturbances and maintain functionality over time, while sustainability encompasses efficient resource use and environmental stewardship for long-term viability. As articulated by Keeney [1], sustainable agriculture integrates plant and animal production to meet human needs, enhance environmental quality, and optimize resource use, emphasizing productivity alongside environmental and social well-being.

Historically, Ethiopia's traditional agricultural practices fostered a symbiotic relationship between crops and livestock, nurturing sustainable ecosystems [2]. However, the advent of modern farming methods and industrialization has disrupted this balance, often overlooking the inherent interconnectedness of these systems [3]. This shift, despite technological advances, has had significant impacts on regions like Africa, characterized by small farms, limited land sizes, and a strong imperative to diversify income sources [4].

Sub-Saharan Africa presents unique challenges and opportunities within its smallholder farming systems due to diverse and often remote landscapes. Adaptability, flexibility, and sustainable practices are essential to meet the evolving needs of these systems, necessitating a comprehensive understanding of technology, management practices, and policy frameworks. Agricultural systems are multifaceted, harnessing natural resources for food, fiber, and energy production, with smallholder farming systems dedicated to achieving food sufficiency, environmental sustainability, and socio-economic well-being through integrated approaches.

Mixed farming systems are crucial in developing countries, spanning extensive arable and grassland areas to ensure food security and livelihoods by providing staple crops and essential livestock products to millions. However, these systems face mounting pressures to enhance productivity, combat environmental degradation, and adapt to climate change, driven by global demands for agricultural products. With growing populations and resource constraints, enhancing agricultural productivity while preserving environmental integrity is paramount. Thus, understanding the dynamics within these systems is critical for fostering resilient and inclusive agricultural development.

This paper focuses on examining the resilience, sustainability, and pivotal role of livestock within Ethiopia's smallholder farming systems. It explores intricate interactions among agricultural components—such as land use practices, food production systems, energy generation, environmental stewardship, and socio-cultural dynamics—that collectively shape the resilience and sustainability of these systems. By elucidating these dynamics, this study aims to identify pathways for enhancing the resilience of Ethiopian agriculture amidst evolving socio-economic and environmental challenges.

2. Natural resource: based livestock production

Natural resource–based livestock production involves the utilization of local resources such as land, water, vegetation, and biomass to support livestock rearing. This approach is integral to traditional farming systems in many regions, including SSA, where smallholder farmers rely on the integration of crop and livestock production to sustain their livelihoods.

In SSA, livestock production is intricately linked with the broader food system, playing a crucial role in ensuring food security, income generation, and socio-economic resilience for smallholder farming communities. Recognizing the interconnectedness of crop and livestock systems and adopting integrated agricultural approaches are essential for enhancing productivity and resource use efficiency. Ultimately, these strategies aim to improve the livelihoods of rural populations in the region.

2.1 Spatial framework and resource utilization

Livestock systems are intricate, interconnected human-environment frameworks that integrate livestock with environmental and socio-economic variables through natural resources, market participation, and livelihood strategies [5–7]. The distribution of livestock production is profoundly influenced by resource allocation, which is, in turn, shaped by landscape relief and topography. Key factors such as elevation, slope, and aspect are critical in determining resource distribution, access, and availability of livelihood options. High-altitude areas, for instance, tend to exhibit distinct vegetation types due to cooler temperatures and shorter growing seasons. Slopes affect soil erosion, water runoff, and vegetation cover, while aspect (e.g., north-facing versus south-facing slopes) impacts sunlight exposure, microclimates, and the gradient flow of resources.

Topography significantly affects soil formation processes, with steep slopes often leading to erosion, whereas flat areas tend to accumulate organic matter. Elevation plays a key role in determining temperature, precipitation, and vegetation, subsequently impacting soil nutrient cycling. Furthermore, altitude dictates vegetation zones, such as alpine, subalpine, temperate, and tropical, while slope position influences moisture availability and vegetation composition. Aspect further affects sunlight exposure and vegetation growth.

Spatial variations in rainfall have notable implications for soil moisture, vegetation growth, and nutrient availability, with rain shadow effects on the leeward sides of mountains resulting in drier conditions. Additionally, spatial patterns of grazing, cropping, and other land uses significantly impact soil fertility and vegetation. Terracing on slopes can effectively mitigate erosion and enhance soil fertility, while agroforestry systems, integrating trees with crops and livestock, alter resource distribution dynamics. Bodies of water such as rivers, lakes, and wetlands influence soil moisture, vegetation, and nutrient availability, consequently affecting livestock access to water and their distribution and grazing behavior.

Poccard-Chapuis et al. [8] conducted a study on integrated crop-livestock systems in tropical regions, identifying them as sophisticated and dynamic approaches to land-based livestock production. Their research aimed to unravel the complexities inherent in these systems and evaluate their potential by analyzing their spatial and social configurations within the broader landscape context.

In these systems, community livelihoods are intricately woven into the environmental fabric, relying on factors such as land size holdings, food and feed product availability from agricultural plots, and socio-cultural practices to sustain life throughout the seasons. However, farm operations and labor productivity are often hampered by acute seasonality in tropical climates, where significant disparities exist between wet and dry seasons or periods with and without irrigation water [9]. Thus, the sequence of seasons plays a crucial role in shaping production activities and resource allocations within livelihood systems.

Efforts to integrate these resource-based agricultural production systems and address associated challenges are crucial for regional economic development, necessitating access to adequate information. Natural resource–based livestock production embodies a holistic approach to farming that leverages local ecosystems and traditional knowledge to sustainably meet the needs of rural communities. By integrating livestock and cropping systems within the spatial context of the landscape, farmers can optimize resource use, enhance productivity, and build resilience against environmental and socio-economic challenges.

Achieving sustainable development in integrated crop-livestock systems is a multifaceted challenge, particularly in maintaining the efficiency of circular agricultural production. Agriculture, including livestock production, has both benefits and drawbacks concerning environmental aspects. Challenges associated with livestock husbandry or agricultural farming necessitate research, education, and technology to overcome setbacks, especially within communal business managed by humans.

Drawing upon Darwin's observations in "On the Origin of Species," where he identifies slugs, insects, and native weeds as pivotal in the constant struggle for existence, parallels can be drawn to contemporary agricultural challenges. These challenges, such as vector-borne diseases, wildfires, and the lack of insightful management, underscore the dynamic interplay of evolution and the ongoing need for adaptation and resilience. These challenges reflect broader ecological dynamics and pose significant threats to agricultural sustainability. Addressing these challenges requires concerted efforts in research, education, and technological advancements. For marginalized communities in developing nations, livestock development guided by scientific inquiry can play a pivotal role in mitigating environmental impacts and enhancing the resilience of integrated crop-livestock systems.

2.1.1 Geographical targeting and prioritization

Spatial analysis helps identify areas where interventions are most needed to identify high-potential zones for livestock production (e.g., fertile valleys); marginal areas where sustainable practices are critical; areas prone to degradation (e.g., steep slopes) that require conservation efforts. Prioritizing interventions based on spatial data ensures efficient resource allocation.

2.1.2 Out-scaling research results

Spatial analysis provides insights for scaling up successful practices such as identifying similar landscapes in other regions and adapting context-specific solutions to new areas while ensuring ecological compatibility with local conditions.

Smallholder farmers face constraints in managing integrated crop-livestock production systems, hindering their ability to benefit from research and development efforts. Infrastructural limitations and poor market access further impede farmers' capacity to meet the growing demands for food in SSA and globally.

To address these challenges, research and development interventions should prioritize improving policy and institutional aspects to enable farmers to increase total farm productivity and household income. This involves providing better access to technologies, services, and markets. Additionally, more research and policy instruments are needed to enhance resource use efficiency in integrated farming systems.

A detailed analysis of farmers' circumstances and practices across various regions is essential to understand the specific opportunities and constraints present in different agro-ecological zones and socio-economic settings. Understanding the subsystems within integrated farming systems is crucial for comprehending the bio-economic foundations of rural livelihood systems, which entails accounting for component stocks, resource flows, and interactions.

3. Socio-economic context of smallholder rural agriculture

Livestock play a crucial role in providing diversified income sources for rural communities. By relying on both crops and livestock, smallholders reduce their vulnerability to market fluctuations, climate variability, and other economic shocks. Studies have shown that households engaged in both crop cultivation and livestock rearing are better equipped to withstand adverse events. Livestock products (such as milk, meat, and wool) provide a steady income stream even when crop yields are affected by weather conditions or market volatility [10].

Contrary to the belief that farm size alone determines socio-economic outcomes, it is the interaction effect of livelihood strategies that truly matters. Smallholder families manage a diverse set of livelihood activities, including crop farming, livestock rearing, and off-farm work. Research has found that households combining crop production with livestock rearing achieve better overall well-being [11, 12]. The synergy between these activities allows for risk-sharing, resource utilization, and complementary labor allocation. Understanding these income patterns can help in designing better support mechanisms for rural smallholder farmers, ensuring more balanced and sustainable economic growth across different regions.

Efficient resource use is essential for economic returns. Integrating crop and livestock systems optimizes resource utilization (such as crop residues as livestock feed) and minimizes waste. Smallholders who integrate crop residues into livestock feed achieve higher productivity. For example, using crop residues as fodder reduces the need for external feed sources, leading to cost savings and improved livestock health.

Smallholder systems evolve based on local conditions, including climate, soil, distance to market, and available resources. Farmers adapt their strategies to maximize returns. Case studies across different regions reveal that successful smallholders tailor their practices to suit their specific context. This adaptability ensures long-term sustainability and resilience.

Understanding the intricate web of livelihood strategies in smallholder farming systems is crucial for policymakers. By recognizing the interplay between crops, livestock, and resource utilization, we can design effective agricultural policies that promote economic stability and enhance rural livelihoods.

Tittonell [13] emphasizes the need to stimulate desirable shifts in farming systems by addressing both ends simultaneously. This involves focusing on the knowledge base for the ecological intensification of smallholder landscapes, policy developments, and market dynamics. Additionally, understanding thresholds in specific variables that may indicate the presence of tipping points is critical, although these thresholds are often elusive and site-specific in East African agro-ecosystems.

3.1 Understanding mixed crop-livestock systems

An agricultural system, often referred to as an agro-ecosystem, comprises a multitude of components designed to produce crops and raise livestock to fulfill diverse requirements, including food, fiber, and energy [14]. These systems seamlessly integrate both crop cultivation and livestock rearing on the same land, typically in a mutually advantageous manner. They are prevalent across various agro-ecological zones and have been cultivated for centuries, evolving to adapt to local conditions and requirements. Traditionally, these systems have been managed in an integrated manner, with farmers capitalizing on the synergies between crops and livestock to optimize yields and mitigate risks.

One of the key advantages of mixed crop-livestock systems is their ability to maximize resource utilization. Livestock can utilize crop residues and by-products as feed, while their manure can be used to fertilize crops, creating a closed nutrient cycle. This integration enhances overall productivity and sustainability of the farming system.

Moreover, mixed crop-livestock systems contribute significantly to livelihoods and food security in rural communities. They provide diverse sources of income and nutrition, as well as employment opportunities along the value chain, from farming and animal husbandry to processing and marketing.

Mixed crop-livestock systems represent a dynamic and adaptive approach to agricultural production, rooted in the knowledge and traditions of local communities. Sustaining and enhancing these systems can contribute significantly to achieving food security, promoting rural development, and conserving natural resources in developing countries.

3.1.1 Structural functions of food system

Structural definitions of mixed crop-livestock farming typically classify farms based on the proportional contribution of crops and animals to the farm's total production. The structural function of livestock in mixed system primarily exists in land use function. In smallholder system the plot of the farm is very small [4, 15, 16]. The relationship between farm size and the welfare of smallholder farmers in SSA is complex and multifaceted. While larger farm sizes generally provide advantages such as economies of scale and increased production potential, the reality for most smallholders in SSA is characterized by limited land access. Consequently, a diverse and integrated approach to livelihood activities becomes essential for improving their welfare. The functional integration of farm components is pivotal for the production of food, feed, and fiber, thereby supporting the diverse needs of farm households. Land use management involves a dynamic interplay of annual crops, perennial crops, fallow periods, vegetation covers, and grazing lands.

Food security for smallholders hinges on effectively utilizing their small plots of land throughout the seasons. The storage of food in this mixed system allows for continuity across different times of the year. However, if these components of the food system fail to interact effectively, the size of the farm may not adequately provide food for all family members within the household.

In this context of environmental influence and seasonal variability in food and feed production, farmers focus primarily on cultivating rain-fed annual crops during the primary rainy season. To mitigate seasonal shortages of food and fodder, these activities are complemented by the cultivation of root crops under conditions of moderate rainfall and moisture. The inclusion of perennial crops within the agricultural system serves to stabilize the availability of both food and fodder resources in households.

Research has documented the seasonal dynamics of feed production in the northern part of Ethiopia. Studies by Abegaz [17] and Amejo [18] have identified a distinct seasonal pattern in the availability and quality of forage, which significantly impacts the live weight dynamics of livestock.

As indicated by Abegaz [17], during the late wet season from September to November, cattle graze on natural pasture, resulting in high availability and quality of feed resources, which is reflected in substantial live weight gains. Beginning in December, cattle graze on crop field aftermath, where the reduced availability and quality of feed leads to lower live weight gains. From April onwards, cattle experience a decline in live weight, with the most significant loss occurring in August.

Understanding the structural definitions, functional interactions, and seasonal dynamics within these systems is essential for promoting sustainable agricultural development and improving food security. By recognizing the importance of true integration and leveraging local agricultural practices, stakeholders can enhance productivity, resilience, and livelihoods within mixed crop-livestock farming systems.

However, previous approaches may have overlooked the functional interactions between crops and livestock in mixed crop-livestock systems. While farms categorized under this system may sometimes focus predominantly on either crop or animal production without actual integration, true integration involves functional connections that create mutual benefits for both components. A more nuanced definition proposed by Séré and Steinfeld [19] considers livestock systems where a significant portion of animal feed comes from crop by-products or where non-livestock farming activities contribute substantially to the total production value. This definition acknowledges the intensity of material flow between crops and livestock.

Crop-livestock integration extends beyond material exchange, impacting water flows, soil moisture, and ecological resilience. Interactions within mixed systems can enhance species diversity, soil health, nutrient cycling, and pest management. Synergy between crops and animals, such as livestock providing manure for crops and crop residues feeding animals, contributes to soil fertility, waste reduction, and overall sustainability. The distinction between structural and functional definitions of mixed crop-livestock farming is crucial for understanding environmental performance. True integration, characterized by mutual benefits between crops and animals, is essential for sustainable agricultural systems.

3.1.2 Livestock buffering against risk factors

Livestock also serves as a risk buffer in rural areas. In rural mixed systems, transporting bulk agricultural products to distant markets can be challenging due to a lack of proper infrastructure, transportation facilities, and high transportation costs. However, farm animals can be mobilized on foot, allowing farmers to reach markets even in remote areas. This mobility facilitates the exchange of goods and contributes to the local economy. Crop-livestock diversification is a potent strategy for farm households, benefiting both market value and household well-being. By optimizing the interaction between crops and livestock, we can create resilient and sustainable agricultural systems.

Livestock production serves as more than just an economic pursuit; it acts as a vital lifeline for marginalized communities grappling with the challenges imposed by climate change. By advocating for sustainable practices, preserving local breeds, and bolstering community resilience, Berhe et al. [20] tried to address the persistent issues plaguing rural areas in East Africa. Across diverse agricultural communities—including pastorals, semi-pastorals, agro-pastorals, and mixed farming—climate-induced threats, particularly drought, cast a long shadow. The intricate interplay of factors, coupled with the timing of extreme events, significantly influences the impact of climate change on food security and poverty [21]. This variability in weather patterns, encompassing fluctuations in temperature, precipitation, and wind, underscores the mounting challenges faced by rural populations in East Africa [22].

3.1.3 Environmental function

In the realm of sustainability, an agro-ecosystem embodies a dynamic fusion of various elements: soil characteristics, past management practices, microbial life, plant roots, genetic diversity, animal and insect interactions, human interventions, atmospheric conditions, and pathogenic influences, all unfolding across time and space. These components engage in intricate interactions, shaping the ecosystem in unpredictable ways. Fragmenting the system into simpler parts does not necessarily enhance our ability to forecast its dynamics accurately.

Embracing sustainability entails acknowledging that the components of an agro-ecosystem transcend their individual roles, existing in a symbiotic relationship where each element contributes to the whole. This holistic perspective recognizes that sustainable agro-ecosystems are more than just the sum of their parts; they are intricate networks of interdependent entities. Sustainable practices thus require a comprehensive understanding of these relationships and a commitment to nurturing the interconnectedness that sustains the system's vitality and resilience over time.

A classic example come from southwestern Ethiopia from traditional enset-livestock farming systems [23–26]. Traditional agricultural practices intertwine enset (*Ensete ventricosum*) and livestock production with a variety of other perennial, vegetable, and root crops, forming a harmonious integrated system. Within this framework, the dynamic interplay among soil attributes, historical cultivation methods, microbial communities, root systems, crop genetics, animal husbandry practices, insect populations, human interventions, climatic conditions, and disease pressures unfolds across both temporal and spatial dimensions.

The traditional enset-livestock farming system symbolizes efficiency in resource utilization, yielding significant outputs from relatively small parcels of land. This efficiency is underpinned by the system's multifunctionality, which fosters adaptability and resilience. Moreover, these traditional farming practices demonstrate consistency in delivering productive outputs, owing to their inherent ability to perform various functions simultaneously.

Notably, the traditional enset-livestock farming system exhibits remarkable resistance to disturbances, such as drought, and showcases a capacity for recovery, swiftly regaining its original functionality even in the face of adversity [27–29]. This resilience stems from the integrated nature of the system, which facilitates the efficient allocation of resources and the dynamic response to changing environmental conditions.

In the vast agricultural landscapes of Ethiopia, the synergy between enset and livestock farming has been a cornerstone of sustainable agriculture for centuries [30]. This integrated farming system, deeply rooted in indigenous knowledge, embodies a holistic approach that balances tradition, sustainability, and productivity.

Enset, a versatile plant in the banana family, serves a dual purpose as both food for humans and feed for livestock. Livestock reciprocate by providing manure to enset plants, creating a symbiotic relationship that sustains both components of the farming system. Indigenous knowledge has long guided the management practices, ensuring the sustainability of soil fertility and agricultural productivity.

Despite its resilience, the enset-livestock farming system faces existential threats in the modern era [25, 27, 31]. Acculturation, urbanization, and the introduction of new crops are eroding indigenous knowledge and practices. Climate change exacerbates these challenges, leading to emerging pests and declining soil fertility. The decline in enset cultivation and livestock herds not only threatens food security but also jeopardizes the rich biodiversity and soil health inherent in this traditional farming system.

3.1.4 Asset accumulation

The distribution and intensity of livestock holdings in smallholder systems are influenced by agro-ecological conditions, regional functional specialties, and the diverse roles livestock play in farm households. Agro-ecology encompasses the climatic conditions that influence the types of livestock that can be raised in a given area. For instance, in arid and semi-arid regions, drought-resistant livestock such as goats and camels are more prevalent. Conversely, in regions with more abundant rainfall and lush pastures, cattle and sheep may be more common. The quality of soil and the types of vegetation available also affect livestock distribution. Fertile soils that support rich pastures can sustain larger herds and a greater diversity of livestock. In contrast, poorer soils may only support hardier, less demanding animals. Access to water is a critical factor in livestock management. Regions with ample water resources can support larger herds, whereas in water-scarce areas, the intensity of livestock holding must be lower to prevent overgrazing and degradation of resources. The landscape also plays a role in livestock distribution. In mountainous regions, certain types of livestock, such as goats, which are adept at navigating rugged terrain, are more common. Plains and flatter regions might support larger herds of cattle or sheep.

Livestock provide a direct source of food through meat, milk, and eggs. This is particularly important in regions where crop production may be unreliable due to climatic conditions. Livestock can be a significant source of income through the sale of animals, animal products (such as milk, wool, and eggs), and by-products (like manure for fertilizer). In many smallholder systems, livestock are used as draft animals for plowing fields and transporting goods. This reduces the labor burden on farm households and can improve agricultural productivity. Livestock act as a form of insurance for smallholder farmers, providing a buffer against crop failures or economic shocks. During difficult times, livestock can be sold or used to meet immediate household needs. Recognizing these factors is essential for developing tailored interventions that enhance livestock productivity and sustainability, thereby improving the livelihoods of smallholder farmers. Understanding and supporting the unique needs of different agro-ecological zones can lead to more resilient and productive smallholder farming systems.

4. Sustainability

Livestock play a crucial role in nutrient cycling by converting organic matter (such as crop residues, grasses, and by-products) into valuable manure. This process contributes to soil fertility and overall ecosystem health. Crop residues, grasses, and other organic materials are consumed by livestock. Livestock digest these materials, breaking them down into simpler compounds. The resulting manure contains essential nutrients (nitrogen, phosphorus, potassium, etc.) that enrich the soil.

Livestock manure is rich in nutrients. When applied to the soil, it replenishes essential elements necessary for plant growth. Studies have shown that fields where livestock graze or where manure is applied exhibit improved soil fertility compared to areas without livestock influence. Nutrient-rich manure enhances soil organic matter content, microbial activity, and nutrient availability. Typical example is enset-livestock production system in smallholder system.

By recycling nutrients through livestock, farmers can reduce their reliance on synthetic fertilizers. Research indicates that integrating livestock into cropping systems leads to decreased fertilizer requirements. Livestock manure provides a sustainable alternative to chemical fertilizers, reducing environmental impacts.

Livestock contribute to sustainable agricultural practices by maintaining nutrient cycles. Sustainable farming systems that incorporate livestock have been associated with increased crop yields, improved soil structure, and reduced erosion. Enset-livestock farming, specifically, benefits from nutrient cycling, as enset (a drought-tolerant plant) and livestock coexist harmoniously.

Livestock manure introduces diverse microbial communities into the soil. Soil microbiota thrive on organic matter from manure, enhancing nutrient cycling and decomposition processes. These microorganisms contribute to soil health, disease suppression, and nutrient availability.

Integrated agricultural systems exhibit significant interdependencies among soil, crops, livestock, and the atmosphere. Livestock grazing directly affects soil physical properties, nutrient availability, and carbon sequestration. Furthermore, these interactions play a crucial role in influencing the exchange of greenhouse gases such as carbon dioxide, methane, and oxygen in the atmosphere.

Livestock play a pivotal role in agriculture by supplying draft power for plowing, assisting in weed and pest management, and enhancing soil quality through grazing and the deposition of manure. In Ethiopia, draft oxen are indispensable, cultivating over 99% of grains and root crops, with each household utilizing these animals for approximately 125 days annually [15, 32]. This integration significantly supports crop farming, reducing dependency on fossil fuel-powered machinery and promoting more sustainable agricultural practices.

Grazing animals contribute to weed suppression by consuming undesirable plant species, while certain livestock, such as chickens and ducks, play a crucial role in precision agricultural management by actively feeding on insects and pests, thereby mitigating their adverse effects on crops. The integration of livestock with crop production confers multiple advantages, including enhanced resource use efficiency, improved soil health, and reduced environmental impact. By leveraging the synergies between crop and livestock systems, farmers can establish more sustainable and resilient agricultural systems.

The synergy between enset and livestock farming epitomizes a harmonious relationship between humans, animals, and the environment. Yet, the sustainability of this ancient farming system hangs in the balance, threatened by rapid societal changes and environmental degradation. By fostering localized ownership, enhancing awareness, and leveraging technology, we can revitalize enset-livestock farming for future generations. Through concerted research efforts and innovative interventions, we can ensure that Ethiopia's agricultural legacy endures, nourishing both its people and the land they depend on.

To address these challenges and safeguard the future of enset-livestock farming, a multidisciplinary research approach is imperative. Research must focus on understanding the intricate interactions within the agro-ecosystem, optimizing resource utilization, and enhancing productivity. Key areas of research include improving livestock management practices, enhancing feed supplementation using enset by-products, and assessing the impact of livestock on soil fertility and microbial diversity. Furthermore, integrating traditional knowledge with modern agricultural practices can unlock innovative solutions for sustainable intensification and resilience.

5. Role in rural food systems

The role of sustainability and innovation in smallholder and pastoral systems encompasses a broad array of subjects. Livestock is an indispensable asset for smallholder farmers, significantly impacting human capital through nutrition and health. Animal-source foods, including meat, milk, eggs, and offal, provide high-quality protein and essential micronutrients, thereby enhancing the nutritional adequacy of diets that have traditionally relied on nutritionally poor but readily available staple crops.

Livestock holds profound cultural and social significance in many communities across SSA. These animals are integral to traditional ceremonies, rituals, and social exchanges, reinforcing social cohesion and identity. Smallholder farming systems and pastoralism, which have sustained human civilization for centuries, regard livestock as critical assets with multifaceted roles: they provide food, fuel, fiber, medicine, draft power, transport, manure, social status, economic security, increased income, and political power. The cultural importance of livestock adds distinctive value to various societies, enhancing their social fabric and economic resilience.

For instance, in traditional weddings, livestock such as cows, camels, goats, or sheep, along with products like butter and honey, are often exchanged as part of dowries or bride prices, symbolizing prosperity, fertility, and family bonds. Livestock also plays a pivotal role in spiritual ceremonies and rituals, with animal sacrifices during religious festivals or rites believed to appease deities, bring blessings, or cleanse sins.

Livestock and their products are deeply intertwined with the cultural, nutritional, and spiritual lives of rural communities. Their role transcends mere sustenance, shaping traditions, celebrations, and family dynamics. Fresh milk, butter, and other dairy products are often the first foods given to mothers postpartum, providing essential nutrients such as protein, calcium, and vitamins that aid in recovery. Additionally, milk and milk products are crucial for infant nutrition, serving as valuable alternatives when breastfeeding is insufficient or not possible, offering necessary nutrients for healthy growth.

Historical texts, folklore, and oral traditions frequently reference livestock in cultural practices. Ancient texts from various civilizations describe livestock offerings during religious ceremonies, and anthropological studies have documented the cultural significance of livestock across societies. Through field studies, researchers observe how livestock ownership, rituals, and dietary practices intersect in daily life, providing firsthand evidence of their integral role.

Figure 1 illustrates the variety of animal-sourced foods consumed by postpartum mothers and during various festival ceremonies. These foods, including meats, dairy products, and other livestock-derived items, play a critical role in nutrition and cultural practices. They not only provide essential nutrients for maternal recovery and infant health but also hold significant cultural importance, being integral to traditional rituals and celebratory events. The visual representation underscores the multifaceted role of animal-sourced foods in enhancing both health and cultural heritage within rural communities.

Figure 1.
Diverse forms of animal-sourced foods for postpartum mothers and various festival and cultural ceremonies; source: [33].

6. Materials and methods

6.1 Literature review

The literature review synthesized existing knowledge on the role of livestock in enhancing resilience and sustainability in rural food systems, focusing particularly on Ethiopia. Relevant sources were systematically identified through searches of academic databases (e.g., Scopus, PubMed, Web of Science, Google Scholar), reports from international organizations (e.g., FAO, IFAD), and government publications. Included studies provided insights into livestock-crop integration, socio-economic impacts, cultural significance, and policy frameworks.

6.2 Case study selection

Case studies were selected based on their representation of diverse agricultural practices in Ethiopia, including traditional mixed crop-livestock systems. Local experts and agricultural extension officers assisted in identifying regions known for these practices to ensure comprehensive coverage.

6.3 Data collection

6.3.1 Primary data

Farm Surveys: Structured surveys were conducted in selected communities to gather quantitative data on livestock holdings, crop production, land use, and socio-economic indicators. Surveys included questions on livestock species, numbers, management practices, and contributions to household income. Additionally, global positioning system (GPS) tools were utilized to accurately track the coordinate locations of household resident points.

Interviews: Semi-structured interviews were conducted with smallholder farmers, community leaders, and local experts to collect qualitative data. Topics covered perceptions of livestock's role in food security, traditional knowledge in livestock management, challenges faced, and adaptive strategies employed.

Focus Group Discussions: Discussions with farmer groups explored community perceptions, cultural practices related to livestock, and their roles in local livelihood strategies.

6.3.2 Secondary data

National agricultural statistics, reports from research institutions, and publications from non-governmental organizations (NGOs) were reviewed to complement primary data and provide a broader context of trends in Ethiopian agriculture. Additionally, posters and photos were sourced from cultural exhibition centers focusing on traditional animal-source foods. Metadata from various software sources were retrieved and accessed.

6.4 Data analysis

Quantitative Analysis: Descriptive statistics (e.g., means, percentages) were used to analyze data from farm surveys using statistical software (e.g., SPSS). This characterized livestock-crop integration, land use patterns, and socio-economic impacts.

Qualitative Analysis: Thematic analysis was applied to interview transcripts and focus group discussions to identify key themes related to resilience, sustainability, cultural significance, and challenges in livestock management.

7. Results

Table 1 provides a comprehensive overview of smallholder farming systems across various agro-ecological zones in Ethiopia, illustrating the diverse agricultural practices adapted to local environmental conditions. It highlights distinct spatial distributions where farming activities vary significantly based on factors such as elevation, slope, and climatic nuances.

In wet highland areas like Gmra Qema in Tocha district and Losha in Chencha district, the predominant crops include wheat, garlic, and ensat, often with bamboo tree plantation, alongside cattle and sheep rearing. These regions benefit from higher elevations and suitable climates for these crops, reflecting a focus on staple foods and livestock integral to local livelihoods.

Conversely, dry lowland regions like Alga in Mirab Abaya district specialize in maize intercropping with bananas, coupled with cattle and goat husbandry. Here, farming practices are influenced by lower elevations and different climatic conditions, necessitating adaptations that favor drought-tolerant crops and livestock suited to arid environments.

The variations in farming systems are influenced not only by agro-ecology but also by economic factors such as market opportunities in nearby towns, access to extension services, and household land sizes. For instance, households with small plots in wet highland may prioritize cash crops like vegetables and fruits over staple foods due to their immediate income-generating potential. In the dry lowland areas, banana cultivation emerges as the most promising agricultural activity.

Despite these differences, similarities exist between the wet highland areas of Gmra Qema and Losha, where comparable agro-ecological conditions support similar farming patterns. Both regions demonstrate how smallholder farmers integrate crop

Admin zone	Peasant Admin/district	Latitude	Longitude	Elevation, m	Slope (%)	Agro-ecology	Farming system
Dawuro	Gmra Qema/Tocha	7.084530	37.0281908	2561–2714	7.5–20.4	Wet highland	Wheat/garlic-enset/bamboo-cattle/sheep
Gamo	Losha/Chencha	6.16213333	37.57598333	2641–2743	4.83–16.58	Wet highland	Wheat/cabbage-enset/bamboo/apple-cattle/sheep
Dawuro	Myla/Mareka	7.00065	37.0642	1874–2317	2.5–62.9	Upper sub-humid	Wheat/maize/teff-enset/bamboo-ruminant
	Guzza/Issera	6.761195	37.0281199	1843–2231	9.22–101.44		
Gamo	Fishto/Bonke	5.92611111	37.28163889	1609–2223	8.33–63.49	Wet upper lowland	
	Gress Zala/Bonke	5.90112	37.31773	1787–2225	5.41–56.79		
Dawuro	Qcheme Kessi /Tocha	7.095509	37.0287104	1208–1395	2.63–31.76	Wet lowland	Maize/teff-coffee-cattle/goat
	Tarcha Zura/Mareka	7.099924	37.09708	1195–1342	2.36–15.58		Maize/teff/groundnut-coffee-cattle/goat
	Yallo Worbati/Loma	6.76939	37.317276	1085–1272	6.88–25.77		
Gamo	Alga/Mirab Abaya	6.28074	37.78823	1207–1221	0–5.1	Dry lowland	Maize intercropping/banana-cattle/goat
	Ancover/Mirab Abaya	6.26069444	37.74266667	1216–1255	0.88–3.35		
	Furra/Mirab Abaya	6.17642	37.68773	1200–1209	2.07–5.79		
	Para Gossa/Mirab Abaya	6.23224	37.7296	1255–1284	1.74–6.32		

Source: [15, 16].

Table 1.
Features of farming system in smallholder system.

cultivation with livestock rearing to enhance resilience and productivity, adapting their strategies to local environmental constraints and resource availability.

Understanding these spatial patterns and farming dynamics is crucial for devising effective agricultural development strategies. It enables targeted interventions that promote sustainable practices, improve market access, and introduce climate-resilient farming techniques tailored to the specific needs of each agro-ecological zone. Such initiatives aim to enhance food security, livelihoods, and environmental sustainability among smallholder farmers across Ethiopia.

The data presented in **Table 2** highlight the annual average income of farm households across various major livelihood activities within different Peasant Associations (PAs). Livestock and annual crops emerge as primary sources of income for many PAs, underscoring their agricultural importance. However, significant income disparities exist among PAs, with notable examples like Gress Zala and Furra demonstrating higher average incomes compared to others such as Ancover and Losha. This variance may be attributed to factors like larger farm plot sizes in more affluent households.

The contribution of off-farm activities, remittances, and labor to total income varies across PAs but generally constitutes a smaller proportion compared to income derived from agricultural pursuits. This suggests that while non-agricultural activities supplement incomes, agriculture remains the backbone of livelihoods in these rural communities.

The wide range in average total incomes across PAs reflects disparities in agricultural productivity, market accessibility, and potentially climatic or geographic conditions. Livestock-related activities appear pivotal in generating substantial income for numerous associations, complementing earnings from crop production.

Peasant association	Average total(% within and between activities) income €				Overall average (%)
	Annual crops	Plantation and natural capital	Livestock	Off-farm, remittance, labor	
Gmra Qema	272(32,32)	43(5,5)	425(50,50)	101(12,12)	842(99)
Losha	221(35,26)	81(13,10)	104(16,12)	234(36,28)	640(76)
Myla	410(51,48)	24(3,3)	371(46,44)	3.16(0,0)	808(95)
Guzza	279(38,33)	79(11,9)	356(49,42)	16.6(2,2)	731(86)
Fishto	305(66,36)	35(8,4)	114(25,13)	6.37(1,1)	460(54)
Gress Zala	718(65,85)	40(4,5)	340(31,40)	9.96(1,1)	1107(131)
Qchem Kessi	769(74,91)	23(2,3)	239(23,28)	2.42(0,0)	1033(122)
Tarcha Zura	613(63,72)	27(3,3)	311(32,37)	22.4(2,3)	974(115)
Yallo Worbti	338(51,40)	37(6,4)	276(42,33)	7.09(1,1)	658(78)
Alga	297(31,35)	64(7,8)	433(45,51)	167(17,20)	961(113)
Ancover	181(21,21)	218(25,26)	476(54,56)	5.05(1,1)	880(104)
Furra	574(48,68)	55(5,6)	509(42,60)	67(6,8)	1204(142)
Para Gossa	356(29,42)	145(12,17)	640(53,76)	68(6,8)	1209(143)
Average* (%, n)	410(46,345)	67(8345)	353(40,352)	55(6345)	885

**Gross average income (%, n) among livelihood strategies.*
Source: [15].

Table 2.
Livelihood activities gross average income (€/year/household) in smallholder farmer.

This nuanced understanding of income dynamics can inform targeted policy interventions aimed at enhancing productivity in specific agricultural sectors or fostering diversification into higher-income activities. Policymakers can leverage these data to tailor support programs that address the unique challenges and opportunities faced by each PA, thereby fostering sustainable agricultural development and improving livelihoods across rural communities.

Table 3 provides insights into the travel time distances from peasant households to potential marketing points, categorized by PA. The travel times vary significantly among PAs, ranging from 51.76 minutes (Ancover) to 330.92 minutes (Guzza). This wide range indicates substantial disparities in geographical access to markets across different rural communities.

Peasant associations like Losha, Alga, and Ancover have relatively shorter travel times, indicating better proximity to marketing points. In contrast, PAs such as Guzza, Myla, and Gmra Qema face longer travel times, indicating greater remoteness. The overall average travel time across all PAs is 141.28 minutes ±8.00 minutes (SE). This average serves as a benchmark, highlighting typical travel conditions for accessing markets for smallholder farmers in the studied area.

The disparities in travel times underscore the logistical challenges faced by smallholder farmers in accessing markets. Longer travel times can lead to increased costs, time inefficiencies, and reduced market opportunities. Addressing these challenges is crucial for enhancing economic opportunities and livelihoods in rural areas.

Policymakers can use these data to prioritize infrastructure development such as road improvements or market accessibility initiatives in areas with longer travel times. This targeted approach can help reduce disparities, improve market access, and stimulate economic growth in rural communities. Formulating effective development

Peasant association	**n***	**Mean ± S.E**
Gmra Qema	59	$254.10(41.20)^{ab}$
Losha	62	$59.92(5.11)^{c}$
Myla	62	$245.42(42.51)^{ab}$
Guzza	62	$330.92(52.23)^{a}$
Fishto	62	$96.08(4.28)^{c}$
Gress Zala	63	$82.14(4.32)^{c}$
Qcheme Kessi	55	$172.64(21.00)^{bc}$
Tarcha Zura	39	$184.46(34.35)^{bc}$
Yallo Worbati	62	$155.40(19.83)^{bc}$
Alga	62	$55.95(7.55)^{c}$
Ancover	62	$51.76(7.76)^{c}$
Furra	36	$62.22(6.95)^{c}$
Para Gossa	49	$61.84(9.33)^{c}$
Total	735	141.28(8.00)

^{abc}*column means of the same superscript are not significant ($p < 0.05$); n*= number of respondents in participatory group and household interview.*
Source: [15].

Table 3.
Overall (mean ± standard error (S.E)) distance (minute) from the study Peasant association to potential market.

strategies aimed at improving rural livelihoods and fostering sustainable economic development could be essential.

Table 4 provides a detailed analysis of travel time distances (in minutes) from peasant households to various market types within the smallholder system. Peasant households enjoy relatively convenient access to agricultural or local markets, with a mean travel time of 43.56 ± 1.17 minutes (SE). These markets primarily facilitate intra-PA transactions among producers, facilitating the exchange of agricultural goods and basic necessities within close proximity.

Travel times to consumer center markets are longer, averaging 126.90 ± 10.88 minutes (SE). These markets serve as pivotal hubs for purchasing consumer goods and may also act as secondary markets for agricultural produce. The extended travel times suggest moderate accessibility challenges for accessing broader consumer markets from rural areas.

Similar to consumer center markets, markets across agro-ecological zones have a mean travel time of 152.08 ± 4.93 minutes (SE). This indicates reasonable accessibility across different ecological zones within the smallholder system, catering to diverse agricultural needs and regional markets.

Travel times to urban center markets with larger populations significantly exceed other market types, averaging 404.21 ± 34.19 minutes (SE). This substantial distance underscores the formidable challenges smallholder farmers face in accessing major urban markets from remote rural areas. Such markets likely offer extensive consumer bases and economic opportunities but require improved infrastructure and transport connectivity to reduce travel times and associated costs.

The result underscores the complex landscape of market accessibility for smallholder farmers, emphasizing the need for strategic interventions to enhance connectivity and support sustainable rural development. Understanding these dynamics is crucial for policymakers aiming to improve market access and livelihood opportunities in rural areas.

Figure 2 illustrates the notable variability of weather variables within the study area. The analysis using the Standardized Precipitation Index (SPI) has pinpointed 19 years marked by significant wet and drought events. This observation is further supported by findings from the Normalized Difference Vegetation Index (NDVI), emphasizing the spatial and temporal variability inherent in climate extremes. These disruptive forces encompass a spectrum from severe wet conditions to droughts, floods, and landslides, posing significant risks to socio-ecological systems, especially impacting marginalized farming communities.

Market type	n*	Mean ± S.E
Agricultural/local market	345	43.56(1.17)^{a}
Consumer center market	130	126.90(10.88)^{b}
Market across agro-ecological zone	130	152.08(4.93)^{b}
Urban center market (>75,000 population)	130	404.21(34.19)^{c}
Total	735	141.28(8.00)

^{abc}*column means of the same superscript are not significant ($p < 0.05$); S.E is standard error.*
n= number of respondent in participatory group and household interview.*
Source: [15].

Table 4.
Travel time distance (minute) from peasant households to various types of markets within the smallholder system.

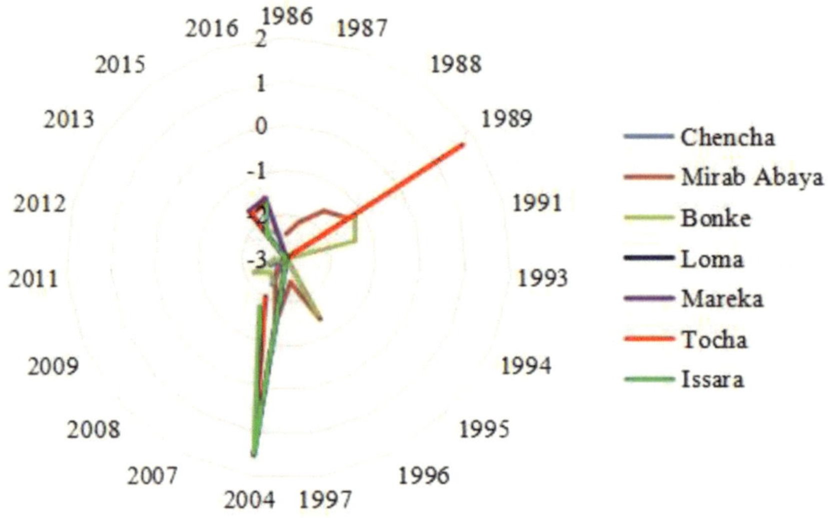

Figure 2.
Standardized precipitation index (SPI) statistical analysis from January 1983–December 2016 of extreme events in districts of the Peasant administration; source: [15, 34].

In specific geographical contexts, such as the Issera district within Guzza PA, Tocha district of Gmra Qema in wet highlands, and Qcheme Kessi PA in wet lowlands, wet stress events associated with climate change are prevalent. Conversely, droughts are frequent in the dry lowlands of Mirb Abaya district affecting Para Gossa, Alga, Furra, and Ancover PAs. These regions are identified as critical areas where the adverse impacts of climate variability are pronounced.

Livestock production emerges as pivotal in these contexts, playing a crucial role as a buffer against the detrimental effects of climatic disturbances. Livestock serve as a resilient asset for farming communities, providing essential support during periods of environmental stress and offering a pathway to enhance resilience and sustainability in the face of ongoing climatic challenges.

This comprehensive understanding underscores the urgency for targeted interventions and adaptive strategies tailored to the diverse climatic conditions and vulnerabilities observed across different PA regions. Such approaches are essential for promoting sustainable agricultural practices and safeguarding livelihoods in vulnerable rural communities.

In response to the severe drought of 2014/15, the American Ambassador to Ethiopia visited the affected districts of Mirab Abaya. However, statistical analyses indicated that during that year, extreme drought conditions, characterized by Standardized Precipitation Index (SPI) values of ≤ -2.2, were actually prevalent in the Bonke and Issara districts. These districts continued to experience prolonged drought conditions, with SPI values persistently ≤ -2.4 through 2016. This underscores the persistent and substantial drought risks faced by these communities, highlighting the need for interventions that go beyond immediate food security measures.

Ethiopia's climate is characterized by four distinct seasons annually, each influencing vegetation cover, rainfall patterns, agricultural production, and livelihoods. Studies, including those by Amejo [35], have confirmed the variability of these seasons using indicators like the Normalized Difference Vegetation Index (NDVI) and rainfall distribution patterns. However, since 2007, there has been a noticeable decline

DOI: http://dx.doi.org/10.5772/intechopen.1006055

in the distinctiveness of these seasonal characteristics, as depicted in **Figure 2**. Prior to 2007, NDVI analyses consistently identified four distinct periods of vegetation cover, a pattern that has since become less discernible.

The magnitude, duration, and frequency of drought events have shown significant variability and an overall increase over time, as highlighted in **Figure 3**. These alterations probably impacted the resilience of the environment, resulting in heightened spatial variability throughout different seasons. Notably, severe events in 2007/08 and 2009/10 transpired across February, July, and December, while droughts specifically in 2012/13 and 2014/15 were predominantly noted in January and February.

This evolving pattern underscores the dynamic nature of drought occurrences in Ethiopia and emphasizes the critical need for adaptive strategies that address both short-term responses and long-term resilience building. Effective interventions should consider the shifting climate patterns and their impacts on agricultural systems and community livelihoods, aiming to mitigate risks and enhance sustainability in vulnerable regions.

Figure 4 provides detailed information on the distribution of livestock in Tropical Livestock Units (TLU) across various PAs within the smallholder system. The overall number of livestock, measured in TLU, varies significantly across different peasant locations, influenced by factors such as agro-ecology, distance, topography, functionality, and adversity of the location. This variability indicates that livestock are integral to the agricultural livelihoods of households in these PAs, serving as a primary source of income, food, and draft power. Additionally, livestock contribute specific functionalities and offer buffering capacity within the farming system, highlighting their critical role in sustaining and enhancing rural livelihoods.

The disparity in distribution among livestock species sheds light on the diverse functions they fulfill within the agricultural landscape. While cattle and small ruminants dominate in terms of sheer numbers and economic contribution, the increased presence of equines in remote regions highlights their unique utility in long-distance transportation tasks.

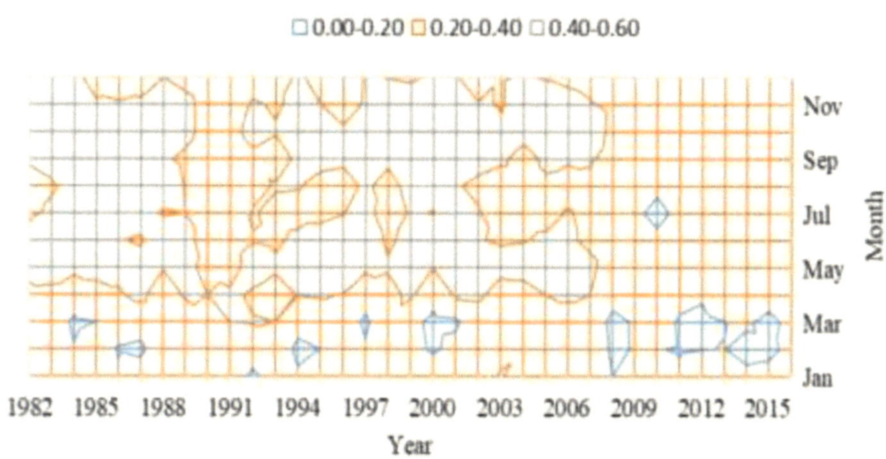

Figure 3.
Temporal variability of NDVI (Blue 0.00–0.20, Red 0.02–0.40 and White 0.4–0.60) in Districts with Peasant Association from January 1982 to December 2016; source: [15, 34].

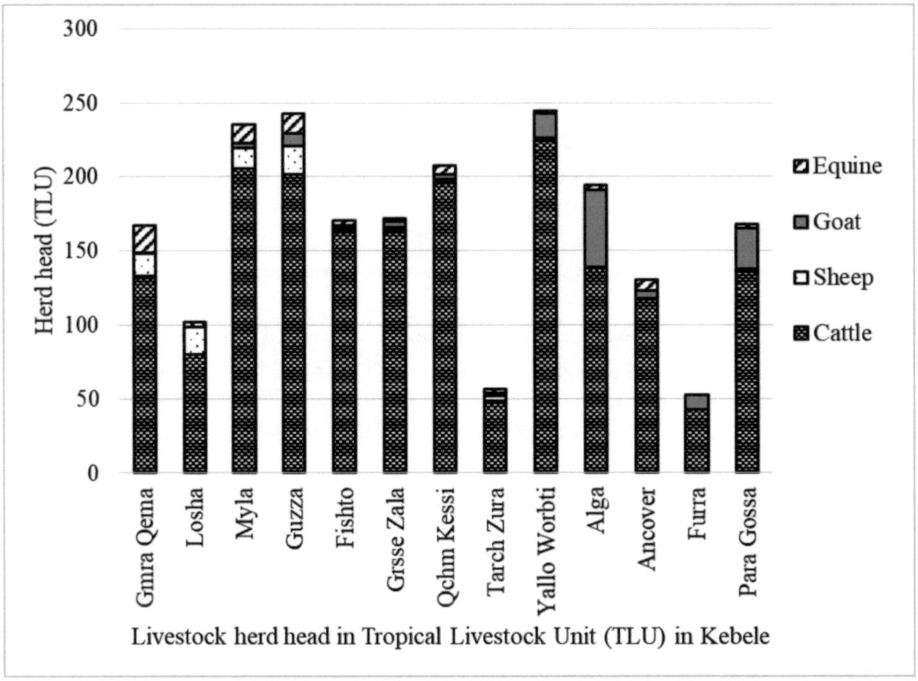

Figure 4.
Distribution of herd head in tropical livestock unit (TLU) in smallholder system; source: [15, 32].

Understanding these nuanced dynamics is paramount for devising targeted interventions aimed at fostering sustainable livestock management practices and optimizing agricultural productivity across different geographical contexts.

8. Comparative analysis

The assessment synthesizes multiple studies focusing on Ethiopia's livestock systems, highlighting the sustainability, resilience, and their implications for rural livelihoods. García de Jalón et al. [36] underscore the influence of different capitals (physical, financial) on adoption of climate-smart agricultural practices (CSAPs) adoption, suggesting targeted policies and investments in practices like improved feed and husbandry techniques. These strategies align with Ethiopia's goals of enhancing productivity while mitigating environmental impacts, critical for sustainable rural development.

Mamphogoro et al. [37] emphasize the significance of sustainable intensification in livestock production by emphasizing improvements in feed quality and forage management. Their approach aims to enhance productivity, mitigate environmental impacts, and bolster food security within agricultural systems. Similarly, Herrero et al. [7] advocate for sustainable intensification in livestock systems and advocate for transitioning towards sustainable food systems. Their framework provides a pathway to integrate livestock into sustainable food systems, addressing environmental challenges and aligning with Ethiopia's agricultural policy goals, which are particularly pertinent to Ethiopia's pasture-based livestock production.

Woodhill et al. [38] call for policy reforms to enhance rural livelihoods, diversify income streams, and capture greater value in food systems. Transforming food systems towards equity, nutrition, and environmental sustainability can bolster Ethiopia's efforts in improving rural well-being and resilience against climate variability.

Khumalo et al. [39] emphasize the transferable principles of CSA from urban to rural contexts, providing insights that can significantly benefit food security and income generation in rural Ethiopia. Their findings underscore the potential for adapting urban agricultural practices to rural settings, thereby enhancing resilience against climate variability and improving livelihoods.

Conversely, Ogisi & Begho [40] examine barriers to the adoption of CSAP practices in SSA. They highlight socio-economic disparities, institutional challenges, and gender inequalities as major obstacles hindering the widespread adoption of CSAPs in the region. Their study underscores the need for targeted interventions and policy frameworks that address these barriers to effectively promote sustainable agricultural practices across SSA, including Ethiopia.

Collectively, these studies underscore the critical importance of adopting existing experiences to enhance resilience against climate variability and ensure sustainability in Ethiopia's livestock systems. They emphasize the necessity of targeted policies that prioritize improvements in feed quality, sustainable intensification of production, and addressing socio-economic and gender disparities for effective implementation of sustainable agriculture practices. By enhancing productivity through sustainable practices, these efforts not only strengthen food security but also support rural livelihoods, aligning closely with the broader development goals of Ethiopia's smallholder farming systems.

The synthesis of these insights provides a solid foundation for developing strategies to navigate challenges and capitalize on opportunities within Ethiopia's livestock sector. Integrating these findings into policy frameworks is crucial for achieving agricultural sustainability and resilience amidst the evolving environmental and socio-economic dynamics of the region. By doing so, Ethiopia can effectively enhance its capacity to adapt to climate change impacts while promoting sustainable development across its agricultural landscape.

9. Conclusions

This chapter emphasizes the crucial role of livestock in bolstering the resilience and sustainability of smallholder farming systems in SSA, with a specific focus on Ethiopia. Traditional agricultural practices in Ethiopia, characterized by integrated crop-livestock systems, exemplify the symbiotic relationship between crop cultivation and livestock rearing. Livestock contribute significantly as draft power and through practices like enset-livestock farming, where animal manure enhances soil fertility and supports crop health and productivity. These integrated approaches optimize resource utilization, improve agricultural output, and provide socio-economic benefits to rural communities, showcasing resilience in the face of environmental and economic uncertainties.

However, contemporary agricultural trends, coupled with climate change and socio-economic shifts, pose challenges to these traditional systems. Fragmented approaches have disrupted the synergistic balance between crops and livestock, undermining their efficiency and sustainability. This case study underscores the importance of holistic agro-ecosystem management, recognizing and reinforcing these interdependencies.

To safeguard and enhance the resilience and sustainability of smallholder farming in SSA, a renewed emphasis on integrated agricultural practices is essential. This entails acknowledging the multifaceted roles of livestock and promoting adaptive management strategies, supported by conducive policies, targeted research, and development initiatives. Interdisciplinary approaches that blend indigenous knowledge with modern scientific advancements are critical for fostering sustainable intensification and resilience.

Policy frameworks should evolve to incentivize and support integrated farming methods, address climate challenges, and ensure equitable access to resources and technologies for smallholder farmers. Capacity-building efforts are vital, encompassing training in sustainable techniques, resource management, and climate adaptation, while fostering community-based approaches to enhance local resilience and knowledge exchange.

The integration of cutting-edge technologies such as spatial analysis, precision agriculture, and digital tools holds promise for optimizing resource efficiency, monitoring environmental conditions, and facilitating informed decision-making within traditional farming systems. This integration, bolstered by robust policies and capacity-building initiatives, is pivotal for advancing the future resilience and sustainability of smallholder farming systems across SSA.

Conflict of interest

The authors declare no conflict of interest.

Acronyms and abbreviations

€	euro
GPS	global positioning system
NDVI	normalized difference vegetation index
PAs	peasant associations
n	sample size
S.E	standard error
SPI	standardized precipitation index
SSA	Sub-Saharan Africa
TLU	tropical livestock units

Author details

Asrat Guja Amejo
Arba Minch University, Arba Minch, Ethiopia

*Address all correspondence to: gujasrat@gmail.com

References

[1] Keeney DR. Toward a sustainable agriculture: Need for clarification of concepts and terminology. American Journal of Alternative Agriculture. 1989;**4**:101-105. DOI: 10.1017/S0889189300002903

[2] Constance DH. Sustainable agriculture in the United States: A critical examination of a contested process. Sustainability. 2009;**2**:48-72. DOI: 10.3390/su2010048

[3] Neo H, Emel J. Geographies of Meat: Politics, Economy and Culture. London: Routledge; 2017. DOI: 10.4324/9781315584386

[4] Hazell P. Is small farm led development still a relevant strategy for Africa and Asia. Chapter. 2013;**8**:192-207

[5] Varijakshapanicker P, Mckune S, Miller L, Hendrickx S, Balehegn M, Dahl GE, et al. Sustainable livestock systems to improve human health, nutrition, and economic status. Animal Frontiers. 2019;**9**:39-50. DOI: 10.1093/af/vfz041

[6] Baiyeri PK, Foleng HN, Machebe NS, Nwobodo CE. Crop-livestock interaction for sustainable agriculture. Innovations in Sustainable Agriculture. 2019:557-582. DOI: 10.1007/978-3-030-23169-9_18

[7] Herrero M, Mason-D'Croz D, Thornton PK, Fanzo J, Rushton J, Godde C, et al. Livestock and sustainable food systems: Status, trends, and priority actions. Science and Innovations for Food Systems Transformation. 2023:375. DOI: 10.1007/978-3-031-15703-5

[8] Poccard-Chapuis R, Alves LN, Grise MM, Bâ A, Coulibaly D, Ferreira LA, et al. Landscape characterization of integrated crop–livestock systems in three case studies of the tropics. Renewable Agriculture and Food Systems. 2014;**29**:218-229. DOI: 10.1017/S174217051400009X

[9] Ruthenberg H. Farming Systems in the Tropics. Oxford: Clarendon; 1971

[10] Thornton PK, Herrero MT, Ericksen PJ. Livestock and climate change. Livestock Exchange Issue Brief. 2011

[11] Ellis F. Household strategies and rural livelihood diversification. The Journal of Development Studies. 1998;**35**:1-38. DOI: 10.1080/00220389808422553

[12] Pender J. Development pathways for hillsides and highlands: Some lessons from Central America and East Africa. Food Policy. 2004;**29**:339-367. DOI: 10.1016/j.foodpol.2004.07.005

[13] Tittonell P. Livelihood strategies, resilience and transformability in African agroecosystems. Agricultural Systems. 2014;**126**:3-14. DOI: 10.1016/j.agsy.2013.10.010

[14] Jones JW, Antle JM, Basso B, Boote KJ, Conant RT, Foster I, et al. Brief history of agricultural systems modeling. Agricultural Systems. 2017;**155**:240-254

[15] Amejo AG, Mekasha DY, Kassa DH, Tana P. Characterization of crop-livestock systems, and the impact of climate change on livestock production in southwestern Ethiopia [doctoral dissertation]. Haramaya University; 2019

[16] Amejo AG, Gebere YM, Kassa H, Tana T. Characterization of smallholder mixed crop-livestock systems in integration with spatial information:

In case Ethiopia. Cogent Food & Agriculture. 2019;**5**:1565299. DOI: 10.1080/23311932.2019.1565299

[17] Abegaz A. Farm Management in Mixed Crop-Livestock Systems in the Northern Highlands of Ethiopia. The Netherland: Wageningen University and Research; 2005

[18] Amejo AG. Estimation of feed dry matter supply in relation to land use and livestock requirements in southwestern Ethiopia. Cogent Food & Agriculture. 2023;**9**:2270276. DOI: 10.1080/23311932.2023.2270276

[19] Seré C, Steinfeld H. World livestock production systems. In: FAO Animal Production and Health Paper 127. Rome, Italy: FAO; 1996

[20] Berhe M, Hoag D, Tesfay G, Tadesse T, Oniki S, Kagatsume M, et al. The effects of adaptation to climate change on income of households in rural Ethiopia. Pastoralism. 2017;**7**:1-5. DOI: 10.1186/s13570-017-0084-2

[21] IPCC. Climate Change 2007: Impacts, Adaptation and Vulnerability. In: Parry ML, Canziani OF, Palutikof JP, van der Linden PJ, Hanson CE, editors. Contribution of Working Group II to the Fourth Assessment Report of the Intergovernmental Panel on Climate Change. Cambridge, UK: Cambridge University Press; 2007. 976 p

[22] Ali A, Erenstein O. Assessing farmer use of climate change adaptation practices and impacts on food security and poverty in Pakistan. Climate Risk Management. 2017;**16**:183-194. DOI: 10.1016/j.crm.2016.12.001

[23] Negash A, Niehof A. The significance of enset culture and biodiversity for rural household food and livelihood security in southwestern Ethiopia. Agriculture and Human Values. 2004;**21**:61-71. DOI: 10.1023/B:AHUM.0000014023.30611.ad

[24] Yemataw Z, Mohamed H, Diro M, Addis T, Blomme G. Enset (Ensete ventricosum) clone selection by farmers and their cultural practices in southern Ethiopia. Genetic Resources and Crop Evolution. 2014;**61**:1091-1104. DOI: 10.1007/s10722-014-0093-6

[25] Borrell JS, Goodwin M, Blomme G, Jacobsen K, Wendawek AM, Gashu D, et al. Enset-based agricultural systems in Ethiopia: A systematic review of production trends, agronomy, processing and the wider food security applications of a neglected banana relative. Plants, People, Planet. 2020;**2**:212-228. DOI: 10.1002/ppp3.10084

[26] Sahle M, Yeshitela K, Saito O. Mapping the supply and demand of Enset crop to improve food security in southern Ethiopia. Agronomy for Sustainable Development. 2018;**38**:1-9. DOI: 10.1007/s13593-017-0484-0

[27] Desalegn R. Resilience and vulnerability: Enset agriculture in southern Ethiopia. Journal of Ethiopian studies. 1995;**28**:23-51

[28] Brandt SA. The 'Tree Against Hunger' Ensete-based Agricultural Systems in Ethiopia. Washington, D.C: American Association for the Advancement of Science; 1997. p. 55

[29] Tsegaye A, Struik PC. Enset (Ensete ventricosum (Welw.) Cheesman) kocho yield under different crop establishment methods as compared to yields of other carbohydrate-rich food crops. NJAS: Wageningen Journal of Life Sciences. 2001;**49**:81-94. DOI: 10.1016/S1573-5214(01)80017-8

[30] Senbeta AF, Sime G, Struik P. Enset farming system–a resilient, climate-robust production system in south and South-Western Ethiopia. Cogent Food & Agriculture. 2022;**8**:2074626. DOI: 10.1080/23311932.2022.2074626

[31] Mellisse BT, Descheemaeker K, Giller KE, Abebe T, van de Ven GW. Are traditional home gardens in southern Ethiopia heading for extinction? Implications for productivity, plant species richness and food security. Agriculture, Ecosystems & Environment. 2018;**252**:1-3. DOI: 10.1016/j.agee.2017.09.026

[32] Amejo AG, Gebere YM, Kassa H, Tana T. Agricultural productivity, land use and draught animal power formula derived from mixed crop-livestock systems in Southwestern Ethiopia. African Journal of Agricultural Research. 2018;**13**(42):2362-2381. DOI: 10.5897/AJAR2018.13258

[33] Dawuro Zone Culture and Tourism Bureau. Cultural Festivals and Photo Exhibition 2017

[34] Amejo AG, Gebere YM, Kassa H, Tana T. Comparative analysis of climate change impact on livestock in relation to biomass base feed availability using standardized precipitation index in South-Western Ethiopia. International Journal of livestock production. 2018;**9**:184-191. DOI: 10.5897/IJLP2018.0478

[35] Amejo AG. Mapping soil terrain resources and descriptions of agro-ecological zone in Dawuro and Gamo Gofa zones in South-Western Ethiopia. Journal of Soil Science and Environmental Management. 2018;**9**:164-179

[36] García de Jalón S, Silvestri S, Barnes AP. The potential for adoption of climate smart agricultural practices in sub-Saharan livestock systems. Regional Environmental Change. 2017;**17**(2):399-410

[37] Mamphogoro TP, Mpanza TD, Mani S. Animal feed production and its contribution to sustainability of livestock systems: African perspective. In: The Marginal Soils of Africa: Rethinking Uses, Management and Reclamation. 2024. pp. 37-54

[38] Woodhill J, Kishore A, Njuki J, Jones K, Hasnain S. Food systems and rural wellbeing: Challenges and opportunities. Food Security. 2022;**14**(5):1099-1121

[39] Khumalo NZ, Sibanda M, Mdoda L. Implications of a climate-smart approach to food and income security for urban sub-Saharan Africa: A systematic review. Sustainability. 2024;**16**(5):1882

[40] Ogisi OD, Begho T. Adoption of climate-smart agricultural practices in sub-Saharan Africa: A review of the progress, barriers, gender differences and recommendations. Farming System. 2023;**1**(2):100019